Charles Seale-Hayne Library
University of Plymouth
(01752) 588 588
LibraryandITenquiries@plymouth.ac.uk

FUNDAMENTAL QUESTIONS IN QUANTUM MECHANICS

Contents

1. W. Bailey	20. S. Cannavo	38. A. O. Barut
2. J. Schumacher	21. D. Albert	39. J. P. Jarrett
3. Z. Fried	22. G. Felming	40. S. Chakravarty
4. L. Ballentine	23. C. Burch	41. R. Koch
5. J. D. Franson	24. S. D'Amato	42. T. Valone
6. R. Baierlein	25. J. Hasbun	43. L. M. Roth
7. P. Pearle	26. I. Pitowsky	44. V. Mansfield
8. R. Reddy	27. A. Cordero	45. D. Page
9. R. Dubisch	28. H. Gomm	46. A. Yasin
10. E. Rogers	29. R. Mallett	47. L. A. Surgalla
11. M. P. Silverman	30. T. Roman	48. D. Mermin
12. J. Dunham	31. R. Chaudhari	49. D. W. Reiss
13. A. Inomata	32. A. Walsled	50. S. Malin
14. D. Leiter	33. P. Gordon	51. A. Shimony
15. K. D. Irani	34. C. C. Gerry	52. R. Griffiths
16. W. Stroup	35. F. Rohrlich	53. D. Henderson
17. W. Wootters	36. S. J. Silverman	54. M. Buettiker
18. D. Park	37. M. Semon	55. D. Greenberger
19. R. Singh		

Not Present: J. Buschman, J. W. Corbett, M. Dresden, T. Jordan, J. Kimball, J. Moskowitz, L. Motz, H. Nebel, A. Ochadlick, J. Onello, W. L. Reese, R. Rockefeller, H. Story, A. Tonomura.

© 1986 by Gordon and Breach Science Publishers S.A.
P.O. Box 161, 1820 Montreux 2, Switzerland. All rights reserved.

Gordon and Breach Science Publishers

P.O. Box 786
Cooper Station
New York, NY 10276
United States of America

P.O. Box 197
London WC2E 9PX
England

58, rue Lhomond
75005 Paris
France

14-9 Okubo 3-chome
Shinjuku-ku, Tokyo 160
Japan

Library of Congress Cataloging-in-Publication Data

Conference on Fundamental Questions in Quantum
    Mechanics (1984: State University of New York)
    Fundamental questions in quantum mechanics.

    1. Quantum theory—Congresses. I. Roth, Laura M.,
1930—   . II. Inomata, Akira, 1931–   .
III. Title.
QC173.96.C65 1984   530.1'2   86-262

FUNDAMENTAL QUESTIONS IN QUANTUM MECHANICS

Proceedings of the Conference on
Fundamental Questions in Quantum Mechanics
held at the State University of New York at Albany,
April 12–14, 1984

Edited by
Laura M. Roth and Akira Inomata
State University of New York at Albany

GORDON AND BREACH SCIENCE PUBLISHERS
New York · London · Paris · Montreux · Tokyo

INTRODUCTION

Despite the great successes of quantum mechanics, there remain
many puzzling and controversial questions about its formulation,
interpretation, and philosophical implications. Currently, new
experiments promise to shed light on some of these questions. The
papers collected in this volume are based on a symposium entitled
"Fundamental Questions in Quantum Mechanics" that was held 12–14
April 1984 at the State University of New York at Albany, under the
auspices of the State University's Conversations in the Disciplines
program. At this conference, a variety of topics relating to fundamental
questions in quantum mechanics were covered. This collection of papers
thus will provide an overview of the field and of the interesting problems
still to be solved in this area. There is a considerable bias for physics over
philosophy; this reflects the interests and concerns of the organizers of
the conference.

We have grouped the papers into various topics in this proceedings,
although it is naturally too difficult to develop a rigid classification of
this work. Section I concerns Bell's theorem, which—according to most
interpretations—enables one to test experimentally between local
realism (that is, local hidden variables) and quantum mechanics. These
recent experiments, most agree, have been decided in favor of quantum
mechanics. Mermin has given a very graphic description of the Bell
theorem arguments by imagining that particles obeying local realism
depart from their interacting state with definite instructions on how
to respond to the two detectors. In his present contribution, he extends
the Bell arguments to apply to arbitrary spin and shows how the
correlations go over into the classical result. Briefly, quantum correlations exist for all spins, but, as the spin increases, their propensity to
be smeared out by experimental uncertainty becomes more pronounced.
Jarrett's short paper propounds a very significant result that
distinguishes two aspects of the locality condition in the Bell theorem
proof, one of which is a weak locality condition that precludes
superluminal signals, which quantum mechanics obeys; the other is a
rather subtle aspect that he calls "completeness". The two together give

a strong locality condition that is the usual factorization of the probabilities assumed under local realism. It is evidently impossible to use a quantum measurement in one place to detect the direction of the analyzer in another place.

Unfortunately, Shimony's most enjoyable talk on hidden variables and Bell's theorem could not be included, but we do have an extended abstract that includes a commentary and an interpretation of Jarrett's result. Barut's paper shines an interesting light on Bell's theorem. He shows that spins that are completely classical and are not forced to give up and down values when detected actually give the same correlations as quantum spins. Perhaps the emperor has no clothes. But, of course, his classical spins would give experimental results different from those observed. In another theoretical experiment, he models spin quantization, or wave packet reduction, by the appearance of an *attractor*.

In Section II, we include papers on the measurement problem, which has been a persistent question in quantum mechanics since Schrödinger's cat. The question is, When, if ever, is a quantum superposition reduced to a definite result or set of classical probabilities? Pearle, in his contribution, propounds a theory of dynamic reduction. By adding an extra term to the Hamiltonian, he shows that the wave packet is reduced by a mechanism similar to the "Gambler's Ruin" game, the fact that, given a finite amount of money, eventually someone wins or loses it all. Pearle presents illuminating insights into the requirements of dynamic reduction theories, especially that of a constant mean value of a property, and suggests ways to test such theories. The neutron interference work does not seem too favorable to his theory. Page introduces a preferred basis of states for considering quantum measurements in which the consistency of information stored in identical subsystems is maximized. Page is of the school that holds with Everett that there is no reduction of the wave packet, and he hopes that his preferred basis of states would enable one to treat the registry of an event in consciousness as a measurement. Ballentine discusses various aspects of the measurement problem and concludes that a reduction of the wave packet is untenable within quantum mechanics. He advocates an interpretation of quantum mechanics put forward by Marganau in 1936, that the wave function does not describe an individual system but an ensemble, in which case no reduction is necessary.

Machida and Namiki were unable to attent the conference, but we have included their paper on the measurement problem. In contrast to Ballentine, they describe the reduction of the wave packet by using the S-

matrix, and show that there is reduction in the limit of infinite time and infinite size of the apparatus. Schumacher applies quantum mechanics to the visual process and uses arguments applied by Bohr and Bohm, as well as Bell, to external quantum measurements, to argue that vision is not an inner show invoked by an outer world, but that inner and outer are aspects of one whole, not analyzable into autonomously existent elements. He gives several examples from sensory physiology to support his views. Albert introduces a quantum mechanical automaton that is aware of its internal states and of the results of a measurement and tells what it observes and predicts future results, correctly, while still a quantum system. With this rather bizarre description of a person he is addressing the same question: Can (or should) self-awareness or cognition of internal states be included in quantum mechanics?

Section III includes papers in macroscopic quantum interference. Greenberger discussed neutron interference experiments that beautifully confirm the quantum mechanical superposition principle. In experiments performed by Rauch in Vienna, spin-polarized neutron beams, weak enough in intensity to correspond to single neutrons, are split by diffraction in a neutron interferometer, and one beam has its spin flipped before the two beams are recombined. Interference was not observed when the analyzer was directed along the original polarization direction, when the measurement could determine which "slit" the neutron had gone through, but was observed in the transverse direction, when there was a combination of superposition and interference. Greenberger's contribution celebrates these results and discusses and dismisses several objections which have been raised. Chakravarty discusses prospects for producing macroscopic quantum coherence effects in superconducting quantum interference devices (SQUID) and, in particular, quantum tunnelling. The prospects do not seem too encouraging, although photoinduced tunnelling seems promising. Buetticker describes calculations of quantum effects in small conducting rings of normal metal, predicting both Josephson-like oscillations in closed rings and a periodicity in the resistance of such rings with the magnetic flux, but with a single instead of a double charge flux quantum, unlike a superconductor. The experimental situation is still not clear. On a controversial note, Franson reports measurements using a large optical interferometer that give a reduction of the visibility of an interference pattern at single photon intensities, which he argues is in disagreement with quantum mechanics but in agreement with local theories.

Section IV, the Aharonov–Bohm effect, begins with Tonamura's paper, which describes experiments giving an impressive demonstration of the magnetic Aharonov–Bohm effect by means of electron holography. An electron beam was sent through tiny ferromagnetic rings, and the interference patterns clearly demonstrated the effect through the shift of the fringes. Several detracting objections were answered by new experimental procedures, such as coating the rings with gold to preclude penetration of the magnet by the electron wave function. Seeing is believing! Silverman discusses the problem of a quantum particle circulating about a confined magnetic flux and the question of whether the canonical or kinetic angular momentum is to be quantized to integer values. He suggests a split beam interference experiment to test whether the rotation generator has eigenvalues dependent on magnetic flux, in which case a spinless particle would manifest a Fermionic rotation property. Semon and Taylor discuss the scalar potential Aharonov–Bohm effect in terms of "modular momentum". Gerry and Inomata discuss the effects of the multiple connectness of space in terms of Feynman path integral theory, the homotopy group, and winding number. They show that the textbook derivation of flux quantization in a superconductor, which is based on singlevaluedness of the wave function, is invalid and would lead to universal flux quantization and, hence, to no observable magnetic Aharonov–Bohm effect. Rather, they show that the flux quantization comes from the minimization of the free energy.

Several new schemes for interpreting or formulating quantum mechanics are described in Section V. Griffiths generalizes the usual formula for transition probabilities to yield conditional probabilities for selected sequences of events or consistent histories, and finds that, within limits, the classical rules of probability apply. His interpretation has no need for a collapse of the wave packet. D'Amato expands on such ideas by considering two-time measurements, i.e. measurements for sums of quantum operators at two different times, and finds that the behavior of a quantum system is sometimes completely determined by the outcome of two-time measurements bracketing the given time. Pitowsky describes a scheme for quantum mechanics in which the operators for position and momentum commute with one another, but the strange structure of the quantum state function leads to all the quantum properties. Dubisch considers how Hawking-type particle production near black holes affects the requirements for a quantum theory of gravity. Interactions are related to transformations of complex structures, which he argues should be regarded as gauge transformations.

Then in Section VI, Interpretations of Quantum Mechanics, we include an interesting set of somewhat miscellaneous topics. To begin with, Park discusses how time enters into classical and quantum mechanics. He shows that the oft-quoted uncertainty relation between time and energy is parasitic on the coordinate and momentum relations and, hence, that time (or proper time, relativistically) is thus far a c-number. He describes a kind of quantum clock. Wooters discusses the possibility of eliminating space as well as time from the descriptions of systems of quantum particles and replacing them with quantum correlations of internal variables such as spin. The time version has been used in connection with quantum gravity. If, for example, two spins are correlated, one can be interpreted as a clock, in which case the other is observed to precess.

Dresden highlights the superposition of states as a key feature of quantum mechanics. After a discussion of how the superposition of particle states manifests itself in the production and decay of K mesons, he describes a statistical model of a minimum quantum system, i.e. one that has this key feature of superposition and little else, and shows interesting effects, such as memory effects, which do not seem to be obtainable in a classical model.

Irani puts very clearly the questions that quantum mechanics raises in the theory of explanation, in that it provides a predictive theory but not a process-descriptive system. He sees in Weiskopf's quantum ladder a relativity of existence of systems depending on the energy of observation. Irani sees the situation as pointing to underlying principles of order which are as yet unknown to us. He concludes that traditional realism is untenable but that a total commitment to instrumentalism is not taking place. Roth introduces the Buddhist Madyamika philosophy of Emptiness, which is best interpreted as interdependence, in the hopes of its providing guidance about quantum mechanics. The Madyamika view shows agreement with the lack of real atomic elements of existence, which quantum mechanics and Bell's theorem imply, but also points to an interdependence between inner mental events and the outer world, a theme shared, interestingly, by a number of contributions to this conference.

We end this volume with Abner Shimony's delightful summary of the conference.

We take this opportunity to thank the Research Foundation, the

University, and the Physics Department, especially Professor John Kimball, Sue Pero, Joann Oroglio, and numerous graduate students, for helping with the conference.

Laura M. Roth
Akira Inomata

GENERALIZATIONS OF BELL'S THEOREM TO HIGHER SPINS AND HIGHER CORRELATIONS

N. David Mermin

Laboratory of Atomic and Solid State Physics, Cornell University, Ithaca, NY 14853

Several recent generalizations of Bell's theorem are reviewed, involving spins larger than one-half, or correlation functions of large numbers of spins. The underlying question of interest is how the strange quantum theoretic correlation functions approach ordinary classical behavior as spin tends to infinity.

I shall begin by describing a relatively new and very simple way to prove a version of Bell's theorem,[1] which applies to a large class of correlation experiments like the one first described by Einstein, Podolsky and Rosen[2] (EPR). This class includes Bohm's spin-$\frac{1}{2}$ version of the experiment,[3] to which Bell's original theorem and most of its subsequent generalizations apply. It also includes the higher spin versions of that experiment up to but not including the classical ($s \to \infty$) version. The point of the exercise is this:

Bell's theorem demonstrates that the EPR reality criterion is incompatible with the quantitative numerical predictions of the quantum theory in the case of Bohm's spin-$\frac{1}{2}$ correlation experiment. On the other hand the reality criterion is explicitly satisfied for the classical ($s \to \infty$) version of that experiment. By examining versions of Bell's theorem valid for the case of general spin-s, one might hope to learn how the strange correlations in the spin-$\frac{1}{2}$ case turn, with increasing s, into the entirely unstrange correlations prevailing in the classical limit. (Many physicists do not find EPR correlations strange. For their benefit I

7

introduce a technical term: correlations will be said to be "strange" if they violate the EPR reality criterion (or some of its later generalizations, characterized by the phrase "local realism") and "unstrange" or "ordinary" if they do not.)

The result of such an exercise[4] is that correlations in the EPR experiment are as strange for general spin-s as they are for spin-$\frac{1}{2}$. The reality criterion is entirely incompatible with all such experiments, whatever the value of the spin. It is the ordinary classical correlations that are atypical.

Although the context in which it arises here is rather novel, this kind of non-uniformity in the appearance of classical behavior is not unusual. No matter how large the quantum number n, for example, the distribution of positions in the various stationary states of a one-dimensional harmonic oscillator is a strongly oscillatory function of position that differs drastically from the simple monotonic classical distribution. It is only the limit in resolution that accompanies any real measurement of position that is responsible for the washing out of the short wavelength quantum oscillations and the restoration of the classical distribution function. This limited resolution has nothing to do with the uncertainty principle. Any given detector will have a limiting resolution: the higher the resolution, the higher n must be, but for any specified resolution there is a quantum number above which the distribution looks classical.

A similar phenomenon holds for local realism. Two aspects of how this comes about will be the subject of the second part of my talk. I will describe a rather unusual way of representing the exact quantum theoretic pair distribution for the spin-s EPR experiment, which reveals explicitly how the classical distribution is buried in the quantum theoretic ones, and why it can be uncovered by appropriate smoothing. I shall then raise some questions (to which I can only give partial answers) about the quantitative way in which increasing lack of resolution in the detectors might obliterate evidence that the EPR reality condition is untenable in the very simple case of spin-$\frac{1}{2}$.

Einstein, Podolsky and Rosen formulated their reality criterion as follows: "If without in any way disturbing a system, we can predict with certainty . . . the value of a physical quantity, then there exists an element of physical reality corresponding to this physical quantity." Bohn pointed out that this has immediate implications for the description of two spin-$\frac{1}{2}$ particles, flying apart from one another in a state of zero total spin. The implications are a direct consequence of angular momentum conservation, and are therefore valid for a pair of particles of any intrinsic spin-s, when the two particle state has zero total spin.

Because the total spin is zero, its component along any direction must vanish, and therefore if the spin of one particle is measured along a direction $\hat{a}$ and yields the value m, a subsequent measurement of the spin of the other along the same direction $\hat{a}$, must yield the value $-m$. Under the "locality" assumption that the spin measurement on the first particle can be made "without in any way disturbing" the second, which can be very far away, we conclude from the reality criterion that "there exists an element of physical reality corresponding to" the spin component of the second particle along the direction $\hat{a}$. But since we could equally well have chosen to measure the spin of the first particle along another direction $\hat{b}$, we must conclude that there is also an element of physical reality corresponding to the spin component of the second particle along the direction $\hat{b}$. By reversing the order of measurements we can make a similar argument about the first particle, and are therefore led to conclude that there must be "elements of reality" corresponding to the spin components of both particles along any directions whatever.

The mathematical embodiment of these "elements of reality" is a function of direction, $\mu(\hat{a})$, assigning to every direction $\hat{a}$ one of $2s + 1$ possible values, thereby specifying the result of measuring the spin of the first particle (or the negative of the result of measuring the spin of the second) along any direction $\hat{a}$ at all. The function μ can differ from one run of the experiment to the next, and the statistics of the spin correlation experiments are determined by the statistical distribution of these functions over many runs of the experiment.

If we limit ourselves to experiments in which both spin components are measured along the same direction, we only reconfirm that the function carried by one particle is the negative of the function carried by the other. We can, however, also consider experiments in which the two particles are subject to spin measurements along different directions $\hat{a}$ and $\hat{b}$. The quantum theory makes a definite prediction for the distribution $q_{ab}(m, m')$ of values obtained in such a pair of measurements, and that function can be interpreted as giving us additional statistical information about the values carried by the function μ in two directions: namely the probability $p_{ab}(m, m')$ of μ having the value m in the direction $\hat{a}$ and the value m' in the direction $\hat{b}$ is just

$$p_{ab}(m, m') = q_{ab}(m, -m'). \tag{1}$$

The right-hand side of Eq. (1) is a perfectly well-defined quantum theoretic distribution, which can be explicitly expressed in terms of the matrix elements of a rotation operator:

$$q_{ab}(m, -m') = |\langle m| \exp(i\theta S_y)|m'\rangle|^2/(2s+1), \qquad (2)$$

where θ is the angle between the directions $\hat{a}$ and $\hat{b}$. (I shall make virtually no use of this detailed form for q in the first part of this talk.) The left side of (1), however, is quantum theoretic nonsense, since it has the meaning of a joint distribution for two incompatible observables—spin components of a single particle along two distinct directions. Nevertheless it is given a supra-quantum theoretic meaning by the EPR reality criterion, and EPR regarded the inability of the quantum theory to give meaning to such an object as a clear indication that the theory was "incomplete". That this is not the correct conclusion to draw can be seen by going one step farther in the argument:

If the functions μ exist they carry more information than is contained in the pair distributions $p_{ab}(m, m')$. We can, for example, inquire into the statistics of μ having the values m, m' and m'' in three distinct directions $\hat{a}$, $\hat{b}$ and $\hat{c}$, as characterized by the distribution $p_{abc}(m, m', m'')$. There is, however, no way to determine the precise form of these third-order distributions, as Eq. (1) gave us the pair distributions. All we can say with assurance is that they must be related to the pair distributions defined in (1) by the marginal relation

$$p_{ab}(m, m') = \sum_{m''} p_{abc}(m, m', m''), \qquad (3)$$

and the analogous two relations for $p_{ac}(m, m'')$ and $p_{bc}(m', m'')$.

It is an elementary exercise to show that the quantum theoretic pair distribution is incompatible with the existence of a third-order distribution satisfying these marginal relations, for *any* three distinct coplanar directions $\hat{a}$, $\hat{b}$ and $\hat{c}$, and *for any value of the spin-s*. The demonstration of this fact is the generalized Bell's theorem.[4]

Before proceeding to the demonstration, pause to note explicitly that the EPR reality criterion *is* satisfied in the classical limit of the spin-s EPR experiment. In that limit the two particles have spins of the same magnitude and opposite senses along a definite direction $\hat{n}$, which is randomly distributed from one run of the experiment to the next. It is convenient to take to the magnitude of that spin to be unity, a convention we can achieve in both the quantum and the classical case by introducing new variables $x = m/s$, $y = m'/s$, $z = m''/s$, etc. In any given run of the classical experiment, the function $\mu(\hat{a})$ (for the particle whose spin is directed along n) is just $\hat{n} \cdot \hat{a}$. The classical limit of the quantum theoretic distribution for spin components x and y must therefore be

$$q^c_{ab}(x, y) = \left\langle \delta(x - \hat{n} \cdot \hat{a})\delta(y + \hat{n} \cdot \hat{b}) \right\rangle, \tag{4}$$

where the brackets indicate a uniform average over all directions $\hat{n}$. The pair distribution characterizing a single particle is thus

$$p^c_{ab}(x, y) = \left\langle \delta(x - \hat{n} \cdot \hat{a})\delta(y - \hat{n} \cdot \hat{b}) \right\rangle, \tag{5}$$

and the third-order distribution required by the EPR reality condition that satisfies all the marginal relations is evidently given by

$$p^c_{abc}(x, y, z) = \left\langle \delta(x - \hat{n} \cdot \hat{a})\delta(y - \hat{n} \cdot \hat{b})\delta(z - \hat{n} \cdot \hat{c}) \right\rangle. \tag{6}$$

I have displayed here these well-defined and meaningful classical distribution functions for several reasons: (1) so we can note which property of the classical distribution invalidates the proof of the generalized spin-s Bell's theorem; (2) to emphasize the trivial validity of the EPR reality criterion and its implications for the existence of higher-order distributions in the classical limit; (3) because I shall be making use of this form in the second part of my talk; and (4) because it ought to be more widely known to forestall attempts one finds in the literature[5,6] to impose subsidiary conditions on hidden variable representations that are manifestly not satisfied by the classical distribution function (for which the hidden variable does exist and is nothing but the spin axis $\hat{n}$).

Having noted the form of the classical limit, I proceed to the proof of the generalized Bell's theorem. It exploits only the following set of properties of the quantum theoretic distribution:

(a) *Bilinearity of the correlation functions.* The mean of the product xy,

$$\left\langle xy \right\rangle_{ab} = \sum_{xy} xy q_{ab}(xy) = (\Psi, (S^1 \cdot \hat{a})(S^2 \cdot \hat{b})\Psi), \tag{7}$$

is explicitly bilinear in the axes $\hat{a}$ and $\hat{b}$ (where Ψ is the zero angular momentum state for two spin-s particles).

(b) *Rotational invariance of the zero angular momentum state.* This requires that the bilinear form in (7) must simply be proportional to $\hat{a} \cdot \hat{b}$:

$$\left\langle xy \right\rangle_{ab} = K_s(\hat{a} \cdot \hat{b}). \tag{8}$$

(The constant K_s is simply $-(1 + 1/s)/3$, but this information is not needed for the argument.)

(c) *Physical locality.* This is the statement that the distribution of results obtained at just one of the detectors should be independent of the *setting* of the far detector:

$$\sum_y q_{ab}(x, y) \qquad \text{is independent of the direction } \hat{b}. \qquad (9)$$

If this were not so, then a change in the setting of one of the detectors would be revealed by a change in the statistical behavior of the particles at the other, and one would have genuine action at a distance. The quantum theoretic distributions do indeed satisfy (9).

(d) *Perfect correlations for parallel axes.* This is the basis for EPR's assignment of "an element of reality" to the result of spin measurements along any directions whatever:

$$q_{aa}(x, y) = \delta_{y, -x}. \qquad (10)$$

(e) *Non-vanishing extremal distributions.* The quantum theoretic distributions (2) satisfy:

$$q_{ab}(x, y) \neq 0 \qquad \text{if } |x| \text{ or } |y| = 1 \quad \text{and} \quad \hat{a} \neq \pm\hat{b}. \qquad (11)$$

This is a straightforward consequence of the well-known form assumed by the matrix elements of a rotation operator when m or m' is $\pm s$. It can be viewed as a manifestation of the uncertainty principle: even in the maximally aligned state there is a non-vanishing probability of finding the particle with any other spin component along any other direction. *This condition is the only one of the properties that is not maintained in the classical limit.* (If a classical spinning particle is known to be perfectly aligned along a given direction, then its spin component along any other direction is entirely determined.)

There is a very simple proof that the quantum theoretic distributions q are incompatible with the existence of third-order distributions p_{abc} satisfying (1) and (3), for any three distinct coplanar directions $\hat{a}$, $\hat{b}$ and $\hat{c}$, and for any value of the spin-s:

Since the directions are coplanar and distinct there are non-zero numbers A, B and C such that

$$A\hat{a} + B\hat{b} + C\hat{c} = 0, \qquad (12)$$

and we can label the axes so that $|A| \geqslant |B| \geqslant |C|$. If a third-order distribution exists then we can define a non-negative quantity f_{abc}:

$$f_{abc} = \sum_{xyz} (Ax + By + Cz)^2 p_{abc}(xyz). \qquad (13)$$

If the squared trinomial in (13) is explicitly multiplied out then each of the resulting nine terms contains at most two of the three variables x, y and z. Consequently in each such term the third-order distribution p can be reduced back to one of the quantum theoretic distributions q because of the marginal relations (3) and the relation (1) between the pair distributions p and the quantum theoretic distributions. Terms in which the two variables are distinct can then be evaluated directly from (8), while the other terms (for example the one in x^2) can be converted to forms given by (8) by using (9) to set the two axes equal, and (10) to replace x^2 by $-xy$.

The nine terms that result can be recombined to give simply

$$f_{abc} = -K_s(A\hat{a} + B\hat{b} + C\hat{c})^2,   \tag{14}$$

a quantity that vanishes by virtue of the defining condition (12) satisfied by A, B and C.

Note that for the classical third-order distribution (6) this vanishing is, in fact, a transparent consequence of the condition (12), and the definition (13) of f_{abc}. In the quantum theoretic case, however, it is impossible for f_{abc} to vanish:

Since the sum (13) is term by term non-negative it is bounded below by any partial sum. In particular,

$$f_{abc} \geqslant \sum_z (|A| + |B| + Cz)^2 p_{abc}(\text{sgn}(A), \text{sgn}(B), z).   \tag{15}$$

If we replace the squared trinomial by its lower bound at $z = -\text{sgn}(C)$, then the sum on z gives a pair distribution which can be expressed in terms of the quantum theoretic distribution q through (1), and we have the lower bound

$$f_{abc} \geqslant (|A| + |B| - |C|)^2 q_{ab}(\text{sgn}(A), -\text{sgn}(B)).   \tag{16}$$

But property (e) of the quantum theoretic distributions guarantees that the right side of (16) must exceed zero. Consequently the third-order distribution required by the EPR reality criterion cannot exist.

That the quantum theory violates the EPR reality criterion in the spin correlation experiment is thus a very general feature of such experiments, not at all peculiar to the spin-$\frac{1}{2}$ case. The question then arises of how the classical distribution (5), which manifestly does satisfy the reality criterion, is to be recovered from the quantum theoretic distributions (2) in the limit of large s.[7] The connection is revealed by the following derivation of a rather unusual representation for the square of the matrix element of the rotation operator appearing in the quantum theoretic distribution (2):[8]

For given s, consider the set of functions of m given by

$$\langle sm| Y_{l0}(S)|sm\rangle \tag{17}$$

for $l = 0, 1, \ldots, 2s$, where $Y_{l0}(S)$ is the tensor operator constructed out of the components of the spin operator S. Since Y_{l0} is a polynomial of degree l in S_z, the quantity defined in (17) is a polynomial in m of degree l.

Consider next the quantity obtained by rotating the tensor operator in (17) through an angle θ about the y-axis:

$$\langle sm| \exp(i\theta S_y) Y_{l0}(S) \exp(-i\theta S_y)|sm\rangle. \tag{18}$$

We can rewrite (18) in two different ways:

(i) Since Y_{l0} is a tensor operator it transforms under a rotation like a spherical harmonic. Since, however, $\langle sm| Y_{lm'}(S)|sm\rangle$ vanishes unless $m' = 0$, the only rotation matrix element $D^l_{m'0}$ that appears in the evaluation of (18) is that with $m' = 0$. Since that matrix element is just a Legendre polynomial, the quantity appearing in (18) is simply

$$P_l(\cos\theta)\langle sm| Y_{l0}(S)|sm\rangle. \tag{19}$$

(ii) We can, on the other hand, expand (18) in intermediate states. Since $\langle sm| Y_{l0}(S)|sm'\rangle$ vanishes unless $m = m'$, that expansion simplifies to

$$\sum_{m'} |\langle sm| \exp(i\theta S_y)|sm'\rangle|^2 \langle sm'| Y_{l0}(S)|sm'\rangle. \tag{20}$$

The identity of (19) and (20) means that the eigenvalues of the real symmetric $(2s + 1)$ dimensional matrix $|\langle sm| \exp(i\theta S_y|sm'\rangle|^2$ are simply the Legendre polynomials $P_l(\cos\theta)$, with the corresponding eigenvectors being just the diagonal matrix elements $\langle sm| Y_{l0}(S)|sm\rangle$. Considered as functions of m these vectors are polynomials of degree l; being non-degenerate eigenvectors of a real symmetric matrix they are also orthogonal, and taken together these two conditions determine them completely.

To make comparison with the classical limit it is convenient to regard these eigenvectors as functions of the discrete variable $x = m/s$ that ranges in $2s + 1$ discrete steps from -1 to 1. Since the functions are orthogonal polynomials in x, their defining properties are simply the discretized versions of the defining properties of the ordinary Legendre polynomials. To emphasize this resemblance we denote them by $P_l^s(x)$, and normalize them in a way that goes over to the usual normalization convention for the Legendre polynomials as $s \to \infty$:

$$[1/(2s + 1)] \sum_x [P_l^s(x)]^2 = 1/(2l + 1). \tag{21}$$

By writing the standard representation of the real symmetric matrix $|\langle sm| \exp(i\theta S_y)|sm'\rangle|^2$ in terms of its eigenvalues and eigenvectors, we thus arrive at the following representation for the quantum theoretic distribution (2):

$$q_{ab}(m, -m') = p_{ab}(m, m') = (2s + 1)^{-1} |\langle sm| \exp(i\theta S_y)|sm'\rangle|^2$$

$$= (2s + 1)^{-2} \sum_l (2l + 1) P_l(a \cdot b) P_l^s(x) P_l^s(y). \tag{22}$$

To make the connection with the classical distribution (5) complete and explicit, I define a function $\delta^s(x, z)$ of a discrete variable x and a continuous variable z by:

$$\delta^s(x, z) = (2s + 1)^{-1} \sum_l (2l + 1) P_l^s(x) P_l(z). \tag{23}$$

Since

$$P_l(\hat{a} \cdot \hat{b}) = (2l + 1)\langle P_l(\hat{a} \cdot \hat{n}) P_l(\hat{b} \cdot \hat{n})\rangle, \tag{24}$$

where the angular brackets denote a uniform average over all directions $\hat{n}$, we can write the quantum theoretic distribution (22) in a form analogous to the classical distribution (5):

$$q_{ab}(x, -y) = p_{ab}(x, y) = \langle \delta^s(x, \hat{n} \cdot \hat{a})\delta^s(y, \hat{n} \cdot \hat{b})\rangle. \tag{25}$$

Since the delta function on the continuous interval $(-1, 1)$ can be represented in the form

$$\delta(x - x') = (1/2) \sum_l (2l + 1) P_l(x) P_l(x'), \tag{26}$$

a comparison of (23) with (26) reveals that as $s \to \infty$, the δ^s-functions are becoming more and more like ordinary δ-functions. On the other hand they cannot, for any value of s, share with ordinary δ-functions the property of being everywhere non-negative, for if they were, one could define, as in (6), a non-negative quantum theoretic third-order distribution,

$$p_{abc}(x, y, z) = \langle \delta^s(x, \hat{n} \cdot \hat{a})\delta^s(y, \hat{n} \cdot \hat{b})\delta^s(z, \hat{n} \cdot \hat{c})\rangle, \tag{27}$$

whose existence is prohibited by the generalized Bell's theorem.

The $\delta^s(x, \hat{n} \cdot \hat{a})$ are thus rather curious objects. If interpreted as conditional probabilities that the measurement of a spin-s particle along a direction $\hat{a}$ will yield the value $m = xs$, given that the particle is actually spinning about the direction $\hat{n}$, then they perfectly reproduce the quantum theoretic spin-s EPR correlations. But there is no way to give meaning to the phrase "spinning about the direction $\hat{n}$", and the interpretation as conditional probabilities suffers from the appearance of negative "probabilities". The functions are thus reminiscent of Wigner's phase space "distribution functions".

The generalized Bell's theorem establishes that no matter how large s, the representation of the quantum theoretic spin-s distribution in the form (25) lacks the essential non-negativity of the analogous representation (5) of the classical distribution, that makes possible a simple interpretation in terms of conditional distributions. One can eliminate this defect by smoothing the quantum theoretic distributions, folding them with functions representing the inability of the detectors to resolve nearby spin components. With increasing smoothing, the δ^s functions should increasingly resemble the correspondingly smoothed δ functions, since differences due both to the discreteness of the P_l^s and the termination of the sum in (23) at $l = 2s$ should wash out.

Of course once one deals with smoothed distributions one can no longer apply the EPR reality criterion in its original form, since two spin measurements along the same direction will no longer yield opposite values with unit probability, and therefore the identification of $p_{ab}(xy)$ with $q_{ab}(x, -y)$ no longer has even a supra-quantum theoretic meaning as a joint distribution for a single particle, when q is the smoothed quantum theoretic distribution.

Under these more general conditions, we adopt as the criterion for local realism the widely studied requirement[9] that there should exist a "conditionally independent" representation of the smoothed distribution q, such as (25) would give if the δ^s-functions were non-negative. In its most general form such a conditionally independent representation would give the smoothed distribution in the form

$$q_{ab}(xy) = \langle f_a(x, \lambda) g_b(y, \lambda) \rangle, \tag{28}$$

where f and g are non-negative distributions in x and y, and the average is over any conceivable set of supplementary parameters ("hidden variables") such as the unit vector $\hat{n}$ in (25), where the distribution of these supplementary parameters is, of course, required to be independent of the settings $\hat{a}$ and $\hat{b}$ of the two detectors.

If a representation such as (28) exists then one can construct non-negative functions that behave like higher-order joint distributions for two or more values of both the x- and the y-variables. For example the fourth-order function

$$q_{aa',bb'}(xx', yy') = \langle\, f_a(x, \lambda)f_{a'}(x', \lambda)g_b(y, \lambda)g_{b'}(y'\lambda)\,\rangle,\qquad (29)$$

when summed over any one of the x- and any one of the y-variables, returns the correct pair distribution in the two unsummed variables.

The inequalities of Clauser and Horne[10] (which are the theoretical underpinning of the recent experiments by Aspect et al.[11]) are necessary conditions for the existence of a non-negative fourth-order function satisfying (29), for a given set of pair distributions, q_{ab}, $q_{a'b}$, $q_{ab'}$, $q_{a'b'}$, when the x- and y-variables assume only two values.

It is interesting to illustrate this procedure in the spin-$\frac{1}{2}$ case. Here the quantum theoretic distribution $q_{ab}(xy)$ is simply

$$q_{ab}(xy) = \tfrac{1}{4}[1 - (\hat{a}\cdot\hat{b})xy],\qquad (30)$$

which can be represented in the form (25) with the δ^s functions given by

$$\delta^s(x, z) = \tfrac{1}{2}[1 + 3^{1/2}xz].\qquad (31)$$

Suppose that the error rate at each detector is r: i.e. that a fraction r of the time a detector signals the wrong value (up instead of down or vice versa) for the spin. Then the distribution q inferred from the detector signals will not be (30), but

$$q_{ab}^r(xy) = \tfrac{1}{4}[1 - (1 - 2r)^2(\hat{a}\cdot\hat{b})xy].\qquad (32)$$

To see this note that the coefficient of xy in (30) is the mean of the product xy. If the error rate is r, then one will record the correct value $\hat{a}\cdot\hat{b}$ for the product xy if none or both of the detectors err, but the negative of the correct value, if precisely one detector errs. Hence the mean of the product xy is reduced by detector error by the factor

$$(1 - r)^2 + r^2 - 2r(1 - r) = (1 - 2r)^2.\qquad (33)$$

The smoothed distribution (32) can be represented in the form (25) if we replace the δ^s-functions (31) by the smoothed functions

$$\delta_r^s(x, z) = \tfrac{1}{2}[1 + 3^{1/2}(1 - 2r)xz].\qquad (34)$$

If $1 - 2r$ is less than $3^{-1/2}$, i.e. if

$$r \geq \tfrac{1}{2}(1 - 3^{-1/2}) = 21.1\%,\qquad (35)$$

then the smoothed δ^s-functions will be everywhere non-negative, and we can construct a conditionally independent representation.[8] No new generalized Bell inequality, applying to data from any correlation experiment, no matter how intricate, will be violated.

Does this error rate have any real significance, or is it an artifact of the particular way we chose to represent the smoothed distribution in terms of smoothed δ^s functions? The answer is that it is an artifact. Anupam Garg[12] has recently pointed out that the smoothed distribution q_{ab}^r can also be represented in the form (25) provided we use for the x-variable the smoothed δ^s-function

$$\delta_r^s(x, z) = \tfrac{1}{2}[1 + 2^{1/2}(1 - 2r)xz],\tag{36}$$

but for the y-variable (which is associated with data collected at the other detector), the alternative form

$$\delta_r^s(y, z) = \tfrac{1}{2}[1 + 2^{1/2}(1 - 2r)xz/|z|].\tag{37}$$

These will both be non-negative if

$$r \geqslant \tfrac{1}{2}(1 - 2^{-1/2}) = 14.6\%.\tag{38}$$

Whether a conditionally independent representation can be found for an even lower error rate is an open question; at the moment 14.6% is the lowest known upper bound to the critical error rate r_c. One can also get a lower bound on r_c from the Clauser–Horne inequalities. These give as a necessary condition for the existence of a conditionally independent representation for the smoothed data collected from four sets of correlation experiments (two a-axes and two b-axes),

$$|\langle xy \rangle_{ab} \pm \langle xy \rangle_{a'b}| + |\langle xy \rangle_{ab'} \mp \langle xy \rangle_{a'b'}| \leqslant 2,\tag{39}$$

where $\langle xy \rangle_{ab}$ is the mean of the product xy in the runs where one detector was along $\hat{a}$ and the other along $\hat{b}$. For the quantum theoretic pair distributions this quantity is just the inner product $-\hat{a} \cdot \hat{b}$, and for suitable choice of axes $\hat{a}$, $\hat{a}'$, $\hat{b}$ and $\hat{b}'$ the left side of (39) can be made as large as $2^{3/2}$.

If the error rate is r, however, the $\langle xy \rangle$ are reduced from their quantum theoretic values by the factor $(1 - 2r)^2$. The Clauser–Horne inequalities (39) will continue to be violated for suitable choices of the four directions unless r is large enough that $2^{3/2}(1 - 2r)^2$ no longer exceeds 2. This happens when

$$r = \tfrac{1}{2}(1 - 2^{-1/4}) = 7.96\%.\tag{40}$$

The critical error rate r_c therefore exceeds 7.96%.

It has been shown[13] that the Clauser–Horne inequalities are not only necessary, but also sufficient for the existence of a non-negative joint "distribution" $p_{aa'bb'}(xx'yy')$. Thus if r exceeds 7.96% no data gathered from a 2×2 experiment (one with two choices for the axis along which each spin is measured) can violate local realism. If r exceeds 14.6% no data gathered from an $N \times N$ experiment can violate local realism whatever the value of N. What happens when r is inbetween?

An intriguing possibility is that the critical error rate for given N increases with increasing N to some limiting value in the range between 7.96% and 14.6%. Garg[12] has shown that it remains at 7.96% in the 3×3 case. I have given examples[14] of distribution functions where it does increase with increasing N, but these examples do not have the quantum theoretic structure. There exist simple computer algorithms that will determine the critical error rate for any value of N,[15] but in practice they may be prohibitively expensive to execute. I offer this as an open question the answer to which might be accessible to a little more ingenuity than I have been able to muster. I find it an appealing idea that the manner in which the strange correlations disappear with decreasing resolution might be by retreating to successively more and more intricate interconnections among the pair correlations, requiring data from more and more pairs of detector orientations to reveal them.

Whether these points are worth pursuing is itself a difficult question. I personally enjoy the pursuit for its own sake, but I am also encouraged by some recent words of a great philosopher,[16] which I have taken the liberty of setting as the poem they are, suggesting that perhaps there still is something to be learned from the enterprise:

> We have always had a great deal of difficulty
> understanding the world view
> that quantum mechanics represents.
>
> At least I do,
> because I'm an old enough man
> that I haven't got to the point
> that this stuff is obvious to me.
>
> Okay, I still get nervous with it. . . .
>
> You know how it always is,
> every new idea,
> it takes a generation or two
> until it becomes obvious
> that there's no real problem.

I cannot define the real problem
therefore I suspect there's no real problem,
but I'm not sure
there's no real problem.

Acknowledgment

This work was supported in part by the National Science Foundation through Grant No. DMR-83-14625.

References

1. J. S. Bell, *Physics* (*N.Y.*) **1**, 195 (1964).
2. A. Einstein, B. Podolsky and N. Rosen, *Phys. Rev.* **47**, 777 (1935).
3. D. Bohm, *Quantum Theory* (Prentice-Hall, Englewood Cliffs, N.J., 1951), pp. 614–619.
4. A. Garg and N. D. Mermin, *Phys. Rev. Lett.* **49**, 901 (1982). Early weaker spin-s generalizations of Bell's theorem (N. D. Mermin, *Phys. Rev.* **D22**, 356 (1980); Mats Ogren, *Phys. Rev.* **D27**, 339 (1983)) did not point toward this result, but suggested that violations of a generalized Bell inequality would become harder to find with increasing s.
5. P. Suppes and M. Zanotti, in *Studies in the Foundations of Quantum Mechanics*, ed. P. Suppes (Philosophy of Science Association, East Lansing, Mich., 1980), pp. 173–179.
6. T. D. Angelidis, *Phys. Rev. Lett.* **51**, 819 (1983).
7. Of course the positive lower bound in (16) becomes exceedingly small as s becomes large. I suspect one can construct spin-s Bell inequalities where the magnitude of the violation does not diminish with increasing s. Whether or not this is possible, however, my concern here is with how the *structure* of the classical distribution emerges from the quantum theoretic form as s becomes large.
8. N. D. Mermin and G. M. Schwarz, *Found. Phys.* **12**, 101 (1982).
9. A good discussion of this representation is given by J. S. Bell, *J. Phys.* (*Paris*) **43**, C2–41 (1981).
10. J. F. Clauser and M. A. Horne, *Phys. Rev.* **D10**, 526 (1974).
11. A. Aspect, J. Dalibard and G. Roger, *Phys. Rev. Lett.* **49**, 1804 (1982).
12. A. Garg, *Phys. Rev.* **D28**, 785 (1983).
13. A. Fine, *Phys. Rev. Lett.* **48**, 291 (1982).
14. N. D. Mermin, *Phil. Sci.* **50**, 359 (1983).
15. A. Garg and N. D. Mermin, *Found. Phys.* **1**, 1 (1984).
16. R. P. Feynman, *Int. J. Theor. Phys.* **21**, 471 (1982).

AN ANALYSIS OF THE LOCALITY ASSUMPTION IN THE BELL ARGUMENTS

Jon P. Jarrett

Department of Philosophy, Harvard University, Cambridge, MA 02138

The strong locality condition employed in Bell's Theorem has been the subject of considerable controversy. This paper is devoted to the analysis of the physical content of this strong locality condition and its precise relationship to the relativistic locality constraint. The discussion focuses on a "decomposition theorem". It can be shown that this strong locality condition is logically equivalent to the conjunction of two weaker conditions: the relativistic locality constraint and another condition which, at least from a certain "classical" perspective, may be regarded as having to do with the "completeness" of the state descriptions allowed by the theory. This is an abbreviated version of a paper to appear in Noûs **18** (*1984*) *under the title "On the Physical Significance of the Locality Conditions in the Bell Arguments".*

In a now-famous *Physical Review* article appearing almost a half-century ago, Einstein, Podolsky and Rosen[1] introduced a Gedankenexperiment for the purpose of exposing what they called the "incompleteness" of quantum mechanics. Recent years have seen modified versions of that very Gedankenexperiment adapted for actual laboratory settings and performed. Ironically, the results of those experiments are now generally taken to constitute strong evidence for the view that even if one were to acknowledge a sense in which quantum mechanics may be said to be "incomplete", genuine features of the world would preclude the "comple*tion*" of quantum mechanics in any way which Einstein would have found acceptable.

In the context of such experiments, Bell's Theorem establishes the incompatibility of quantum mechanics and certain so-called "local realistic" assumptions. Any theory which satisfies these assumptions must make predictions which are constrained by the theorem to satisfy the Bell inequalities. Since the actual experimental results, in good agreement with the predictions of quantum mechanics, do violate the Bell inequalities, there is good evidence that one or more of the "local realistic" premises of the theorem is false.

In fact there are different versions of Bell's Theorem, corresponding to different sets of "local realistic" assumptions. However, each version of the theorem employs as one such assumption some "locality" condition; i.e. a constraint which is to be regarded as a prohibition against "action-at-a-distance" in some sense of that phrase.

There are versions of Bell's Theorem which employ determinism as a premise. In these versions of the theorem, the derivation of a Bell inequality proceeds by adopting as the locality condition a requirement which, subject to minor qualifications, is warranted by relativity theory. This locality condition prohibits superluminal signals. In view of the independent successes of relativity theory, these versions of the theorem are commonly regarded as a serious indictment of determinism.

Somewhat more general versions of Bell's Theorem drop the assumption of determinism and employ a more stringent locality condition. The physical content of this condition and the precise relationship between the relativistic locality condition and this stronger locality condition are topics about which there appears to be some degree of confusion. If we are to understand the full significance of the Bell results, it is essential to know just what kinds of theories these results tell against.

It is hoped that some light may be shed on these matters by calling attention to a certain "decomposition theorem". It can be demonstrated that the stronger locality condition is logically equivalent to the conjunction of two weaker conditions, one of which is the relativistic "no superluminal signals" locality condition and the other of which, at least from a certain "classical" perspective, may be interpreted as having to do with the "completeness" of the state descriptions admitted by the theory.

In the summary which follows, precise statements of these conditions will be given. For a detailed discussion of the physical interpretation of these conditions and for a proof of the decomposition theorem, see Jarrett.[2] The present account will be given in the context of the standard Bohm–Bell–EPR setup consisting of pairs of spin-$\frac{1}{2}$ particles and Stern–Gerlach devices, although the associated photon polarization experiments could just as easily have been used.

In what has become a fairly standard notation (see, for example Clauser and Horne[3]), the locality condition from which the generalized Bell-type inequalities are derived may be expressed as follows:

$$p_{12}(\lambda, a, b) = p_1(\lambda, a) \cdot p_2(\lambda, b),$$

where $p_1(\lambda, a)$ is the probability of a spin up outcome of a measurement of the a-component of the spin of particle 1, $p_2(\lambda, b)$ is the probability of a spin up outcome of a measurement of the b-component of the spin of particle 2, and $p_{12}(\lambda, a, b)$ is the probability of joint spin up outcomes when both measurements are performed, all for two-particle systems in the state λ. (Here and throughout, the obvious universal quantification clauses which belong in the full statement of such conditions are suppressed.) Further specification of the states of the measuring devices is suppressed in this formalism, but whatever additional information is deemed appropriate may also be encoded in the parameters a and b.

By way of the assumption that the p_1 and p_2 functions are well defined, this notation builds in a locality condition of a particular type. For purposes of the present exposition, a notation is needed which permits the formulation of a constraint ("completeness"—to be introduced below) which is independent of this locality condition. This requires a departure from the customary, more wieldy formalism above.

In what follows, the basic probabilities will have this form: $\pi_\lambda(d_L, x_L; d_R, x_R)$. The labels L and R will be used to refer to the left-hand member and right-hand member, respectively, of the pair of measuring devices, of a pair of particles, of a pair of measuring device single-particle subsystems, etc. $\pi_\lambda(d_L, x_L; d_R, x_R)$ is to be interpreted as the probability that the outcomes of a measurement of the d_L-component of the spin of particle L and a measurement of the d_R-component of the spin of particle R, for a pair of particles in the state λ, are x_L and x_R respectively. x_L and x_R may take on the values $+1$ and -1, corresponding to the up and down outcomes respectively. d_L and d_R may be taken to include all relevant information regarding the measuring device states.

It will also be useful to allow d_L and d_R to assume the value 0 for cases in which one of the measuring devices performs no measurement. The associated "outcome" will be expressed by the value 0 as well. For example, $\pi_\lambda(d_L, x_L; 0, 0)$ is the probability for the outcome x_L of a measurement of the d_L-component of the spin of particle L of a pair of particles in state λ, when no measurement is performed on particle R.

The conditions of interest can now be expressed as follows:

LOCALITY

$$\pi_\lambda(d_L, x_L; 0, 0) = \sum_{x_R = \pm 1} \pi_\lambda(d_L, x_L; d_R, x_R)$$

and

$$\pi_\lambda(0, 0; d_R, x_R) = \sum_{x_L = \pm 1} \pi_\lambda(d_L, x_L; d_R, x_R).$$

COMPLETENESS

$$\pi_\lambda(d_L, x_L; d_R, x_R) = \sum_{x'_R = \pm 1} \pi_\lambda(d_L, x_L; d_R, x'_R) \cdot \sum_{x'_L = \pm 1} \pi_\lambda(d_L, x'_L; d_R, x_R).$$

STRONG LOCALITY

$$\pi_\lambda(d_L, x_L; d_R, x_R) = \pi_\lambda(d_L, x_L; 0, 0) \cdot \pi_\lambda(0, 0; d_R, x_R).$$

Locality requires that the probability for each possible outcome of spin measurements be determined "locally" in the sense that it may depend only on the state of the two-particle system and on the state of the measuring device which participates in the measurement. In particular, that probability must be independent of which (if any) component of spin the distant measuring device is prepared to measure.

It is important to note that locality does not exclude the possibility of acquiring information bearing on the outcome of a subsequent measurement on one particle by performing a measurement on the distant particle. Since the state λ of the two-particle system presumably contains information about both particles, a measurement on one particle, from the outcome of which partial knowledge of λ may be inferred, may therefore yield information about the other particle as well. Strictly correlated pairs of measurement outcomes represent the extreme situation of this sort. Such correlations are entirely compatible with locality.

Locality, however, does exclude the possibility that the preparation of either measuring device in any particular state can exert an influence on the distant subsystem so as to affect the probabilities for the possible outcomes of measurements performed at the distant site. It can be shown that this condition expresses the requirement, warranted by relativity theory, that spin measurements in Bell-type setups not be capable of exploitation for the purpose of sending superluminal signals.

Although completeness applies to indeterministic (as well as deterministic) theories, it is a constraint which expresses certain "classical" intuitions. One who wishes to reserve the name "completeness" for some more general condition which is not based on these intuitions may substitute the phrase "classical-completeness" for the term "completeness".

Completeness requires that, given the state of the two-particle system and both measuring device states, the two measurement outcomes are stochastically independent. This constraint may be seen as a plausible requirement for theories which represent observable phenomena as the effects of interacting (but otherwise independently existing) entities whose physical state is exhaustively characterized by the specification of some (not necessarily unique) set of definite, well-defined properties.

For theories which are complete in this sense, the probability for a specified outcome of a specified spin measurement is determined by the state of the two-particle system and the states of the two measuring devices. In particular, that probability is independent of the outcome of any spin measurement at the distant site; i.e. that probability is invariant under conditionalization on the outcome of the distant measurement. For complete theories, however, unlike local throries, the probability for a specified outcome of a specified spin measurement may very well exhibit a dependence on the state of the distant measuring device. The "classical" intuition underlying completeness is just this: By including in the state descriptions of the two-particle system and the measuring devices all of those properties of the systems (i.e. precise numerical values of whatever physical quantities happen to be appropriate for describing whatever entities are posited by the theory) in virtue of which measurement interactions yield each possible outcome with its assigned probability, such state descriptions automatically "screen off" any correlation of the pairs of measurement outcomes which might arise from the omission from the state descriptions of any predictively relevant information.

Of course, knowledge of the outcome x_R, let us say, of an R measurement may (depending on the character of the laws governing the interactions) ground inferences about the properties of particle L. In this way, one perhaps could learn from the outcome x_R that particle L is characterized by such-and-such values of such-and-such quantities with such-and-such probabilities; but the very most that one could infer in this way from the outcome x_R is that particle L definitely possessed certain properties. If all such information about particle L is antecedently included in the state description, then knowledge of the outcome x_R provides only redundant information about particle L. It then follows that the conditional probability of the outcome x_L of the L measurement given the outcome of the R measurement must be equal to the unconditioned probability for x_L. In other words, the two outcomes are stochastically independent.

It should be mentioned that certain complications arise for the interpretation of completeness put forward here when time-dependent states are considered. These complications cannot be addressed here, but there is reason to believe that the resulting problems are not insurmountable (see Jarrett[4]).

Shimony[5] has proposed a rather different (but compatible) interpretation of the completeness condition. He distinguishes two types of non-locality, "controllable" and "uncontrollable", which correspond, respectively, to violations of locality and completeness, as defined here. Whereas, as can be shown, the violation of locality entails the in-principle possibility of a "controllable" action-at-a-distance in the form of superluminal communication, violations of completeness represent a non-locality which cannot be exploited in any such way. Shimony goes on to prove that the non-locality associated with quantum phenomena in general (i.e. not only for Bell-type experiments) is of this "uncontrollable" variety. Shimony's term for this type of non-locality is "passion-at-a-distance".

Don Howard has recently asked whether completeness does not entail the separability of the two-particle state. Separable states are ones which could be decomposed into two parts, each of which is associated with one of the two particles. This remains an open question.

It is a straightforward exercise to demonstrate that strong locality is logically equivalent to the conjunction of locality and completeness. Different presentations employ different variants of strong locality, but they all do essentially the same work. Strong locality, in whatever formulation, is the assumption used in the derivation of generalized Bell-type inequalities.

The interpretation that strong locality inherits from locality and completeness may be exploited to help to understand the significance of the Bell arguments. Those arguments, together with the relevant experimental results, are generally taken to provide excellent evidence that strong locality cannot be satisfied by any empirically adequate theory. Since locality is contravened only on pain of a serious conflict with relativity theory (which is extraordinarily well confirmed independently), it is appropriate to assign the blame to the completeness condition.

In terms of the interpretation of completeness presented here, one must conclude that certain phenomena simply cannot be adequately represented by any theory which ascribes properties to the entities it posits in such a way that no measurement on the system may yield information which is both non-redundant (not deducible from the state descriptions) and predictively relevant for distant measurements. That "information" is just not contained in the state description, however "complete" that description may be.

Despite the general acceptance long ago of the "orthodox" interpretation of quantum mechanics, it is probably fair to say that only since the Bell-type experiments of recent years has there been solid evidence that the world-view which underlies completeness must be relinquished. The predictions of quantum mechanics, in good agreement with the experimental results, do satisfy locality, but violate completeness (and so, too, violate strong locality). Investigation of the quantum-mechanical violation of completeness leads immediately to deep, unresolved mysteries at the foundations of quantum mechanics (including the infamous "measurement problem").

While this analysis provides no resolution of the quantum mysteries, it is hoped that it may help to sharpen the focus on our view of those mysteries. By separating out the relativistic component of the strong locality condition (thereby exposing strong locality as the relativistic locality condition for complete theories), there emerges a clarification of that class of theories excluded by the Bell arguments: the class of theories which satisfy completeness. Although the term "incompleteness" may suggest a defect (as if incomplete theories ought to be "completed"), incomplete theories (e.g. quantum mechanics) are by no means *ipso facto* defective. On the contrary, when the results of Bell-type experiments are taken into account, the truly remarkable implication of Bell's Theorem is that the failure of this kind of "classical" completeness reflects a genuine feature of the world itself.

References

1. A. Einstein, B. Podolsky and N. Rosen, *Phys. Rev.* **47**, 777 (1935).
2. J. P. Jarrett, *Noûs* **18** (1984).

3. J. F. Clauser and M. A. Horne, *Phys. Rev.* **D10**, 526 (1974).
4. J. P. Jarrett, "Bell's Theorem, Quantum Mechanics, and Local Realism", Ph.D. thesis, The University of Chicago (1983), unpublished.
5. A. Shimony, in *Proceedings of the International Symposium on the Foundations of Quantum Mechanics*, ed. S. Kamefuchi *et al.* (Phys. Soc. Japan, Tokyo, 1983).

THE SIGNIFICANCE OF JARRETT'S "COMPLETENESS" CONDITION

Abner Shimony

Departments of Philosophy and Physics, Boston University, Boston, MA 02215

A Bayesian analysis shows that Jarrett's "completeness" condition is actually a kind of locality condition. The failure of this condition, however, would not permit superluminal communication.

Jarrett's remarkable Decomposition Theorem[1] exhibits the equivalence of what he calls "strong locality" (which is the condition used in one guise or another to derive Bell-type inequalities) to the conjunction of "locality" and "completeness".

Concerning locality I have nothing to add to what Jarrett says, except to give references[2-4] to several demonstrations that standard quantum mechanics satisfies this condition. These references do not use his terminology and notation, but what they demonstrate is that quantum mechanical correlations cannot be used to send a message faster than light, and Jarrett himself has proved that this condition is equivalent to his locality.

Jarrett's "completeness" condition is

$$\pi_\lambda(d_L, x_L; d_R, x_R) = \sum_{X=\pm 1} \pi_\lambda(d_L, x_L; d_R, X) \cdot \sum_{Y=\pm 1} \pi_\lambda(d_L, Y; d_R, x_R), \quad (1)$$

which we can rewrite as

$$\pi_\lambda(x_L, x_R) = \sum_{X=\pm 1} \pi_\lambda(x_L, X) \cdot \sum_{Y=\pm 1} \pi_\lambda(Y, x_R), \quad (2)$$

29

since the parameters d_L and d_R are fixed in Eq. (1) and appear on both sides of the equation. Using Bayesian methods and standard conditional notation

$$\pi_\lambda(x_L, x_R) = \pi_\lambda(x_L) \cdot \pi_\lambda(x_R|x_L) = \sum_X \pi_\lambda(x_L, X) \cdot \frac{\pi_\lambda(x_R) \cdot \pi_\lambda(x_L|x_R)}{\pi_\lambda(x_L)}$$

$$= \sum_X \pi_\lambda(x_L, X) \cdot \sum_Y \pi_\lambda(Y, x_R) \cdot \frac{\pi_\lambda(x_L|x_R)}{\pi_\lambda(x_L)}. \tag{3}$$

Hence Eq. (2), and therefore the completeness condition, is equivalent to

$$\pi_\lambda(x_L|x_R) = \pi_\lambda(x_L). \tag{4}$$

It is clear from Eq. (4) that what Jarrett calls "completeness" is actually a type of locality, for Eq. (4) asserts that the occurrence of the event x_R (i.e. the passage or non-passage of the right-going particle through its analyzer) does not affect the probability of the occurrence of the event x_L. If λ is interpreted to be the complete state of the right- and left-going particles, then it does not seem appropriate to interpret Eq. (4) epistemically, as saying that the occurrence of x_R does not reveal information which is probabilistically relevant to x_L, but rather causally: the occurrence of x_R does not affect the propensity of x_L. Consequently, I strongly suggest a change of Jarrett's terminology.

Suppose that Eq. (4) fails. Could one use this failure to send a message? The answer is yes. Prepare an ensemble of pairs in the complete state λ, and perform an analysis of the right-going particle of each pair. Now make the following binary decision:

(i) Intercept none of the left-going particles, regardless of the value of x_R;

(ii) Intercept all the left-going particles such that $x_R = +1$, allowing only the left-going particles such that their partners exhibit $x_R = -1$ to travel towards their analyzer.

If decision (i) is taken, then the statistics exhibited at the L-analyzer are governed by $\pi_\lambda(x_L)$. If decision (ii) is taken, the statistics are governed by $\pi_\lambda(x_L|-1)$. The failure of Eq. (4) implies that the statistics resulting from the two decisions would be different, and therefore a message can be sent.

Is the message communicable in principle faster than light? Clearly no, because it is sent no faster than the beam of left-going particles themselves, which can go no faster than light. Consequently, even though quantum mechanics violates Eq. (4) (and Eqs. (1) and (2)) under certain circumstances, the violation does not permit superluminal communication.

References

1. J. P. Jarrett, "An Analysis of the Locality Assumption in the Bell Arguments", a paper included in this Proceedings.
2. G. C. Ghirardi, A. Rimini and T. Weber, *Lett. Nuovo Cim.* **27**, 263 (1980).
3. P. Eberhard, *Nuovo Cim.* **46B**, 392 (1978).
4. D. Page, *Phys. Lett.* **91A**, 57 (1982).

THEORETICAL EXPERIMENTS ON THE FOUNDATIONS OF QUANTUM THEORY

A. O. Barut

Department of Physics, University of Colorado, Boulder, CO 80309

I present here three theoretical experiments on the foundations of quantum theory. By theoretical experiments I mean simple models which can be analyzed precisely or calculated exactly. The first is an explicit calculation of correlations in a classical model of hidden variables which gives the same correlation function as quantum theory, and I discuss its implications to Bell's inequalities and conclusions drawn from it. The second is on the questions raised by EPR-experiments and their resolution by an explicit solution in which the two emitted particles do not represent two but only one quantum state. The third theoretical calculation is a dynamical nonlinear model for the motion of a classical spin which shows how spin in principle could be quantized, and how the reduction of the wave packet could be understood. Simple and exact models can be more valuable than many long abstract discussions.

I. ON SPIN CORRELATIONS

As is well known the quantum mechanical spin correlations in an EPR-situation, *together* with Bell inequalities, are used to draw far-reaching conclusions of philosophical nature, such as the nonexistence of objective reality. There is no doubt that quantum mechanics is correct and nobody expects the experiments to give different results than the one predicted by the rules of quantum theory. But Bell inequalities as applied to this situation are supposed to eliminate any "local-realistic" hidden variable theory. They are based on two assumptions, and the purpose of

33

the first theoretical experiment is to test these assumptions, not quantum mechanics. The explicit model we calculate elucidates in a simple and intuitive way the concepts entering the Bell's inequalities and tells us what it is really that Bell inequalities do or do not eliminate.

We shall compare the quantum mechanical correlation function $E_Q(a, b)$ of two spin projections coming out from a singlet state ψ

$$E_Q(a, b) = \langle \psi | \boldsymbol{\sigma}_1 \cdot \mathbf{a} \boldsymbol{\sigma}_2 \cdot \mathbf{b} | \psi \rangle / \langle \psi | \psi \rangle, \qquad (1)$$

evaluated completely quantum mechanically, with the correlation function of two classical spins

$$E_C(a, b) = \langle | \mathbf{S}_1 \cdot \mathbf{a} \mathbf{S}_2 \cdot \mathbf{b} | \rangle / \langle | \rangle, \qquad (2)$$

calculated completely classically. Both (1) and (2) can be written as

$$E(a, b) = \langle AB \rangle / [\langle A^2 \rangle \langle B^2 \rangle]^{1/2}$$

where $A = \frac{1}{2}\boldsymbol{\sigma}_1 \cdot \hat{\mathbf{a}}$ and $B = \frac{1}{2}\boldsymbol{\sigma}_2 \cdot \hat{\mathbf{b}}$ for the quantum case and $A = \mathbf{S}_1 \cdot \mathbf{a}$ and $B = \mathbf{S}_2 \cdot \mathbf{b}$ for the classical case.

Eq. (2) refers to the following experiment.[1] Two spin vectors are emitted in opposite directions $\mathbf{S}_1 = -\mathbf{S}_2$. Let the angles of $\mathbf{S}_1$ relative to the origin be (θ, ϕ). Consider two detector directions $\mathbf{a}$ with angles $(0, 0)$ and $\mathbf{b}$ angles $(\theta_0, 0)$. We measure $A = \mathbf{S}_1 \cdot \mathbf{a}$ and $B = \mathbf{S}_2 \cdot \mathbf{b} = -\mathbf{S}_1 \cdot \mathbf{b}$. We repeat this experiment many times assuming that $\mathbf{S}_1$ (hence $\mathbf{S}_2$) comes out uniformly in all possible directions.

Clearly the expectation values of A and B are zero. *Since classical spin can have any magnitude*, the only way to normalize the random variables A and B is to divide them by $[\langle A^2 \rangle]^{1/2}$ and $[\langle B^2 \rangle]^{1/2}$. Thus, Eq. (2), more explicitly, is

$$E_C(a, b) = \int d\lambda \, A(a, \lambda) B(b, \lambda) / [\langle A^2 \rangle \langle B^2 \rangle]^{1/2}, \qquad (3)$$

where $d\lambda = (4\pi)^{-1} \sin\theta \, d\theta \, d\phi$, $\langle A^2 \rangle = \int d\lambda (\mathbf{S}_1 \cdot \mathbf{a})^2$ and $\langle B^2 \rangle = \int d\lambda (\mathbf{S}_2 \cdot \mathbf{b})^2$, or,

$$E_C(a, b) = \int d\lambda \, \tilde{A}(a, \lambda) \tilde{B}(b, \lambda) \qquad (3')$$

with $\tilde{A} = A / [\langle A^2 \rangle]^{1/2}$ and $\tilde{B} = B / [\langle B^2 \rangle]^{1/2}$.

Thus, our classical experiment refers to a "local" situation: Eq. (3') is one of the assumptions to derive Bell's inequalities which our model fulfills.

A simple calculation of Eq. (3) gives

$$E_C(a, b) = -\cos(\mathbf{a} \cdot \mathbf{b}) \tag{4}$$

which agrees exactly with the evaluation of Eq. (1). Hence

$$E_C(a, b) = E_Q(a, b). \tag{5}$$

This equality does not mean that classical spin is the same as the quantum spin, but only that the particular spin correlation is the same. It says, however, also that firstly a hidden variable model exists (λ is the set of hidden variables) to give the observed correlation function, and secondly that the classical model, like quantum, also violates Bell's inequalities. So what is going on? Bell's inequalities are mathematically correctly derived. But they make use of a second additional assumption that we did not need to use. This second assumption asserts that the densities $\tilde{A}$, $\tilde{B}$ in Eq. (3′) are bounded, by 1 for photons, or by $\frac{1}{2}$ for spin-$\frac{1}{2}$:

$$|\tilde{A}| \leqslant \tfrac{1}{2}, \qquad |\tilde{B}| \leqslant \tfrac{1}{2}. \tag{6}$$

This second assumption is true for quantum spin, but not true for classical spin. Now Eq. (3′) is supposed to be a classical expression in the hidden variable space λ, hence we cannot impose a quantum condition on a classical situation and then say that the correlation function eliminates an underlying classical way of calculating the expectation values. Bell's inequalities are thus based half on quantum and half on classical arguments. A hidden variable model must be calculated completely classically, and that what this simple model has allowed us to do.

One might argue that quantum measurements are "yes" or "no", discrete propositions, whereas our model involves continuous values of spin components. Of course there is a difference between quantized and continuous spins. But we can group classical events suitable into spin "up", and spin "down" and discretize our model. More precisely, we define the new random variables, again for both quantum and classical case,

$$X_\pm = \tfrac{1}{2}(1 \pm \tilde{A}), \qquad Y_\pm = \tfrac{1}{2}(1 \pm \tilde{B}), \tag{7}$$

and calculate

$$P_{ij}(a, b) = \int d\lambda\, X_i Y_j; \qquad i, j = +, - \tag{8}$$

and obtain

$$P_{ij} = \tfrac{1}{4}[1 \pm \cos(\hat{a} \cdot \hat{b})] \tag{9}$$

where $+$ if $i = j$ and $-$ if $i \neq j$, and

$$\int d\lambda \, X_j = \int d\lambda \, Y_j = \tfrac{1}{2}. \tag{10}$$

The relation between $E(a, b)$ and the P_{ij} is

$$E(a, b) = P_{++}(a, b) + P_{--}(a, b) - P_{+-}(a, b) - P_{-+}(a, b). \tag{11}$$

The Eqs. (8–11) are true both for quantum theory and for our classical model and all agree with experiment. Now Bell's inequalities make the *additional* assumption that $X_i(a, \lambda)$, $Y_j(b, \lambda)$ are probabilities themselves, hence positive and bounded by one, and show that such models do not agree with quantum theory. This assumption is traced back to the assumption (6).

In quantum theory each event contributes to one of the $P_{ij}(a, b)$ only; in classical experiment each event contributes to all $P_{ij}(a, b)$. In order to discretize the classical measurement we call an event "up" or $+$, if $X_+ > X_-$. This happens when $-\pi/2 \leqslant \phi \leqslant +\pi/2$. (Upper half plane of spin direction relative to the vector **a** at the north pole.) Similarly we call the second spin "up" or $+$, if $Y_+ > Y_-$. And this will happen for $\theta - \pi/2 \leqslant \phi \leqslant \pi/2$ (i.e. upper half plane of spin direction with vector **b** at the nord pole). Similarly for spin "down" or $-$. We choose the normalization for counting at angle ϕ to be $N \cos \phi$. Then

$$P_+ = \int_{-\pi/2}^{\pi/2} N \cos \phi \, d\phi = \tfrac{1}{2} \quad \text{with} \quad N = \tfrac{1}{4}, \tag{12}$$

and for the joint probability we find

$$P_{+-} = \int_{\theta-\pi/2}^{\pi/2} N \cos \phi \, d\phi = \tfrac{1}{2}(1 + \cos \theta), \tag{13}$$

similarly for P_{++}, etc.

This discretization of the classical spin is in agreement with the model I shall discuss in Section III.

II. ON EPR-PROBLEM ITSELF

The questions raised by the EPR-Gedankenexperiment are independent of the correlation experiment and Bell inequalities, although they also originate from the correlations contained in a quantum mechanical coherent state, e.g. singlet combination of two spin-$\frac{1}{2}$ states. A coherent state is split into two parts. One performs a measurement on one. This seems to put the whole system into a statistical mixture instead of the original coherent pure state, and moreover, certain noncommuting quantities appear to be simultaneously measurable. I think these questions are related to and solved together with the following one: When a system emits two particles (e.g. pair production) do we still have a single quantum state, or really a state of *two* particles, in other words, does the pair production take place instantaneously, (as in the picture of second quantization), or does it take place gradually or dynamically? To my knowledge, the problem has not been discussed from this point of view, so I present an explicit model.

Consider the nonlinear wave equation[2]

$$u_t - u_x = -2u[1 - u^2]^{1/2} = F(u) \tag{14}$$

or, with $u = \sin^2 \phi$, $\phi_t - \phi_x = -\sin \phi$. This equation has a solution of the form

$$u = \cosh^{-2}[(t - x) + f(t + x)], \tag{15}$$

where f is an arbitrary function of its argument. Taking f to be a quadratic function, we can arrange the solution so that it represents a wave shape growing, then splitting, and then the two parts going away from each *other* as two distinct particles (Fig. 1). Yet the whole development is a single solution, a single wave or state, with two correlated parts, and *not* two particles. Different parts of a single wave function evolving in time are of course necessarily correlated, and the problems of EPR disappear.

We can generalize the model to both spin and position variables. The two component equation, for example,

$$\psi_t + \boldsymbol{\sigma} \cdot \nabla \psi = -2\psi[1 - \psi^2]^{1/2} \tag{16}$$

or

$$\left(\frac{\partial}{\partial t} + \frac{\partial}{\partial x}\right)\psi_1 = F(\psi_1); \qquad \left(\frac{\partial}{\partial t} - \frac{\partial}{\partial x}\right)\psi_2 = F(\psi_2) \tag{16'}$$

has the solution

$$\psi = \begin{pmatrix} \psi_1 \\ \psi_2 \end{pmatrix} = \begin{pmatrix} \cosh^{-2}[(t + x) + f_1(t - x)] \\ \cosh^{-2}[(t - x) + f_2(t + x)] \end{pmatrix} \tag{17}$$

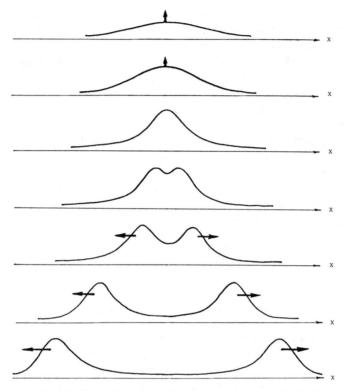

Figure 1   Schematic presentation of a wave growing and splitting.

and one can arrange f_1 and f_2 so that in Fig. 1, if one of the bumps has spin up, $\begin{pmatrix} \psi_1 \\ 0 \end{pmatrix}$, the other will be spin down, $\begin{pmatrix} 0 \\ \psi_2 \end{pmatrix}$ at large separation.

III. QUANTIZATION OF SPIN AND WAVE FUNCTION REDUCTION (OR COLLAPSE)

In this "experiment" we shall calculate the motion of a classical velocity or spin vector obeying a nonlinear equation due to the radiation reaction. The equation of motion of the Lorentz–Dirac electron in proper time is

$$\dot{v}_\mu = (e/m)F_{\mu\nu}v^\nu + \lambda(\ddot{v}_\mu + \dot{v}^2 v_\mu) \qquad (18)$$

where $\lambda = \frac{2}{3}(e^2/m)$ $(c = 1)$ is the characteristic time (or length) of self-energy. We assume that the spin $\mathbf{S}$ also satisfies a similar equation in an external field:

$$\dot{\mathbf{S}} = \mathbf{S} \times \mathbf{B} + \lambda(\ddot{\mathbf{S}} + \dot{S}^2\mathbf{S}). \qquad (19)$$

For a spin vector of constant magnitude

$$\mathbf{S} = S_0(\cos\phi\sin\theta, \sin\phi\sin\theta, \cos\theta), \qquad (20)$$

$\dot{S}^2$ simplifies to

$$\dot{S}^2 = -\dot{\mathbf{S}}\cdot\dot{\mathbf{S}} = S_0^2(\sin^2\theta\,\dot{\phi}^2 + \dot{\theta}^2). \qquad (21)$$

Hence the z-component of Eq. (19) is

$$-\sin\theta\,\dot{\theta} = \lambda[-\cos\theta\,\dot{\theta}^2 - \sin\theta\,\ddot{\theta} + S_0^2(\sin^2\theta\,\dot{\phi}^2 + \dot{\theta}^2)\cos\theta],$$

which for $S_0^2 = 1$ becomes

$$\dot{\theta} = \lambda(\ddot{\theta} - \lambda\sin\theta\cos\theta\,\dot{\phi}^2). \qquad (22)$$

It can be verified that this nonlinear equation has a special solution for which $\dot{\phi}^2 = \omega^2 = $ const. and

$$\tan\theta = \tan\theta_0\, e^{-t/\lambda}. \qquad (23)$$

If the initial spin direction is in the upper half plane, the final spin direction will be at the north pole, and if the initial spin points to the lower half plane, it will tend towards the south pole[3]. The two spin directions, north and south pole, relative to the direction of the magnetic field B, are attractors. Thus the classical spin ends up to be "quantized", and the wave function reduced.

Nonlinear effects were essential in the last two models. That the nonlinearities really can be the cause of quantum effects is still an hypothesis[4]. We need to know more about the nonlinear effects, in particular those in the theory of electrons and photons[4], e.g. Eq. (18).

IV. CONCLUSIONS

All three conclusions reached in this essay may be quite unorthodox: (1) there exists a classical hidden variable model which gives the same spin correlation function as quantum mechanics, hence Bell's inequalities are not sufficient to rule out such models, (2) in the EPR-experiment, the "two" space-like separated particles must be considered a single quantum state, and (3) there is a possible nonlinear dynamical mechanism which models the quantization of spin (and the reduction of wave packet) in the form of appearance of attractors.

References

1. A. O. Barut and P. Meystre, *Phys. Lett.* **105A**, 458 (1984).
2. A. O. Barut, *Phys. Lett.* **67A**, 257 (1978).
3. A. F. Ranada and M. F. Ranada, *J. Phys.* *A12*, 1419 (1979).
4. A. O. Barut, "Relativistic Electron Theory and Foundations of Quantum Theory", in *Proceedings of International Symposium on Foundations of Quantum Mechanics*, ed. S. Kamefuchi *et al.* (Phys. Soc. Japan, Tokyo, 1983), pp. 321–326.

STATEVECTOR REDUCTION AS A DYNAMICAL PROCESS

Philip Pearle

Department of Physics, Hamilton College, Clinton, NY 13323

In a dynamical reduction theory, the statevector evolves with time from a superposition of macroscopically distinguishable states to one of those states. A review is given of ideas leading to dynamical equations which describe this process. These equations depend upon randomly fluctuating matrix elements whose particular time dependence determines the outcome of each experiment. The probabilistic behavior of nature is attributed to lack of experimental control of these matrix elements. In previous work we considered these matrix elements to be functions of white noise. Here that restriction is relaxed to arbitrary functions of short, but not necessarily zero, correlation time.

I. INTRODUCTION

This is a brief review of some work that has been published over an eight year period.[1-5] What I have done is add an extra term to the Schrödinger equation that reduces the statevector.

What this means is the following. When a measurement is completed at $t = 0$, the statevector according to quantum theory has the form

$$|\psi, t\rangle = \sum_n a_n(t)|\varphi_n, t\rangle \qquad t \geqslant 0 \tag{1}$$

where the states $|\varphi_n, t\rangle$ describe the various possible macroscopically distinct outcomes of the experiment (the "pointer positions" as they are sometimes called). The squared amplitudes $|a_n(t)|^2$, which are interpreted as the probabilities of the various outcomes, are constant in time according to the description by the usual Schrödinger equation. However, the extra term I add to the Schrödinger equation rapidly drives all these amplitudes to 0, except one which is driven to 1, so that soon the statevector looks like

$$|\psi, t\rangle = 1 \cdot |\varphi_0, t\rangle \tag{2}$$

for example. A statevector which has the form (2) is said to be reduced.

Moreover, each time the modified Schrödinger equation is solved it can produce a different reduced result. This is not because it has been solved incorrectly, but because the equation depends on randomly fluctuating matrix elements. The particular fluctuations of these matrix elements determine the particular solution. The physical implication is that there is a randomly fluctuating interaction between different "pointer positions" (macroscopically different states in a superposition) that determines which "pointer position" will actually be observed. In this way, the randomness observed in experimental outcomes is explained. Naturally, the extra term must produce the predictions of quantum theory for most experiments. This means, for example, that a fraction $|a_6(0)|^2$ of the total number of solutions of the modified Schrödinger equation must reduce to the form (2).

It turns out that there are many different modified Schrödinger equations that will behave in this way. Here I will only discuss the one I like best. So as not to keep up the suspense, let me present the equation now, and justify it later.

The modified Schrödinger equation is

$$i \frac{d}{dt} |\psi, t\rangle = H|\psi, t\rangle + G|\psi, t\rangle. \tag{3}$$

It contains the usual Hamiltonian operator

$$H = \sum_{n,m} |\varphi_n, t\rangle H_{nm} \langle \varphi_m, t|. \tag{4}$$

The matrix elements $H_{nm} = \langle \varphi_n, t|H|\varphi_m, t\rangle$ do not depend upon the statevector $|\psi, t\rangle$, so the first term of (3) is linear in $|\psi, t\rangle$. We shall take the time dependence of $|\varphi_n, t\rangle$ to be

$$|\varphi_n, t\rangle = e^{-iHt}|\varphi_n, 0\rangle \tag{5}$$

so that the macroscopic states $|\varphi_n, t\rangle$ continue to evolve according to the usual quantum mechanics even during reduction. Then the matrix elements $H_{nm} = \langle \varphi_n, 0|H|\varphi_m, 0\rangle$ are time independent. Because of its bilinear construction, H is also a basis-independent operator.

The "unusual" Hamilton operator G depends upon the statevector $|\psi, t\rangle$, so the second term of (3) is nonlinear in $|\psi, t\rangle$:

$$G \equiv \sum_{n,m} |\varphi_n, t\rangle (e^{2i\theta_n} A_{nm} e^{-2i\theta_m}) \langle \varphi_m, t| \qquad (6a)$$

$$e^{2i\theta_n} \equiv \langle \varphi_n, t|\psi, t\rangle / \langle \varphi_n, t|\psi, t\rangle^*. \qquad (6b)$$

In addition, the matrix elements $A_{nm} = \langle \varphi_n, t|A|\varphi_m, t\rangle$ are time dependent. They are supposed to fluctuate randomly, with a short correlation time. (This may be totally due to the complicated fluctuations in the macroscopic states $|\varphi_n, t\rangle$,[4] or one may need to make the operator A explicitly time dependent. Actually, as will be seen, only the phases and not the magnitudes of the matrix elements A_{nm} need fluctuate.)

The matrix elements A_{nm} represent a new kind of interaction in physics, a nonlocal interaction between different macroscopic states in a superposition. The magnitudes $|A_{nm}|$ are supposed to be negligibly small between states which only differ microscopically, but as the difference between the states $|\varphi_n, t\rangle$, $|\varphi_m, t\rangle$ involves increasing mass or growing spatial separation or growing number of correlated particles, the magnitudes $|A_{nm}|$ are supposed to grow and cause the reduction term to dominate the statevector dynamics. No explicit dependence of A on the dynamical variables will be given here, nor will we discuss the numerical magnitude of A_{nm} which determines the scale of the reduction time (as will be seen).

Furthermore, A is a basis dependent operator. Indeed, the nonlinear part of the Schrödinger equation only has this form in what we call the "preferred" basis, the basis in which the statevectors $|\varphi_n, t\rangle$ describe what we actually observe in Nature. We will not have anything to say here about preferred bases, although this is a most important problem.[3,6-9]

By substituting $|\psi, t\rangle$ from (1) into the Schrödinger equation (3), one obtains a Schrödinger equation for the amplitudes $a_n(t)$:

$$i\, da_n/dt = \sum_n A_{nm} a_m^* a_n / a_n^* \qquad (7)$$

(the Hamiltonian H has disappeared because of Eq. (5), i.e. the $a_n(t)$ are interaction picture amplitudes). We claim that the solutions of Eq. (7) have the following properties:

Property 0. $\sum_n |a_n(t)|^2 = 1$.

Property 1. For each individual solution (corresponding to a particular sample $A_{nm}(t)$) all the $|a_n|^2$'s end up equal to 0 except one which goes to 1.

Property 2. From the complete ensemble of solutions, a fraction $|a_n(0)|^2$ of these end up with $|a_n|^2 = 1$ $(|a_{\neq n}|^2 = 0)$.

Property 0 follows easily from Eq. (7). It is properties 1 and 2 that will require some proof.

We close this introduction by mentioning an interesting similarity between the nonlinear Schrödinger equation (7) and the ordinary linear Schrödinger equation

$$i\, da_n/dt = \sum_n A_{nm} a_m \tag{8}$$

when expressed in "action-angle" variables x_n, θ_n

$$a_n \equiv x_n^{1/2} \exp(i\theta_n). \tag{9}$$

The equations for x_n, θ_n corresponding to Eqs. (7) and (8) are remarkably similar:

$$dx_n/dt = \pm 2 \sum_m (x_n x_m)^{1/2} |A_{nm}| \sin(\theta_n - \theta_m + \alpha_{nm}) \tag{10a}$$

$$d\theta_n/dt = -\sum_m (x_m/x_n)^{1/2} |A_{nm}| \cos(\theta_n - \theta_m + \alpha_{nm}) \tag{10b}$$

where

$$A_{nm} \equiv |A_{nm}| \exp(i\alpha_{nm}). \tag{10c}$$

The positive sign in Eq. (10a) corresponds to Eq. (7); the negative sign corresponds to Eq. (8). It will be seen that this difference in sign is responsible for the strikingly different behavior of the solutions of Eqs. (7) and (8).

II. THE CONSTANT MEAN HYPOTHESIS

It is remarkable that properties 1 and 2 can be ensured by requiring a simple ensemble average behavior of the second and first moments of the x_n's:

$$\langle x_n(t) x_m(t) \rangle \xrightarrow[t \to \infty]{} 0 \qquad \text{for all} \quad n \neq m \Rightarrow \text{property 1} \tag{11a}$$

$$\langle x_n(t) \rangle \xrightarrow[t \to \infty]{} x_n(0) \qquad \Rightarrow \text{property 2.} \tag{11b}$$

If, for any sample solution,[10] a pair x_n and x_m are both nonzero, then the ensemble average $\langle x_n x_m \rangle$ will not vanish, contradicting the above statement of property 1. Therefore each sample solution must have all x_n's except possibly one vanish, and because of property 0 that nonvanishing x_n must equal 1.

Now, because of property 1, each x_n will have either the value 0 or 1 at time infinity, so the average value of x_n is computed simply as 1 multiplied by the fraction of solutions for which x_n achieves the value 1. If this equals $x_n(0)$, the probability of the nth outcome as predicted by quantum theory, then we have achieved property 2. The behavior expressed by (11b) is so important, we have called it the "fundamental property of dynamical reduction theories". For example, the first theory of this type due to Bohm and Bub[11] satisfies this property.

At time 0 all $|a_n(0)|^2 = x_n(0)$ in the ensemble have the same value. Thus we can write $x_n(0) = \langle x_n(0) \rangle$ with no loss of generality, and rewrite (11b) in the symmetrical form:

$$\langle x_n(t) \rangle \xrightarrow[t \to \infty]{} \langle x_n(0) \rangle. \qquad (12)$$

Eq. (12) suggests that property 2 can be achieved by the simple dynamical *constant mean hypothesis*

$$d\langle x_n(t) \rangle /dt = 0. \qquad (13)$$

That is, I am proposing a physical law: reduction takes place dynamically in such a way that Eq. (13) is always satisfied. The dynamical reduction theory of Bohm and Bub does not obey Eq. (13).

Before talking about a more detailed dynamics, one of whose features is that it satisfies Eq. (13), it is worthwhile looking at some consequences of the constant mean hypothesis alone.

An experimental test of dynamical reduction theories, it has always been supposed, is to conduct two consecutive experiments, spaced by a time interval shorter than the reduction time. The idea is that the amplitudes during the reduction following the first experiment are quite different from those of quantum theory. Therefore the probabilities of the outcomes of the second experiment evolving out of those "wrong" amplitudes will be "wrong", where "wrong" here means different from the probabilities predicted by quantum theory.

However, when one analyzes two (or more) consecutive experiments with arbitrary time separation, one finds that the probabilities predicted by quantum theory and the probabilities predicted by a dynamical reduction theory obeying the constant mean hypothesis are identical[5] provided that the states corresponding to different outcomes of the first experiment are not made to interfere in the second experiment. (This is overwhelmingly the usual case, since each experiment generally involves a macroscopic registration of the experiment's outcome, which prevents such a subsequent interference.) At whatever time the constant mean hypothesis is not obeyed, if the second experiment is performed at that time, the predictions will differ from those of quantum theory. (This is the case with the Bohm–Bub theory, where the constant mean hypothesis is not obeyed at any finite time.)

Then, how could one experimentally test a dynamical reduction theory obeying the constant mean hypothesis? If one believes that spontaneous reductions take place, the theory can be tested against quantum theory. For example, consider a neutron incident on a double slit. Its wave packet will split into two and we may hypothesize that these packets are each two preferred basis states which start to undergo reduction dynamics. That is, we may think of the two packets as being like two small "pointers", separated as they are by a small but macroscopic distance and possessing a mass equivalent to that of two thousand electrons. One would not expect the reduction to proceed rapidly in this case (the matrix elements $|A_{mn}|$ would be small), but perhaps one packet might grow slightly at the expense of the other before they hit a detector placed a fair distance away. Under these circumstances, one would expect the two-slit interference pattern to be a bit "washed out". Such an experiment was done by A. Zeilinger et al.[12] The neutrons travelled for about 5 meters for a time interval of about .05 seconds, and the interference pattern agreed with that predicted by quantum theory to slightly better than 1%. Therefore, Zeilinger (private communication) reports a lower limit on the reduction time in this experiment of about 8 seconds.

The point here is that the first "experiment" (separation of the packets at the double slit) which is hypothesized to initiate the reduction does not involve a macroscopic apparatus, so it is possible for the second experiment (detection of the particle) to measure interference between states corresponding to different outcomes of the first experiment. In general, experiments which measure interference of macroscopic systems, such as Leggett's[13] proposed experiment to test coherent flux tunnelling in a SQUID, can most likely be interpreted as testing dynamical reduction theories.

In passing, I may mention that I believe it should be possible to argue that if spontaneous reductions take place in a macroscopic system in thermal equilibrium such as a gas, then the second law will be violated unless the constant mean hypothesis is obeyed. (The entropy will fluctuate up and down on a time scale characterized by the reduction time, and will not just remain constant or increase.) However, I do not as yet have a satisfactory argument.

The point of this discussion is to persuade you that if the statevector does reduce dynamically, then it ought to obey the constant mean hypothesis. I now turn to the description of a "game" that bears a close analogy to the reduction process I will be modelling.

III. THE GAMBLER'S RUIN GAME

Consider two gamblers, G_1 and G_2, who have $100 between them. Say G_1 starts with $40 and G_2 with $60. They toss a fair coin: heads means G_1 win $1 from G_2, tails means the opposite. They play until one gambler loses all his money.

This is analogous to a two-state vector reduction. Let $x_1(t)$ equal the fraction of the total amount of money possessed by G_1 at time t, and likewise define x_2. Then

Property 0. $x_1 + x_2 = 1$.

Property 1. The game ends (either $x_1 \rightarrow 1$, $x_2 \rightarrow 0$ or $x_1 \rightarrow 0$, $x_2 \rightarrow 1$).

Property 2. G_1 wins with probability .4, G_2 wins with probability .6.

(Property 2 is easy to prove.[14,3,4]) Of course, analogous results hold regardless of the sums of money G_1 and G_2 start with.

The constant mean hypothesis is also obeyed. After one toss, half the games have G_1 with $39, half the games have G_1 with $41, etc., so $\langle x_1 \rangle = .4$. (If a gambler loses, all the games in which this occurs must be correctly weighed in the average, where correct weighing means that each string of infinite coin tosses must be counted as a game even though the game is over after a finite number of tosses.)

Thus one can think of the statevectors during reduction as playing the gambler's ruin game with each other until one statevector wins.

The gambler's ruin game can be generalized to N players who play in pairs, and who play continuously in time, i.e. who have a probability $\sim dt$ that they will play in the time interval dt. We shall make this generalization, and further, suppose that the rate of plays slows down in a particular way as a gambler gets close to losing his money, so that the

$n - m$th rate of play is proportional to $x_n x_m$. Then one can write down the Chapman–Kolmogorov equation for the ensemble of games, and pass to the continuous limit in the well-known way,[15] obtaining the diffusion equation describing the ensemble of games:[3]

$$\frac{\partial \rho}{\partial t}(\mathbf{x}, t) = \sum_{n,m=1}^{N} \sigma_{nm}^2 \left(\frac{\partial}{\partial x_n} - \frac{\partial}{\partial x_m} \right)^2 x_n x_m \rho(\mathbf{x}, t). \qquad (14)$$

Here, $\rho(\mathbf{x}, t)\, dx_1 \cdots dx_N$ is the fraction of games at time t with $\mathbf{x}$ values in the volume $dx_1 \cdots dx_N$, and σ_{nm}^2 are constants characterizing the rate of play of the $n - m$th pair of gamblers.

As one expects because of the way it was derived, it is a consequence of Eq. (14) that

Property 0. $\rho(\mathbf{x}, t) \sim \delta(1 - x_1 - \cdots - x_N)$.

Property 1. $d/dt\langle x_n x_m \rangle = -2\sigma_{nm}^2 \langle x_n x_m \rangle$ so

$$\langle x_n x_m \rangle_t = \langle x_n x_m \rangle_0 \, e^{-2\sigma_{nm}^2 t} \xrightarrow[t \to \infty]{} 0.$$

Property 2. $d/dt\langle x_n \rangle = 0$.

Property 0 follows from the $(\partial/\partial x_n - \partial/\partial x_m)$ form of the differential operator. Properties 1 and 2 follow from multiplying both sides by $x_n x_m$ or x_n, integrating over all x, and integrations by parts.[1-3]

To summarize, if one wants to construct a dynamical reduction equation, one may select it so that the ensemble of its solutions obey the diffusion equation (14), since then the solutions will possess properties 0, 1, 2 which are all that one needs. In the last section we will show that indeed the ensemble of solutions of the nonlinear Schrödinger equation (7) is described by the diffusion equation (14), so that Eq. (7) satisfactorily describes the dynamical reduction of the statevector.

IV. STOCHASTIC DIFFERENTIAL EQUATION

Because the matrix elements $A_{nm}(t)$ are presumed to be random functions of time, it is best (and most practicable) to solve Eq. (7) by finding out how the ensemble of solutions behaves, and not to try to see how an individual sample solution behaves. There are a number of different schemes for doing this.

In past work,[2] for simplicity, I have chosen to use the powerful methods of Itô.[16] This requires one to restrict $A_{nm}(t)$ to having zero correlation time, i.e. to being a Hermitian matrix of white noise functions. However, there is no need to be so restrictive. If $A_{nm}(t)$ has a "short" but nonzero correlation time τ_0, that is, if

$$\langle A_{nm}(t)A_{rs}(t+\tau)\rangle - \langle A_{nm}(t)\rangle\langle A_{rs}(t+\tau)\rangle = 0 \qquad \text{for} \quad \tau > \tau_c \qquad (15a)$$

$$0 < |A_{nm}|\tau_c \ll 1 \qquad (15b)$$

one cannot use Itô's method but other methods are available, for example Projection Operator techniques.[17]

Here we will use a result of van Kampen.[18] van Kampen shows that the set of stochastic differential equations

$$\frac{du_n}{dt} = F_n(\mathbf{u}, t) \qquad (16)$$

(the functions F_n depend on a random variable which we do not explicitly write down) has an ensemble of solutions which obeys the diffusion equation

$$\frac{\partial \rho(\mathbf{u}, t)}{\partial t} = \sum_{n,m} \frac{\partial}{\partial u_n} \int_0^\infty d\tau \left\langle F_n(\mathbf{u}, t) \frac{\partial}{\partial u_m} F_m(\mathbf{u}, t - \tau) \right\rangle \rho(\mathbf{u}, t). \qquad (17)$$

In Eq. (17) terms of order $|F_n|\tau_c$ are neglected, i.e. the functions F_n are presumed to have a short correlation time as in Eqs. (15). The average is over the random variable.

We can rewrite Eq. (17) to separately display the "drift" and "diffusion" parts:

$$\frac{\partial \rho(\mathbf{u}, t)}{\partial t} = \sum_{n,m} \frac{\partial^2}{\partial u_n \partial u_m} \rho(\mathbf{u}, t) \int_0^\infty \langle F_m(\mathbf{u}, -\tau)F_n(\mathbf{u}, 0)\rangle \, d\tau$$

$$- \sum_n \frac{\partial}{\partial u_n} \rho(\mathbf{u}, t) \sum_m \int_0^\infty \left\langle F_m(\mathbf{u}, -\tau) \frac{\partial}{\partial u_m} F_n(\mathbf{u}, 0) \right\rangle d\tau. \qquad (18)$$

(We have used in Eq. (18) an assumption that F_n describes a stationary stochastic process so that the expectation values are independent of t.)

We will use Eq. (18) to obtain the diffusion equation describing the ensemble of solutions of both the nonlinear and linear Schrödinger equations (7) and (8). This is most easily done using the differential equations (10a,b) for the action-angle variables x_n, θ_n. We set $u_1, \ldots, u_N$ equal to $x_1, \ldots, x_N$ and $u_{N+1}, \ldots, u_{2N}$ equal to $\theta_1, \ldots, \theta_N$, so the right-hand side of Eq. (10a) is F_n ($n = 1, \ldots, N$) and the right-hand side of Eq. (10b) is F_n ($n = N + 1, \ldots, 2N$).

It is only the phases α_{nm} (Eq. (10c)) of the amplitudes A_{nm} that need to behave stochastically, so we will choose the magnitudes $|A_{nm}|$ as constants. Some statistical properties of these phases need to be assumed. First, α_{nm} with different indices are statistically independent (apart from the symmetry $\alpha_{nm} = -\alpha_{mn}$ which follows from the Hermiticity of A_{nm}). Next, the $\alpha_{nm}(t)$ are uniformly distributed on the interval 0 to 2π, as they would be with an initial random phase assumption and if they performed random walk on the unit circle thereafter. Third, the correlations between $\alpha_{nm}(0)$ and $\alpha_{nm}(\tau)$ are short (in the sense of Eqs. (15)), and the conditional probability that $\alpha_{nm}(0)$ takes on a value given the value of $\alpha_{nm}(-\tau)$ is a symmetric function of the difference $\alpha_{nm}(0) - \alpha_{nm}(-\tau)$ (i.e. no angle or direction of motion on the unit circle is favored). For simplicity we will assume that this conditional probability is independent of the indices n, m (all angles behave similarly).

Then it is straightforward to calculate the expectation values appearing in Eq. (18). Because of trigonometric identities we only need to calculate, for any $\alpha(t) \equiv \alpha_{nm}$,

$$\langle \sin(\alpha(0) - \alpha(-\tau)) \rangle = \langle \sin(\alpha(0) + \alpha(-\tau)) \rangle$$
$$= \langle \cos(\alpha(0) + \alpha(-\tau)) \rangle = 0 \qquad (19a)$$

$$\int_0^\infty d\tau \langle \cos(\alpha(0) - \alpha(-\tau)) \rangle \equiv \sigma^2 \tau_c \qquad (19b)$$

where σ^2 is a number of order of magnitude 1 (the integral in (19b) actually is only over the range $0 \leqslant \tau \leqslant \tau_c$ as the expectation value vanishes for $\tau > \tau_c$).

The nonzero diffusion coefficients are

$$\int_0^\infty \langle F_m(\mathbf{u}, -\tau) F_n(\mathbf{u}, 0) \rangle \, d\tau =$$

$$1 \leqslant n, \quad m \leqslant N, \quad m = n: \qquad 2\sigma^2 \tau_c x_n \sum_k x_k |A_{nk}|^2 \qquad (20a)$$

$$m \neq n: \qquad -2\sigma^2 \tau_c x_n x_m |A_{nm}|^2 \qquad (20b)$$

$$N + 1 \leqslant n, \quad m \leqslant 2N, \quad m = n: \qquad \tfrac{1}{2}\sigma^2 \tau_c x_n^{-1} \sum_k x_k |A_{nk}|^2 \quad (20c)$$

$$m \neq n: \qquad \tfrac{1}{2}\sigma^2 \tau_c |A_{nm}|^2. \qquad (20d)$$

Note that these coefficients are identical for both the nonlinear and linear Schrödinger equations.

The drift coefficients for the phase angles all vanish. It is in the calculation of the drift coefficients for the x_n that the sign difference in Eq. (10a) between the nonlinear and linear Schrödinger equations plays its important role, since

$$\sum_{m=1}^{N} \left\langle F_m \frac{\partial}{\partial x_m} F_n \right\rangle = \mp \sum_{m=N+1}^{2N} \left\langle F_m \frac{\partial}{\partial \theta_{m-N}} F_n \right\rangle \qquad 0 \leqslant n \leqslant N. \quad (21)$$

(The upper and lower signs correspond to the nonlinear and linear Schrödinger equations respectively.) As a result, the drift vanishes for the nonlinear Schrödinger equation, guaranteeing the constant mean hypothesis behavior. The drift for the linear Schrödinger equation is

$$\sum_{m=1}^{2N} \int_0^{\infty} \left\langle F_m(\mathbf{u}, -\tau) \frac{\partial}{\partial u_m} F_n(\mathbf{u}, 0) \right\rangle d\tau = 2\sigma^2 \tau_c \sum_k |A_{nk}|^2 (x_k - x_n). \quad (22)$$

Putting these results into the diffusion equation (18), we obtain for the nonlinear Schrödinger equation

$$\frac{\partial \rho}{\partial t} = \sigma^2 \tau_c \sum_{n,m} |A_{nm}|^2 \left(\frac{\partial}{\partial x_n} - \frac{\partial}{\partial x_m} \right)^2 x_n x_m \rho$$

$$+ \tfrac{1}{2} \sigma^2 \tau_c \sum_{n,m} |A_{nm}|^2 \left(x_n^{-1} x_m \frac{\partial^2}{\partial \theta_n^2} \rho + \frac{\partial^2}{\partial \theta_n \partial \theta_m} \rho \right) \quad (23)$$

and for the linear Schrödinger equation

$$\frac{\partial \rho}{\partial t} = \sigma^2 \tau_c \sum_{n,m} |A_{nm}|^2 \left(\frac{\partial}{\partial x_n} - \frac{\partial}{\partial x_m} \right)^2 x_n x_m \rho - 2\sigma^2 \tau_c \sum_{n,m} |A_{nm}|^2 \frac{\partial}{\partial x_n} (x_m - x_n) \rho$$

$$+ \tfrac{1}{2} \sigma^2 \tau_c \sum_{n,m} |A_{nm}|^2 \left(x_n^{-1} x_m \frac{\partial^2}{\partial \theta_n^2} \rho + \frac{\partial^2}{\partial \theta_n \partial \theta_m} \rho \right) \quad (24)$$

results which we have obtained previously[2] using Itô's methods.

If we integrate over the phase angles in Eq. (23), we obtain the diffusion equation (14) for the gambler's ruin problem, which completes our proof that the nonlinear Schrödinger equation (7) satisfactorily describes the dynamical reduction of the statevector.

If we multiply Eq. (24) by x_n and integrate over all variables, we obtain

$$d\langle x_n \rangle / dt = 2\sigma^2 \tau_c \sum_m |A_{nm}|^2 (\langle x_m \rangle - \langle x_n \rangle).$$

This shows that the drift term causes the expectation values to obey Pauli's Master Equation, based upon Fermi's Golden Rule. The effect of the drift term[1] is to cause Ergodic behavior, to asymptotically smear the probability density uniformly over the whole configuration space $0 \leqslant x_n \leqslant 1, \sum x_n = 1$. This is dramatically different from the reduction behavior which causes the probability density to end up at the "corners" $x_n = 1, x_{\neq n} = 0$. Loosely speaking, *the Schrödinger equation without the drift reduces the statevector.*

References

1. P. Pearle, *Phys. Rev.* **D13**, 857 (1976).
2. P. Pearle, *Int. J. Theor. Phys.* **18**, 489 (1979).
3. P. Pearle, *Found. Phys.* **12**, 249 (1982).
4. P. Pearle, in *The Wave-Particle Dualism*, ed. S. Diner *et al.* (Reidel, Dordrecht, 1984).
5. P. Pearle, *Phys. Rev.* **D29**, 235 (1984).
6. O. Kubler and H. Zeh, *Ann. Phys.* (*N.Y.*) **76**, 405 (1973).
7. D. Deutsch, "Quantum Theory as a Universal Physical Theory", Univ. of Texas Center for Theoretical Physics report, 1980.
8. W. H. Zurek, *Phys. Rev.* **D24**, 1516 (1981).
9. D. Page, "Information Basis of States for Quantum Measurements", in this volume.
10. For simplicity, this argument is phrased as if there are a finite number of sample solutions, when actually there are an uncountable infinite number. The argument should really contain the phrase "a nonzero measure set of sample solutions" instead of the phrase "a sample solution". A set of measure zero of sample solutions may not behave properly, but they have no physical significance.
11. D. Bohm and J. Bub, *Rev. Mod. Phys.* **38**, 453 (1966).
12. A. Zeilinger, R. Gaehler, C. G. Shull and W. Treiner, in *Neutron Scattering—1981* (*Argonne*), Proceedings of the Conference on Neutron Scattering, ed. J. Faber, Jr. (AIP, New York, 1982).
13. A. J. Leggett, *Suppl. Prog. Theor. Phys.* **69**, 80 (1980).
14. W. Feller, *An Introduction to Probability Theory and its Applications* (Wiley, New York, 1950), Chapter 14.
15. S. Chandrasekhar, *Rev. Mod. Phys.* **15**, 1 (1943); also in *Selected Papers on Noise and Stochastic Processes*, ed. N. Wax (Dover, New York, 1954).
16. See, e.g., E. Wong, *Stochastic Processes in Information and Dynamical Systems* (McGraw-Hill, New York, 1971); L. Arnold, *Stochastic Differential Equations: Theory and Applications* (Wiley, New York, 1974); Z. Schuss, *Theory and Applications of Stochastic Differential Equations* (Wiley, New York, 1980).
17. H. Grabert, *Projection Operator Techniques in Nonequilibrium Statistical Mechanics*, Springer Tracts in Modern Physics 95 (Springer-Verlag, New York, 1982); R. Mazo, in *Stochastic Processes in Nonequilibrium Systems*, Lecture Notes in Physics 84, ed. L. Garrido, P. Seglar and P. J. Shepherd (Springer-Verlag, New York, 1978).
18. N. G. van Kampen, *Phys. Rep.* **24**, 171 (1976). See Eq. (19.6) there. I am choosing van Kampen's approach in response to a suggestion by Max Dresden.

INFORMATION BASIS OF STATES FOR QUANTUM MEASUREMENTS

Don N. Page†

Center for Theoretical Physics, The University of Texas at Austin, Austin, TX 78712 and Department of Physics,‡ The Pennsylvania State University, University Park, PA 16802

A preferred basis of states for analyzing the results of a quantum measurement is described. This information basis is the basis in which the data recorded in the memory units of a measuring instrument are the most consistent. Such a basis is useful for determining what observable has been best recorded by a measurement.

If quantum mechanics is a complete theory of physical reality, it can in principle describe the process of measurement: a measuring apparatus is dynamically coupled to a system to be measured so that the state of the apparatus becomes dependent upon the state of the system. The unitary evolution of the combination of apparatus, measured system, and environment (assuming these together form an isolated system, e.g. the universe) in general leads to a mixed state of the apparatus in which all or nearly all observables do not have precise values. Yet common experience suggests that measurements lead to fairly definite results. This apparent conflict between the predictions of the quantum mechanical formalism and ordinary experience is the measurement problem of quantum mechanics.

---

† Alfred P. Sloan Research Fellow.

‡ Permanent address.

The conventional view is that pure quantum mechanics (which gives unitary evolution of isolated systems by dynamical equations of motion such as the Schrödinger equation) must be supplemented by a projection hypothesis.[1] According to this hypothesis, the state of a system changes randomly to one of the eigenstates of an observable being measured. This suggestion leaves the enormous problem of when and how this nonunitary "collapse of the wavefunction" occurs. A number of proposals have been made,[1-4] but it is hard to see how to make any of them precise, as would befit a physical theory. In practice it might be very difficult to determine when this collapse occurs, but in principle one could find out by looking for an absence of otherwise predicted interference effects after the collapse.[5] At present such interference experiments have been limited to fairly simple systems, but they have shown no evidence for any collapse of the wavefunction. Thus one might question whether this postulated process occurs at all.

Indeed, Everett and others have shown[6,7] that our experience is perfectly consistent with pure quantum mechanics in which the universal wavefunction has a unitary evolution and never collapses. The reason that a definite measurement result is observed, whereas the wavefunction is a superposition of eigenstates, is ascribed to the fact that each measurement result occurs in a separate branch or component of the superposition of "many worlds" (eigenstates of the measured observable). All branches in the universal wavefunction have physical reality, but one's individual brain state in each branch is only conscious of the observation results in that branch and hence is not directly aware of the other results occurring in the other branches.

It has been objected[8] that the Everett interpretation has "a heavy load of metaphysical baggage" in its "infinitely many unobservable worlds". This is somewhat analogous to the metaphysical baggage in classical general relativity of ascribing reality to the interiors of all black holes we shall never enter, since they are causally disjoint regions that can never influence us. One might thus prefer to restrict spacetime to be the causal past of our wordline, so that the real universe would only contain points we could in principle eventually see. But from another viewpoint one might find it metaphysically more objectionable to suppose the dynamical evolution of spacetime is cut off at the boundary of what we can eventually see. Analogously, one may find the infinitely many unobservable worlds of the unitarily evolved universal wavefunction metaphysically less objectionable than the ad hoc proposal that the wavefunction continually collapses so that only our branch has reality.

As d'Espagnat[9] summarizes the choice, "Whether we should economize preferably on universes or on principles is likely to remain a matter in which each of us can follow his own preferences". In this paper I explicitly assume pure quantum mechanics, with unitary evolution and no collapse of the universal wavefunction (or density matrix or C^*-algebra state, whichever is the assumed quantum mechanical description of the state of the entire universe).

The universal wavefunction (assuming henceforth for inessential simplicity that the universe as a whole is in a pure state and can be described by a wavefunction) is in general a superposition of many branches. In each branch separately a measurement result can have a definite value, though not in the entire superposition. In principle this implies that there is no measurement problem, but for the analysis of specific measurements there remains the ambiguity of which basis of states is to be used to expand the wavefunction.[5,10,11] That is, what is the measured observable, whose eigenstates are the branches in each of which the measurement has a definite outcome?

It would be particularly crucial to resolve this ambiguity of bases in the conventional view, because in it the wavefunction is supposed to collapse to one of the basis states. In the view adopted here in which the wavefunction does not collapse, I would argue (though Deutsch[5] and Zurek[11] disagree) that there is no *fundamental* necessity for choosing any one particular basis for expanding the wavefunction and analyzing the quantum measurement process. The situation appears to be analogous to the choice of a coordinate basis in either special or general relativity: the theory is invariant under a change of basis.

However, in relativity it is often convenient to choose a coordinate basis that is related to the situation being described, such as having the laboratory or other bulk matter such as galaxies be at rest (i.e. stay at fixed spatial coordinates). Similarly, the analysis of specific measurements in quantum mechanics may be more convenient in certain bases than in others. For example, to describe what a person is aware of after an observation, it would be useful to expand the wavefunction in a basis of components in each of which the person's awareness is of a fairly definite character, rather than in a basis of superpositions of greatly different conscious states. Each of the former basis states could then be interpreted as one of the branches of the "many worlds", with the state of the person in each branch being aware only of the fairly definite observations in that branch.

Thus one tool in the analysis of a measurement would be the choice of a basis of states in which the measurement results can be simply described. Two proposals for such a basis have already been made,[5,11] and in this paper I suggest a third. Each of these bases is necessarily somewhat ad hoc, since the quantum mechanical reality is independent of bases, but each has certain advantages in analyzing different aspects of this reality.

The ambiguity of bases has two parts. The first is the division of the universe into subsystems representing the measuring apparatus, measured system, and environment. Deutsch[5] has proposed a method for doing this based upon both the Hamiltonian and the state of the universe, assuming known finite dimensionalities for the Hilbert spaces of the subsystems. For the case of a division into two subsystems, his proposal is to choose the subsystems so that if joint correlations in their combined density matrix were dropped, the resulting product density matrix would remain a product under an infinitesimal evolution by the joint Hamiltonian. This is an interesting proposal that should be examined in more detail, but it appears likely that this rigidly determined product structure would generally lead to a very complicated and impractical division of the system into subsystems. For example, it is not at all clear that Deutsch's proposed division is consistent with the usual externally imposed requirement that "the subsystem in the white coat is the observer".[5]

The division of the universe into subsystems is a matter of convenience which depends in a complicated and perhaps subjective way upon the analysis being considered. Presumably one would want a product structure in which the mutual interactions of the subsystems are in some sense small compared with their self-interactions, so that the subsystems can be considered to have a certain degree of autonomy. On the other hand, the mutual interactions should not be eliminated or else there would be no measurement at all. Because of these difficulties, I do not wish to specify here any particular choice of the product structure. However, I will assume that a choice is made, so that the Hilbert space H for the universe is written as a product of Hilbert spaces H_I for the measuring instrument, H_S for the measured system, and H_E for the environs: $H = H_I \times H_S \times H_E$. (One could read "instrument" as "observer" or "apparatus", except that I wish to avoid the connotations of consciousness in "observer" and reserve the word "apparatus" for the smaller subsystem considered by Zurek.[11] Similarly, "environs" could be read "environment", except that Zurek uses the latter word to include the non-apparatus part of the instrument.)

The second part of the ambiguity of what observable is measured by a measurement is the choice of a basis for H_I. Strictly speaking, this is a basis for the results recorded in the measuring instrument and not for the state of the measured system. But for each instrument basis state one can take the relative state[6,7] for the rest of the universe, trace over the state of the environment, and thus obtain the corresponding state (mixed, in general) for the measured system. (Only if the measured system and the environment are kinematically independent[5]—described by a product density matrix—will this state of the system be pure.) The degree to which the instrument basis state represents what one would like to know about the measured system is a measure of the success of the measurement process. I shall not discuss in detail this question of measurement success but only the information stored in the measuring instrument.

Deutsch proposes, as what he calls the *interpretation basis*,[5] the basis of eigenstates of the density matrix formed by the outer product of the density matrices of the subsystems. In other words, it is the product basis in which each subsystem has a diagonal density matrix in its corresponding subbasis. This is of course quite a natural choice. It is unique (after the division into subsystems has been made) if the density matrix of each subsystem has entirely distinct eigenvalues. If various eigenvalues coincide (e.g. at zero if several allowed measurement results in H_I are inconsistent with the actual evolution so that the corresponding probabilities are zero), then the basis is defined only up to unitary transformations in the subspaces of eigenstates of common eigenvalues. This can be a problem for ideal instruments with the appropriate initial states and couplings so that many states (e.g. those with inconsistent data stored in separate memory locations) have zero probability. Of course, one may say that such instruments form a set of measure zero, so that the degeneracy will never actually arise, but if the ideal degeneracy is only slightly broken, the interpretation basis will be unstable to small changes in the nearly equal eigenvalues.

Another disadvantage of the interpretation basis for the instrument is related to the possible disadvantage of rigidly sticking to Deutsch's proposed product structure. If it is convenient to divide the instrument itself into certain subsystems (e.g. distinct memory units), the interpretation basis for the instrument as a whole will not in general be a product basis in terms of the bases for its subsystems. Again I will not attempt to answer here the difficult question of how to define a convenient division of the instrument into subsystems, but I will assume that a definite choice is made in any particular analysis.

Zurek has proposed an alternate basis, the *pointer basis*.[11-13] He assumes a definite but unspecified division of the universe into subsystems (unlike Deutsch), thereby ignoring the problem of the choice of product structure, which I am also avoiding. He also implicitly assumes an apparatus subsystem Hilbert space H_A which has the same dimensionality as the measured system Hilbert space H_S, and everything else is considered as environment. In my notation, Zurek's apparatus would be a subsystem of the instrument, and his environment would be a joint system of the rest of the instrument plus the environs. (Thus the environs does not include any of the instrument, whereas the environment does, since the instrument plus environs is the same as the apparatus plus environment, i.e. everything in the universe except the measured system itself.)

Once the system, apparatus, and environment are defined, Zurek's pointer basis for the apparatus is the orthonormal set of apparatus states whose corresponding projection operators commute with the apparatus-environment interaction Hamiltonian. Zurek originally assumed these states are also eigenstates of the apparatus Hamiltonian,[11-13] which is a strong assumption that would not be true for a general product structure. The assumption could be satisfied perfectly by choosing the product structure so that the eigenstates of the complete Hamiltonian formed the product basis states, but the diagonal elements of the density matrix would never change in this pointer basis, so it would not be suitable for describing a measurement. However, one may be able to find a more suitable product structure and pointer basis that does not unduly restrict the system-apparatus coupling but which does keep the apparatus Hamiltonian approximately commuting with the apparatus-environment interaction Hamiltonian. Then after the apparatus effectively decouples from the system, the pure apparatus basis states of the pointer basis remain nearly stationary as the apparatus evolves in interaction with the environment.[14] Pure Hilbert-space superpositions of the pointer basis apparatus states in general evolve into mixed apparatus states, as phase information for them is displaced into the environment. Thus the pointer basis is the basis in which information in the apparatus is best preserved. However, it is completely independent of the apparatus-system interaction and the state of the apparatus after the measurement, so it may not be the most convenient basis for describing the actual information stored in the apparatus.

As an alternative to the interpretation basis and the pointer basis, I shall suggest an *information basis* as giving a more convenient basis in which information is stored. This basis formalizes the strategy of redundancy[12,15,16] that utilizes the amplification which occurs during a measurement, a concept stressed by Bohr.[17] It does not, however, insist that the amplification be irreversible, an idealization not in principle achieved by any finite amplification. The idea is that any realistic measuring instrument actually stores its measurement results in more than one subsystem. When essentially the same information is stored in more than one memory unit, one can define the basis in which it is best stored as the basis in which it is the most consistent.

To illustrate this point and develop the motivation for a precise definition of the information basis to be given below, consider the difference between an unamplified and an amplified measurement, say of a spin-$\frac{1}{2}$ system with basis states $|\uparrow_S\rangle$ and $|\downarrow_S\rangle$ for spin either up or down. For an unamplified measurement, suppose the instrument also consists of spin-$\frac{1}{2}$ with basis states $|\uparrow_I\rangle$ and $|\downarrow_I\rangle$. (Since this instrument has the same number of basis states as the system, it is an apparatus in Zurek's sense.) Assume the instrument is initially in the state $|\uparrow_I\rangle$ but that the system-instrument coupling during the measurement induces a perfect correlation between the instrument state and the system state without changing the z-component of the system spin:

$$|\uparrow_I\rangle \, (a|\uparrow_S\rangle + b|\downarrow_S\rangle) \rightarrow a|\uparrow_I\rangle \, |\uparrow_S\rangle + b|\downarrow_I\rangle \, |\downarrow_S\rangle. \tag{1}$$

After the measurement, the mixed state of the instrument is $aa^*|\uparrow_I\rangle\langle\uparrow_I| + bb^*|\downarrow_I\rangle\langle\downarrow_I|$. This has no single direction for the instrument spin but rather represents a combination of spin up and spin down. However, it can be written in any orthonormal basis, in which case it represents a combination of spins in any two opposite directions (though generally with off-diagonal terms), so it is somewhat unclear which direction of spin is recorded. If $|a| \neq |b|$, one can apply Deutsch's criterion to get the up-down spin basis as the basis in which the density matrix is diagonal. But if $|a| = |b|$, all bases are equivalent, and there is no preferred spin direction that the instrument has recorded. If there is a coupling to the environment that obeys Zurek's assumptions, there will be a pointer basis in which the diagonal elements of the density matrix remain unchanged, but the spin direction for this basis need not be related to the z-direction.

On the other hand, if the instrument consists of two spin-$\frac{1}{2}$ subsystems (each an apparatus) with bases $|\uparrow_1\rangle, |\downarrow_1\rangle$ and $|\uparrow_2\rangle, |\downarrow_2\rangle$ respectively, it can make an amplified measurement of the system. For example, suppose that during the measurement both instrument subsystems become perfectly correlated with the z-component of the system spin without altering that component. If the initial system state was $a|\uparrow_S\rangle + b|\downarrow_S\rangle$, the final state for the instrument plus system becomes

$$a|\uparrow_1\rangle |\uparrow_2\rangle |\uparrow_S\rangle + b|\downarrow_1\rangle |\downarrow_2\rangle |\downarrow_S\rangle, \tag{2}$$

where I have assumed the environs remain kinematically independent. Then the instrument mixed state is

$$aa^*|\uparrow_1\rangle |\uparrow_2\rangle \langle\uparrow_2|\langle\uparrow_1| + bb^*|\downarrow_1\rangle |\downarrow_2\rangle \langle\downarrow_2|\langle\downarrow_1|. \tag{3}$$

In this case if one requires consistency between the spins of the two instrument subsystems, one is restricted to the basis used here as a preferred information basis. In no other spin basis (other than a trivial interchange of up and down) do the spins of the two instrument subsystems agree in each term of the density matrix. Note that Deutsch's criterion to get the interpretation basis is ambiguous here, since the density matrix for the instrument has two zero eigenvalues (zero probabilities for $|\uparrow_1\rangle |\downarrow_2\rangle$ and $|\downarrow_1\rangle |\uparrow_2\rangle$). Zurek's pointer basis, if it exists, will depend upon the apparatus-environment coupling and need not be the basis in which the spin directions are the most consistent.

It is clear how to proceed when the instrument consists of a larger number of subsystems, each of which becomes precisely correlated with the same observable of the measured system. But to make a general definition of the information basis (for a given product structure), one must allow for the possibilities of imperfect measurements and/or perturbations by the environs. In both cases the density matrix of the instrument will not show a perfect correlation between observables of the subsystems in any basis. But one can still define a basis in which the correlation is in some sense optimal. The idea is to get the diagonal elements of the density matrix concentrated as strongly as possible in the terms that represent consistent data in the different instrument subsystems (e.g. memory uhits). In order not to presuppose what data are consistent, consider an arbitrary basis for each subsystem and from it form the product basis for the entire instrument. Take the instrument density matrix ρ_{Iij} in such a basis, drop its off-diagonal terms, and calculate the "entropy" of the resulting diagonal matrix:

$$S_I^D = -\sum_i \rho_{Iii} \ln \rho_{Iii}. \tag{4}$$

This is one relative measure of the inconsistency of the data in this basis, so to optimize the correlation, choose the subsystem bases which minimize this diagonal entropy. The result is the *information basis* of the instrument in a given state.

Of course, the information basis (like the pointer basis) depends crucially upon how the instrument is divided into subsystems. Deutsch's criterion[5] might be used, but it is doubtful that it is always the most convenient. In many cases a more natural division might be possible, such as certain spatially disjoint units of the instrument. The most convenient memory units presumably depend upon how the information is to be read out.[12] Since a consideration of this would further complicate the discussion, for now I prefer to leave this division rather arbitrary, so that it can be altered to suit different analyses. Once it is chosen, the information basis is simply the product basis of apparatus subsystems which minimizes the entropy of the total apparatus density matrix with all off-diagonal terms dropped in that basis.

For a measuring instrument that works perfectly in the sense of recording in each memory unit the appropriate eigenvector if the measured system is in an eigenstate of the intended observable, the measurement basis will be the product basis of these appropriate eigenvectors. Any other instrument designed to check the self-consistency of the measuring instrument need simply compare the states of its subsystems. (This can only be a statistical check, since even an inconsistent instrument usually has some nonzero density matrix elements or probabilities giving consistency, but in principle one may make the confidence level arbitrarily high by checking a sufficiently large ensemble of identical instruments. Also, if the information basis is not known *a priori*, one needs an ensemble of different checking instruments to test all possible bases and hence find the information basis statistically.) To further check whether the self-consistency reflects an accurate measurement of the observed system, the checking instrument must also compare the instrument state with the system state. The diagonal entropy used in the definition of the information basis makes no distinction as to whether the measurement is accurate as well as self-consistent.

If the environment of each instrument subsystem (i.e. the external environs plus the other subsystems) has a negligible effect before the measurement is completed, the information basis will initially be independent of it, unlike the pointer basis. But as the interaction with the environment carries off correlations, the information basis will change to reflect the consistent information actually remaining in the instrument. If the pointer basis exists for each instrument subsystem or apparatus, the information basis will asymptotically tend toward it, since the pointer basis tells what information will be preserved indefinitely (assuming a time-independent Hamiltonian after the measurement which commutes with each apparatus Hamiltonian). Thus another criterion for a good measuring instrument, besides self-consistency and accuracy, would be that the product of its apparatus pointer bases coincide closely with its information basis.

With the present definition of the instrument's information basis as the product basis of its memory subsystems which minimizes the "diagonal entropy" S_I^D defined by Eq. (4), one can ask whether specific measurements are indeed simply described in that basis. For example, if the instrument is taken to be a human observer, one can ask whether the person's conscious awareness is of a fairly definite character in each information basis state. The complexity of the problem precludes a complete analysis here, but I shall suggest that the answer may be yes.

The suggestion is that each item of a person's awareness is stored in several memory subsystems of the brain and forms an element of consciousness only if these memory subsystems agree concerning it. The neural connections of the brain are presumably continually cross-checking and processing the information redundantly stored there. Only if the redundant information is consistent does it "make sense" and become part of the person's conscious awareness. If this suggestion is true, then the information basis for a human observer and particularly for his brain should at least roughly coincide with what might be called the eigenstates of consciousness. Then in each diagonal component of the density matrix for the observer in the information basis, the person's conscious awareness would be of a fairly definite character, rather than a superposition of greatly different conscious states.

One might then use the information basis as a convenience for asking the further question of why the eigenstates of consciousness are aware of macroscopic objects' being in fairly definite relative locations rather than in superpositions or mixtures of greatly different locations, even though the relative locations are almost certainly not nearly so sharp in the full state of the universe. This question puzzled Einstein, who wrote to Born that "one would then be very surprised if a star, or a fly, seen for the first time, appeared even to be quasi-localised".[18] The idea suggested here is that the interaction, via large numbers of photons which are focused by the eyes into images upon spatial arrays of receptors, causes information about the relative spatial quasi-location of the objects to be stored consistently in many memory subsystems, whereas information about superpositions or mixtures of greatly different object locations is not. Thus it may be a consequence of how we are constructed and interact with external objects which causes us to see them in fairly definite relative locations.

Of course, much remains to be done to show in detail how quantum mechanics as a universal physical theory can explain ordinary experience. I am suggesting here that the information basis of states defined above may be a useful tool in this enterprise.

Acknowledgments

My ideas on this subject have been motivated and sharpened by discussions with L. E. Ballentine, D. Deutsch, B. d'Espagnat, A. Shimony, J. A. Wheeler, W. K. Wootters, and W. H. Zurek. This work was supported by NSF Grant Nos. PHY-7826592 and PHY-8117464 and by the Center for Theoretical Physics, whose hospitality is greatly appreciated.

References

1. J. von Neumann, *Mathematische Grundlagen der Quantenmechanik* (Springer, Berlin, 1932); *Mathematical Foundations of Quantum Mechanics*, tr. R. T. Beyer (Princeton University Press, Princeton, NJ, 1955).
2. F. London and E. Bauer, *La Théorie de l'Observation en Mécanique Quantique* (Hermann, Paris, 1939); English translation in *Quantum Theory and Measurement*, ed. J. A. Wheeler and W. H. Zurek (Princeton University Press, Princeton, NJ, 1983), pp. 217–259.
3. G. Ludwig, *Z. Phys.* **135**, 483–511 (1953); and in *Werner Heisenberg und die Physik unserer Zeit* (Friedrich Vieweg und Sohn, Braunschweig, 1961), pp. 150–181.
4. E. P. Wigner, in *The Scientist Speculates*, ed. I. J. Good (Heinemann, London, 1961; Basic Books, New York, 1962), p. 284; reprinted in E. P. Wigner, *Symmetries and Reflections* (Indiana University Press, Bloomington and London, 1967), p. 171.
5. D. Deutsch, *Int. J. Theor. Phys.* **24**, 1–41 (1985).
6. H. Everett, III, *Rev. Mod. Phys.* **29**, 454–462 (1957).
7. B. S. DeWitt and N. Graham (eds.), *The Many-Worlds Interpretation of Quantum Mechanics* (Princeton University Press, Princeton, NJ, 1973).
8. J. A. Wheeler, *Frontiers of Time* (North-Holland, Amsterdam, 1979), p. 397.
9. B. d'Espagnat, *Conceptual Foundations of Quantum Mechanics* (W. A. Benjamin, Reading, Massachusetts, 1976), 2nd ed., p. 272.
10. L. E. Ballentine, *Found. Phys.* **3**, 229–240 (1973).
11. W. H. Zurek, *Phys. Rev.* **D24**, 1516–1525 (1981).
12. W. H. Zurek, in *Quantum Optics, Experimental Gravitation, and Measurement Theory*, ed. P. Meystre and M. O. Scully (Plenum, New York, 1983), pp. 87–116.
13. W. H. Zurek, *Phys. Rev.* **D26**, 1862–1880 (1982).
14. W. H. Zurek, *Prog. Theor. Phys.* (in press).

15. S. Machida and M. Namiki, *Prog. Theor. Phys.* **63**, 1457–1473 and 1833–1847 (1980).
16. H. Araki, *Prog. Theor. Phys.* **64**, 719–730 (1980).
17. N. Bohr, *Atomic Physics and Human Knowledge* (Wiley, New York, 1958), pp. 73, 88.
18. A. Einstein, in M. Born, *The Born–Einstein Letters*, tr. I. Born (Walker, New York, 1971), p. 208.

WHAT IS THE POINT OF THE QUANTUM THEORY OF MEASUREMENT?

L. E. Ballentine

Department of Physics, Simon Fraser University, Burnaby, B.C., Canada V5A 1S6

There are several uses of the quantum theory of measurement: (1) to criticize and clarify the interpretation of quantum mechanics (QM); (2) to examine the relation of QM to other subjects such as irreversibility; (3) to determine practial limits on precise measurements. Concerning (1) it is argued that any interpretation of QM that requires a "reduction" of the state during measurement is untenable and must be abandoned. Concerning (2) it is suggested that the theories of measurement and of irreversibility may illuminate each other through methodological analogies, but that earlier suggestions of a more direct relationship were mistaken.

I. INTRODUCTION

The literature on the quantum theory of measurement (QTM) is both voluminous and diverse, with different authors adopting different assumptions and sometimes apparently having different objectives. Wheeler and Zurek,[1] in the introduction to their recently published reprint collection, emphasize the lack of consensus about the subject. In an attempt to reduce that perceived chaos, I shall offer a preliminary classification of the literature according to its objectives, and shall emphasize some firm conclusions that have been obtained.

Although any paper that develops the possibility of a comparison between quantum mechanics (QM) and experiment could be claimed to be a part of QTM, I shall restrict QTM to the explicit analysis of the measurement process, that is to the interaction between the measurer (apparatus + observer) and the measured. Papers on QTM may be divided into three categories:

(1) Those which analyze the measurement process with the intent of criticizing or clarifying the interpretation of QM (Ex: Heisenberg's uncertainty principle (1927); Schrödinger's cat paradox (1935); numerous papers on state-vector reduction);

(2) Those which use QTM to study the interface between QM and other branches of physics (Ex: those which attempt to use QTM to explain irreversibility, or which invoke irreversibility to explain QTM);

(3) Those concerned with practical limitations on precise measurements (Ex: quantum nondemolition measurements).

Clearly one needs to understand the interpretation of QM before one can apply it reliably to other problems, and I shall have nothing to say about class (3) except that the conclusions of class (1) must be taken into account. This paper will deal primarily with (1), and very briefly with (2).

II. RIVAL INTERPRETATIONS OF QM

QTM provides a forum in which one can evaluate the relative merits of the two principal interpretations of the quantum *state* concept (or rather, the two principal classes of interpretations, since many detailed variations are possible):

(A) A pure state $|\Psi\rangle$ provides a complete and exhaustive description of an *individual* system. A dynamical variable represented by the operator Q has a value (q, say) if and only if

$$Q|\Psi\rangle = q|\Psi\rangle.$$

(B) A pure (or mixed) state describes the statistical properties of an *ensemble* of similarly prepared systems.

Elsewhere[2] I have discussed these interpretations in more detail and have expounded the positive virtues of (B). Here I must confine myself to critical arguments. I shall argue that the results of QTM make (A) untenable.

III. ANALYSIS OF MEASUREMENT

Measurement involves an *object* (I) and an *apparatus* (II). We wish to measure some dynamical variable R belonging to the object. The operator corresponding to R possesses a complete set of eigenvectors,

$$R|r\rangle_I = r|r\rangle_I. \tag{1}$$

A general state vector for I can be expanded in terms of these eigenvectors,

$$|\psi\rangle_I = \sum_r c_r |r\rangle_I, \tag{2}$$

and $|c_r|^2$ is the probability that dynamical variable R takes on the value r when the state is ψ.

The apparatus (II) has a "pointer position" observable A, and a complete set of eigenvectors,

$$A|\alpha, m\rangle_{II} = \alpha|\alpha, m\rangle_{II}. \tag{3}$$

Here α is the "pointer position" eigenvalue, and m labels all the many other quantum numbers needed to specify an eigenvector.

The apparatus is prepared in an initial pre-measurement state, $|0, m\rangle_{II}$, with $\alpha = 0$. One then introduces an interaction between I and II which must produce a unique correlation between the r value of I and the pointer position α_r of II. The properties required of the interaction are specified implicitly by placing constraints on the effect of the time development operator U.

If one requires that the measurement should not change the quantity being measured, then one must have

$$U|r\rangle_I|0, m\rangle_{II} = |r\rangle_I|\alpha_r, m'\rangle_{II}. \tag{4a}$$

Except for quantum nondemolition measurements, this is unnecessarily restrictive. In Ref. 2 I proposed

$$U|r\rangle_I|0, m\rangle_{II} = |\phi_r\rangle_I|\alpha_r, m'\rangle_{II}, \tag{4b}$$

with the final state of I being arbitrary. The purpose of this generalization was to include destructive measurements, which comprise a large fraction of real measurements. However, (4b) is still unnecessarily restrictive, in that it requires the final state to factor. The most general possibility is

$$U|r\rangle_{\text{I}}|0, m\rangle_{\text{II}} = \sum_{r', m'} u^{r;m}_{r', m'}|r'\rangle_{\text{I}}|\alpha_r, m'\rangle_{\text{II}}$$

$$= |\alpha_r; (r, m)\rangle, \qquad \text{say.}^3 \qquad (4c)$$

The final state vector of (4c) is labelled by the "pointer position" eigenvalue α_r. The other labels (r, m) are not eigenvalues, but merely labels indicating where this vector came from via the unitary transformation U. This is the most general possibility, preserving only the one essential requirement for a successful measurement, the establishment of a unique correspondence between the initial value of r and the final pointer position α_r. The values of α_r corresponding to different r values (assumed to be discrete) should be clearly distinguishable by eye. I shall refer to these as *macroscopically distinct* values.

We now consider a general initial state (2) for the object I, which is not an eigenvector of the dynamical variable R that is being measured. Using (4c) and the linearity of the time development operator U, we obtain a final state

$$U|\psi\rangle_{\text{I}}|0, m\rangle_{\text{II}} = \sum_r c_r|\alpha_r, (r, m)\rangle$$

$$= |\Psi^f_m\rangle, \qquad \text{say.} \qquad (5)$$

The final state is a *coherent superposition* of macroscopically distinct "pointer position" eigenvectors.

This fact causes no difficulties for interpretation (B), according to which the initial state describes an ensemble of similarly prepared systems[2] with $|c_r|^2$ being the probability distribution for the r values of object I, and the final state describes a correlated ensemble with $|c_r|^2$ being the probability distribution for the pointer position α_r of the apparatus II. But according to interpretation (A), it follows that since (5) is not an eigenvector of the "pointer position" operator A, the pointer has no well-defined position at the end of the measurement interaction. This is, of course, contrary to observation. So an advocate of (A) must postulate a further process, the so-called "reduction of the state vector",

$$|\Psi^f_m\rangle \rightarrow |\alpha_{r_0}; (r_0, m)\rangle, \qquad (6)$$

where r_0 is the observed value. The classic "problem of measurement" in QM is to explain this process of reduction (more accurately, projection) of the state vector.

IV. EXCUSES FOR "REDUCTION OF THE STATE VECTOR"

There have been many attempts to explain how the process of projection (6) might come about. I shall argue, not only that none of them is satisfactory, but that no such explanation is possible because the projection (6) is incompatible with the linear equation of motion. The explanations, or excuses, include the following.

4.1 The projection is caused by an unpredictable and uncontrollable disturbance of the object by the measuring apparatus.[4]

Any interaction between I and II that might cause such a disturbance is included in the interaction Hamiltonian from which the time development operator U is constructed. If the interaction satisfies the minimal condition (4c) for a successful measurement, then it must necessarily lead to the superposition (5), and not the reduced state (6).

4.2 The observer causes the reduction when he reads the result of the measurement from the apparatus.

One can avoid discussing the strangely psychic character of this proposal by simply including both apparatus and observer in the definition of II, and applying the above analysis unchanged. This method can be used to rule out any alleged external influence as a possible mechanism for state reduction.

4.3 The initial state of the apparatus II is, in practice, not fully specified as a definite pure state, as the notation $|0, m\rangle_{II}$ would imply. Perhaps this imprecision in the state preparation, and the resultant uncertainty in the initial state vector, is sufficient to destroy the coherent superposition in (5).

This kind of imprecision only affects the value of m. The argument is not significantly changed if the initial state vector in (5) is replaced by an arbitrary linear combination of vectors having different m values.

4.4 In view of the imprecise specification of the initial state mentioned in 4.3, one should describe the initial state by means of a mixed state operator, and not by a pure state vector of any kind.

This is the only one of the proposed explanations which requires a non-trivial response. But before doing so in the next section, we should question its relevance. If, in accordance with interpretation (A) (which is the only reason for having considered the projection (6) in the first place), we regard a pure state vector as describing a single system (rather than an ensemble), then the use of a mixed state operator is necessitated only by our ignorance of the state vector. But as long as the state vector *exists*, even if it is unknown, the analysis of Section III applies, along with its conclusion that the final state is a coherent superposition of macroscopically distinct "pointer position" eigenvectors.

Although the "ignorance" interpretation of mixed states is commonly associated with interpretation (A), it may not be a necessary part of it, so we shall go on to an analysis of QTM with mixed states.

V. QTM FOR MIXED STATES

Instead of the initial pure state vector assumed in (5),

$$|\Psi_m^i\rangle = |\psi\rangle_I |0, m\rangle_{II}$$

we now assume an initial mixed state for the system I + II,

$$\rho^i = \sum_m w_m |\Psi_m^i\rangle\langle\Psi_m^i|. \tag{7}$$

Here w_m can be regarded as the probability associated with each of the microscopic states labelled by m (which represents many quantum numbers of the apparatus).

The hope of an advocate of interpretation (A) would now be that the final state would be a mixture of "pointer position" eigenstates, perhaps of the form

$$\rho^d = \sum_r |c_r|^2 \sum_m v_m |\alpha_r; (r, m)\rangle\langle\alpha_r; (r, m)|,$$

but certainly diagonal with respect to α_r. (Any terms that are non-diagonal in α_r would correspond to coherent superpositions of different "pointer position" eigenvectors.) A bit of reflection leads one to realize that the search for a projection of the post-measurement state of the form (6) was doomed to fail because it would have led to a unique measurement result α_{r_0}. But it is universally agreed that QM can make only probabilistic predictions, and ρ^d represents (within the confines of interpretation (A)) the prediction that the result will be α_r with probability $|c_r|^2$. Since the time development is unitary, it is impossible for an initial pure state to lead to a mixed state like ρ^d. But the use of an initial mixed state (7) appears, superficially, to offer renewed hope for interpretation (A).

That hope is very quickly dashed. The actual final state is

$$\rho^f = U\rho^i U^+ = \sum_m w_m |\Psi^f_m\rangle \langle \Psi^f_m| \tag{8}$$

where $|\Psi^f_m\rangle = U|\Psi^i_m\rangle$. Using Eq. (5), we obtain

$$\rho^f = \sum_{r_1}\sum_{r_2} c^*_{r_1} c_{r_2} \sum_m w_m |\alpha_{r_1}; (r_1, m)\rangle \langle \alpha_{r_2}; (r_2, m)|. \tag{9}$$

The terms with $\alpha_{r_1} \neq \alpha_{r_2}$ indicate *coherent superposition* of macroscopically distinct "pointer position" eigenvectors, just as was the case in (5). Since $w_m \geqslant 0$, there is no possibility that these non-diagonal (in α_r) terms can cancel out. If the minimal condition (4c) for a successful measurement is satisfied, then a coherent superposition of r-eigenvectors (2) for the object (I) must inevitably lead to a similar coherent superposition of α_r-eigenvectors for the system of object + apparatus (I + II).[5]

We have therefore proven the *conclusion*:

5.1 "Reduction" of the state of the entire system (I + II) to an incoherent mixture of "pointer position" eigenvectors is incompatible with the linear equation of motion of QM;

and its corollary,

5.2 Any interpretation of the type (A), which requires such a "reduction", is untenable.

VI. RELATION OF QTM TO IRREVERSIBILITY

There is an analogy between the QM descriptions of measurement and irreversibility, which may be schematically indicated as:

Measurement

$$\rho^i \to U\rho^i U^+ \neq \rho^d. \tag{10}$$

An initial state that is non-diagonal with respect to r (the value of the dynamical variable being measured) cannot evolve into a state that is diagonal in the apparatus "pointer position" α_r;

Irreversibility

$$\rho^{\text{non-eq}} \rightarrow U\rho^{\text{non-eq}}U^{+} \neq \rho^{\text{eq}} = e^{-\beta H}/Z. \tag{11}$$

The eigenvalues of the equilibrium state operator, ρ^{eq}, yield the Boltzmann energy distribution. Since the eigenvalues of ρ are unchanged by time development (a unitary transformation), the overwhelming majority of non-equilibrium states cannot evolve into ρ^{eq}.

It is reasonable to hope that insight into one of these problems might illuminate the other. Unfortunately it has sometimes happened that confusion about one problem has obscured the other.

Obfuscation (i): The "reduction" or "projection" of the state upon measurement would, if it occurred, be an irreversible process, since the inverse of a projection operator does not exist. Hence it is suggested that measurement or observation might be the cause of irreversibility.[6] This suggestion leads to the unacceptable conclusion that entropy would not increase (i.e. coffee would not cool, smoke would not diffuse, etc.) unless there was someone to measure it. In the light of the conclusion of Section V, we see that this suggestion and its unacceptable consequences result from the use of an unsound interpretation of QM.

Obfuscation (ii): It is asserted that equilibrium must somehow be reached by macroscopic systems. The apparatus II is macroscopic, so by invoking a transition $\rho^{\text{non-eq}} \rightarrow \rho^{\text{eq}}$ for II the non-diagonal ($\alpha_{r_1} \neq \alpha_{r_2}$) terms in the final measurement state (9) are eliminated.[7] This type of argument is the converse of (i), an imperfect understanding of irreversibility now being used to mask an inadequate theory of measurement.

A useful transfer of insight from one subject to the other may be obtained, not because the same or similar physical processes are involved in measurement and in irreversibility, but rather by means of a methodological analogy. The solution of the classic "problem of measurement" in QM, obtained by the rejection of interpretation (A) in favor of (B), did not involve any change in the mathematical formalism of QM, but rather a change in the *correspondence rules* that relate the mathematical formalism to empirical reality. Specifically, it involved a change in the interpretation of a state vector. Similarly, the conundrum of (11), which *seems* to say (but actually does not) that the attainment of equilibrium is incompatible with QM, will be resolved by a more careful analysis of the relation between the statistical state operator ρ and empirical reality. But that is another story for another day.[8]

Another methodological analogy, this time from irreversibility to QTM, is obtained from the observation that the approach to equilibrium is usually studied, not for an isolated system, but for a system in contact with a heat reservoir. If we allow the system and the reservoir to interact, and then take the trace over reservoir variables to form a partial state operator for the original system, then it may be possible to obtain

$$\text{Tr}_{(\text{res})} U \rho_{\text{sys}}^{\text{non-eq}} \otimes \rho_{\text{res}} U^+ = \rho_{\text{sys}}^{\text{eq}},$$

where U is the time evolution operator for the system and reservoir in interaction. This result has not been proven, but at least it is not incompatible with (11).

In our analysis of measurement we have already included all relevant external influences with the apparatus II, so there is no possibility of adding an analogue of the heat reservoir. But we may calculate partial state operators for the object I and for the apparatus II by taking partial traces of the final state operator (9) of the total system:

$$\rho_{\text{I}}^f = \text{Tr}_{(\text{II})} \rho^f, \qquad \rho_{\text{II}}^f = \text{Tr}_{(\text{I})} \rho^f.$$

These partial state operators may indeed be mixed states, diagonal in r and in α_r, respectively, even if the initial state is a pure state.[9] (One needs some additional properties of the coefficients $u_{r',m'}^{r,m}$ in (4c) to prove this.) But the partial state operators ρ_{I}^f and ρ_{II}^f provide only an incomplete description of the system $\text{I} + \text{II}$, since they do not describe the correlations between I and II. Therefore, this result is of only practical computational significance, and has no bearing on the fundamental interpretation of QM.

VII. CONCLUSIONS

The major conclusions of this paper are those stated at the end of Section V. The interaction between the object of measurement I and the apparatus II leads, in general, to a state of the system $\text{I} + \text{II}$ that is a coherent superposition of macroscopically distinct "pointer position" eigenstates. The so-called "reduction" of the state to an eigenstate or an incoherent mixture of eigenstates is impossible within the (linear) mathematical formalism of QM. Hence any interpretation of the formalism of type (A), Section II, which regards a state vector as the complete description of an *individual* system, and so requires a "reduction", is untenable and must be discarded.

Several years[2,10] ago I advocated a *statistical ensemble* interpretation of QM that is more conservative than (A), in that it retains all the useful predictive power of QM but avoids paradoxes like the classic "problem of measurement". This interpretation also avoids the so-called QM Zeno Paradox, that a continuously observed system cannot change its state,[11] a paradox that can arise only in interpretation (A). Variations and innovations regarding the interpretation of QM are, no doubt, still possible, but whatever form they take, they must not include the fictitious process of state reduction during measurement.[12]

Notes and References

1. J. A. Wheeler and W. H. Zurek (eds.), *Quantum Theory and Measurement* (Princeton University Press, Princeton, N.J., 1983). This book is a reprint collection and large annotated bibliography.
2. L. E. Ballentine, *Rev. Mod. Phys.* **42**, 358–381 (1970).
3. Vectors without a subscript refer to the state space of the entire system I + II, whereas subscripted vectors belong to the state space of only one component, I or II.
4. This "disturbance" theory is no longer widely held. Yet a book as modern as A. Messiah, *Quantum Mechanics* (North-Holland, Amsterdam, 1964) still employs it (see p. 140).
5. This conclusion has also been reached by many other authors, by means of analyses of varying degrees of generality. Probably the most general of these is by A. Fine, *Phys. Rev. D2*, 2783 (1970).
6. This suggestion seems to have its origin in J. Von Neumann, *Mathematical Foundations of Quantum Mechanics* (Princeton University Press, Princeton, N.J., 1955).
7. The theory of A. Daneri, A. Loinger and G. M. Prosperi, *Nuovo Cim.* **44B**, 119 (1966) is of this type. J. Bub, *Nuovo Cim.* **57B**, 503 (1968) has given a detailed critique of their theory.
8. The necessary ideas are presented in the context of classical statistical mechanics by E. T. Jaynes, *Am. J. Phys.* **33**, 391 (1965). A translation of them into the language of quantum statistical mechanics is not difficult. It should be pointed out that Jaynes' occasional use of a subjective interpretation of probability is not essential and may be avoided.
9. H. Margenau, *Ann. Phys.* (*N.Y.*) **23**, 469 (1963).
10. Much of the credit for the ideas in Ref. 2 belongs to A. Einstein. See his "Reply to Criticisms" in *Albert Einstein: Philosopher-Scientist*, ed. P. A. Schilpp (Library of the Living Philosophers, Evanston, Ill.; and Harper-Row, New York, 1949). A

historical study of Einstein's interpretation of QM was given by L. E. Ballentine, *Am. J. Phys.* **40**, 1763 (1972).

11. J. R. Fox, "The Quantum Zeno Paradox Resolved" (preprint, 1983).

12. The first writer to stress the need to eliminate the "reduction" or "projection" postulate from QM was H. Margenau, *Phys. Rev.* **49**, 240 (1936). Some other papers that contain interpretations of QM meeting this requirement are: P. Pearle, *Am. J. Phys.* **35**, 742 (1967); R. G. Newton, *Am. J. Phys.* **48**, 1029 (1980); and J. R. Fox, *Am. J. Phys.* **51**, 49 (1983). The "Many-Worlds" interpretation meets this requirement, but fails on other grounds; see L. E. Ballentine, *Found. Phys.* **3**, 229 (1973).

MANY-HILBERT-SPACE DESCRIPTION OF MEASURING APPARATUS AND REDUCTION OF THE WAVE PACKET

Sigeru Machida

Department of Physics, Kyoto University, Kyoto 606, Japan

Mikio Namiki

Department of Physics, Waseda University, Tokyo 160, Japan

The macroscopic nature of a measuring apparatus is formulated in a continuous direct sum of many Hilbert spaces. On a mathematical basis it is explicitly shown that the reduction of wave packet takes place even in the negative-result-measurement case, keeping the unitarity of the S-matrix of the elementary interaction processes between the object and the apparatus system. Some exactly soluble models of measuring apparati are also presented.

I. INTRODUCTION

The central issue in the theory of measurement is the so-called wave packet reduction problem. The question is whether or not the reduction of the wave packet at the moment of measurement can be described by quantum mechanics itself. Among many papers published,[1] some insist on an affirmative answer to the question, while others argue adversely. Especially, we know the rather recent controversy between the von

Neumann–Wigner theory[2] (a negative view) and the ergodic amplification theory[3] (a positive view). The former school has criticized the latter for the reason that it not only conflicts with the Wigner theorem but also fails to explain the reduction in the negative-result-measurement case.

The Wigner theorem[4] states that the total system can never reach the reduction via the unitary time evolution if it starts from an initial state for which the object system is in a superposed state and the apparatus system in a mixed state. The negative-result-measurement paradox[5] implies that thermal irreversible processes such as discharge phenomena in counters are not necessarily the only cause for the reduction. In this paper, we describe a new version of theory proposed by the present authors[6] and by H. Araki,[7] which is characterized by the macroscopic nature of a measuring apparatus (especially, of its local system) represented by a continuous direct sum of many Hilbert spaces. The theory enables us to overcome the above criticism and paradox against the ergodic amplification theory and then to derive the reduction of wave packet in an explicit form even in the negative-result-measurement case. Our conclusion is that quantum mechanics provides a framework in which the reduction process can be consistently described.

II. REDUCTION OF WAVE PACKET AND ITS DESCRIPTION

Consider a measurement of an observable $\hat{F}$ for a quantum-mechanical object system Q in a state $\psi^Q = \sum_i c_i u_i$, where u_i is the ith eigenstate of $\hat{F}$ and $c_i = (u_i, \psi^Q)$. The reduction of the wave packet (in a measurement of the first kind) is often described only in terms of the Q-states as

$$\hat{\rho}_I^Q = |\psi^Q\rangle\langle\psi^Q| \rightarrow \hat{\rho}_F^Q = \sum_i |c_i|^2 \hat{\xi}(u_i) \tag{1}$$

where $\hat{\xi}(u_i) = |u_i\rangle\langle u_i|$. However, as has been pointed out by Watanabe,[8] this is not satisfactory for describing the reduction of the wave packet. The process (1) leads us to a contradiction for certain cases. For example, consider a case in which $\hat{F}$ has only two eigenstates and $c_1 = c_2 = 2^{-1/2}$. Then we have

$$\hat{\rho}_F^Q = \tfrac{1}{2}[\hat{\xi}(u_1) + \hat{\xi}(u_2)] \tag{2a}$$

or setting $u_\pm = 2^{-1/2}(u_1 \pm u_2)$,

$$\hat\rho_F^Q = \tfrac{1}{2}[\hat{\xi}(u_+) + \hat{\xi}(u_-)] \tag{2b}$$

where $\hat{\xi}(u_\pm) = |u_\pm\rangle\langle u_\pm|$. Eq. (2a) describes the measurement of $\hat{F}$, but (2b) must correspond to the measurement of another operator, say $\hat{G}$, with eigenstates $u_\pm$. In general, $[\hat{F}, \hat{G}] \neq 0$. We are therefore led to a contradiction that (1) describes two mutually incompatible measurements at the same time.

In order to circumvent such a problem, it is necessary to modify (1). What we propose here is to bring the A-states (apparatus states) into the reduction process in the following way:

$$\hat\Xi_I^{\mathrm{tot}} = \hat\rho_I^Q \otimes \hat\sigma_I^A \rightarrow \hat\Xi_F^{\mathrm{tot}} = \sum_i |c_i|^2 \hat\xi^Q(i_i) \otimes \hat\sigma_{F(i)}^A. \tag{3}$$

Here $\hat\Xi^{\mathrm{tot}}$ and $\hat\sigma^A$ are the statistical operator of the total system and that of the apparatus system, respectively, and the subscripts I and F refer to the initial and final states, respectively. The inconsistency seen in (1) does not occur in (3) because of the presence of $\hat\sigma^A$. In fact, (3) gives a complete description for the reduction of wave packet (in a measurement of the first kind). Watanabe's remark is important from an epistemological point of view as well. Not only the Q-states but also the A-states are necessarily involved in expressing the quantum-mechanical measurements. This would lead us to a denial of naive realism in quantum mechanics. Now, we ask the question: Does quantum mechanics accommodate process (3)?

To study this question, we apply quantum mechanics to the total system. Then we erase the phase correlations among u_i's by taking account of interactions between the systems Q and A. As is well known, the von Neumann–Wigner theory never gives us the reduction as a physical process, but merely describes the measurement by

$$\Psi = \psi^Q \otimes \Phi^A = \sum_i c_i u_i \otimes \Phi^A \rightarrow \tilde\Psi = \sum_i c_i u_i \otimes \Phi_i. \tag{4}$$

Here Φ^A stands for the initial state wave function of the apparatus system A. The system A is considered to have an observable, say $\hat{F}$, whose ith eigenstates Φ_i is so designed as to keep a one-to-one correspondence to u_i. Since $\tilde\Psi$ still keeps phase correlations among the u_i, it is obvious that (4) never gives rise to the reduction of the wave packet. Therefore, to understand the reduction, the "abstractes Ich" or the "consciousness" must be invoked. Such an outcome is mainly based on the strong requirement that the superposition principle or equivalently the unitary time evolution should be strictly maintained over the whole process of measurement. In contrast, the ergodic amplification theory gives up the unitarity of the time evolution operator to derive the reduction of wave packet via the dynamical statistical mechanics of thermal irreversible processes. Its proponents[3] have identified the reduction with the thermal irreversible processes, such as discharge phenomena in counters, which are used to amplify a microscopic input up to a macroscopic output. Jauch, Wigner and Yanase[9] criticized the ergodic amplification theory on the basis of the Wigner theorem and the negative-result-measurement.[10] In this paper we try to get around the Wigner theorem and derive explicitly the reduction of wave packet even in the negative-result-measurement case.

III. MACROSCOPIC NATURE OF THE DETECTING APPARATUS

We now wish to explain the basic ideas of our approach to the problem. As has previously been discussed in detail by using the perfect mirror model,[6] the amplification processes to supply energy are not essential to the reduction of wave packet. Therefore, the reduction process should not be identified with a thermal irreversible process such as a discharge in the counter. The two processes are distinct in nature from each other. A "real firing" in the counter will register a result of measurement and so will a "no firing". With this understanding, we analyze the quantum-mechanical measuring process. First of all, we have to point out that the system Q interacts with a *local system* of the apparatus A rather than the whole A. The local system which is to be a "genuine" apparatus is still a macroscopic system with a finite size which is macroscopically very small but microscopically very large. It is then necessary to describe such a local system of A quantum-mechanically.

A macroscopic state variable of the system A observed in a measurement is not an eigenvalue of a quantum-mechanical observable, as in the case of the von Neumann–Wigner theory, but is a kind of average over microscopic variables, analogous to a thermodynamical state variable. The averaging procedure provides a micro-macro scale transformation, through which we can introduce the macroscopic nature of the apparatus system into quantum mechanics. In conventional quantum mechanics, such a notion of the apparatus system has not been given in an explicit form. Taking the finite size effect of the local system of A into account, we apply dynamical scattering theory to the elementary interaction processes between Q and the local system. We never destroy the unitarity of the S-matrix in dealing with the interactions.

The local system of A is still macroscopic and has no definite energy or particle number. Generally speaking, we cannot sharply determine the energy or particle number of a macroscopic system, isolated or not, in a period of time which is long enough for any desirable measurement but much shorter that Poincaré's recurrence period. Thus, we represent the local system of A by the following statistical operator,

$$\hat{\rho}^A = \sum_{N \in I} W_N \, \hat{\rho}_N^A, \tag{5a}$$

$$\hat{\rho}_N^A = \sum_n |\Phi_n^N\rangle \, w_n^N \langle \Phi_n^N|. \tag{5b}$$

In (5a), I stands for an interval ΔN of particle number N about N_0 and W_N for a positive weight factor normalized as $\sum_N W_N = 1$. In (5b), Φ_n^N denotes the nth eigenstate of the Hamiltonian H_N for the N particle system, and w_n^N is the Boltzmann factor for the nth state. The statistical operator $\hat{\rho}^A$ represents a local equilibrium state which depends on a local temperature and other thermodynamical variables. The macroscopic nature of the apparatus also requires us to take the limit $N_0 \to \infty$ and $\Delta N \to \infty$ keeping $\Delta N/N$ very small, so that we can replace the discrete sum in (5a) by

$$\hat{\sigma}^A = \lim_{N_0 \to \infty} \hat{\rho}^A = \int dl \, W(l) \hat{\rho}^A(l) \equiv \omega \cdot \hat{\rho}^A(l). \tag{6}$$

Here, $l = aN$ is a size parameter of the local system in the limit of $N \to \infty$ and $a \to 0$ (a is a characteristic length of the order of the atomic size) and $W(l)$ is a continuous positive function normalized as $\int dl \, W(l) = 1$ and distributed about $L = \lim aN_0$ with width $\Delta L = \lim a \, \Delta N$. $\hat{\rho}^A(l)$ is the statistical operator representing a local system with a sharp size value l, and $\omega \cdot$ is a symbolic expression of the averaging procedure with the weight function $W(l)$.

It is reasonable to assume that the local system interacting with Q has a dimension at least of the same order as that of the spread of the wave packet, i.e. $L \gtrsim (\hbar/\delta p)$, where δp is the momentum uncertainty of Q. Note that

$$p \gg \Delta P = \hbar/\Delta L = \lim(\hbar/a)(\Delta N)^{-1}, \qquad (7)$$

where p is the particle's momentum and hence $\delta p/\Delta P = \Delta L/L \ll 1$. The averaging procedure (6) of the statistical operators is nothing other than the mathematical description of the macroscopic nature of the apparatus which we wish to introduce into the theory of measurement.

It appears that a new postulate is introduced via (6) into quantum mechanics. In solid state physics, however, we often consider a macroscopically small but microscopically large space-time region over which a microscopic variable must be averaged to give a macroscopic quantity at a macroscopic space-time point. The averaging procedure (6) provides an explicit formulation of the implicit notion widely accepted in practical applications of quantum mechanics.

Using the procedure (6), we can now write down the time-dependent statistical operator which describes the entire process of measurement, and also its asymptotic form, as

$$\hat{\Xi}^{\text{tot}}(t) = \omega \cdot \exp(-i\hat{H}t/\hbar)\hat{\rho}^Q \otimes \hat{\rho}^A \exp(i\hat{H}t/\hbar) \qquad (8a)$$

$$\xrightarrow[t \to \infty]{} \omega \cdot \exp(-i\hat{H}_0 t/\hbar)\hat{S}\hat{\rho}^Q \otimes \hat{\rho}^A S^+ \exp(i\hat{H}_0 t/\hbar). \qquad (8b)$$

Here $\hat{\rho}$ corresponds to the initial state before the measurement. In (8b), we have used the definition of the S-matrix,

$$\exp(-i\hat{H}t/\hbar) \xrightarrow[t \to \infty]{} \exp(-i\hat{H}_0 t/\hbar)\hat{S}$$

where $\hat{H}$ and $\hat{H}_0$ are the total and the free Hamiltonian of $Q + A$, respectively.

IV. MATHEMATICAL BACKGROUND AND ITS PHYSICAL IMPLICATION

Before getting into the derivation of the reduction of wave packet, we wish briefly to summarize the mathematical background for our formulation of the macroscopic nature of the measuring apparatus.

Mathematically,[7] von Neumann rings for physical quantities can be represented in a "large" Hilbert space given by the following direct sum of "small" Hilbert spaces:

$$\mathcal{H} = \mathcal{H}_1 \oplus \mathcal{H}_2 \oplus \cdots \oplus \int d\mu(\zeta)\mathcal{H}(\zeta), \tag{9}$$

where $\mu(\zeta)$ is a smooth function of a continuous parameters ζ. We may identify the subscripts $1, 2, 3, \ldots$ with the particle numbers and the continuous parameter ζ with the size parameter l. In this "large" scheme, a state vector, the inner product of two state vectors and an operator are given, respectively, by

$$\psi = \psi_1 \oplus \psi_2 \oplus \cdots \oplus \int d\mu(\zeta)\psi(\zeta) \tag{10a}$$

$$(\varphi.\psi) = \sum_i (\phi_i, \psi_i) + \int d\mu(\zeta)(\varphi(\zeta), \psi(\zeta)) \tag{10b}$$

$$\hat{F}\psi = \hat{F}_1\psi_1 \oplus \hat{F}_2\psi_2 \oplus \cdots \oplus \int d\mu(\zeta)\hat{F}(\zeta)\psi(\zeta) \tag{10c}$$

in terms of the corresponding "small" schemes with subscript i and parameter ζ.

We can also introduce a "large" statistical operator to represent a "large" state by the following direct sum of "small" statistical operators:

$$\hat{\rho} = \hat{\rho}_1 \oplus \hat{\rho}_2 \oplus \cdots \oplus \int d\mu(\zeta)\hat{\rho}(\zeta) \tag{11}$$

satisfying the normalization condition,

$$\text{tr} \, \hat{\rho} = \sum_i \text{tr} \, \hat{\rho}_i + \int d\mu(\zeta)\hat{\rho}(\zeta) = 1. \tag{12}$$

We call the integral part of (9) a "continuous" superselection-rule space (CSRS). The remaining part of (9) is an ordinary ("discrete") superselection-rule space (DSRS). The two spaces are connected to each other through a limiting process as follows:

$$\lim_{N_0 \to \infty} \sum_{N \in I} \oplus \mathcal{H}_N \subset \int d\mu(\zeta)\mathcal{H}(\zeta) \tag{13a}$$

or

$$\lim_{N \to \infty} \mathcal{H}_N \subset \int d\mu(\zeta)\mathcal{H}(\zeta) \tag{13b}$$

in an appropriate topology. From this mathematical consideration, it is evident that our statistical operator (6) belongs to CSRS. Mathematics also tells us that CSRS has a "center" to which classical and macroscopic observables belong. This is a desirable property of this mathematical framework for describing the macroscopic nature of the apparatus system.

Here we make some remarks on the "large" Hilbert space formulation. One could argue that a statistical operator based on (5) has no phase correlation among the states corresponding to different particle numbers and hence that the present theory cannot deal with the case of a superfluid system. However, we first point out that the present theory is not based on (5) but on (6). Since the particle number tends to infinity, $\hat{\sigma}^A$ belongs to CSRS rather than DSRS. Even though $\hat{\rho}^A$ given by (5) is diagonal in the number representation, $\hat{\sigma}^A$ is not necessarily diagonal because the particle number is not definite in CSRS. The continuous space appears only in the limit of an infinite number of degrees of freedom. As will be shown later, the reduction of wave packet takes place only in this limit. This situation is similar to that in the theory of phase transitions. Rigorously speaking, phase transitions occur only in the limit of an infinite number of particles. Nevertheless, we shall see that if we demand $W(l) = \delta(l - l_0)$ or any other condition free from (7) then we can maintain the phase correlations. In this case, the reduction of wave packet will not take place. Our theory is indeed able to describe macroscopic quantum effects such as the case of a superfluid. We also remark that a macroscopic system is not always qualified to be a measuring apparatus. Every measuring apparatus must be "prepared", for instance, to satisfy the condition (7). We do not think that a superfluid system can be adequately prepared as a measuring apparatus.

V. DERIVATION OF THE REDUCTION OF WAVE PACKET

A typical process of measurement consists of two steps, a spectral decomposition of some sort, and a detection process. For example, let us consider a measurement of a particle's momentum by means of an analyzing magnet, as sketched in Fig. 1. In this example, the ith particle of momentum $\mathbf{p}_i$ in the incident beam is first deflected by the magnet in the direction corresponding to its momentum $\mathbf{q}_i$ and is then detected by the detector D_i. The analyzing magnet which changes the momentum

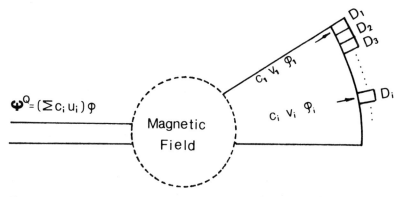

$$\psi^Q = (\textstyle\sum c_i u_i)\,\varphi$$

Figure 1 Momentum analyzer as a preparation step before detection: Object particles with different momenta are spatially separated from one another by a macroscopic distance in the order corresponding to their momenta at the line of detectors.

from $\mathbf{p}_i$ to $\mathbf{q}_i$ gives us the step of spectral decomposition,

$$\psi^Q = \left(\sum_i c_i u_i\right)\varphi \rightarrow \psi^Q_0 = \sum_i c_i v_i \varphi_i. \tag{14}$$

Here u_i and v_i are the eigenstates belonging to momentum $\mathbf{p}_i$ (discretized) and $\mathbf{q}_i$, respectively. φ stands for the envelope function of the wave packet entering the magnetic field and φ_i for that going to the detector D_i. It is obvious that the particle's momentum is determined by the step of detection but not by the step of spectral decomposition. The reduction of the wave packet occurs in the step of detection but not in the step of spectral decomposition† even though the latter step is practically important to improve the experimental accuracy by making a spatial separation of the particle beam.

To describe the detection step explicitly in (8), we have to specify ω, ρ^A and ρ^Q. We put $\omega = \prod_i \omega_i$ and $\hat{\rho}^A_I = \prod_i \otimes \hat{\rho}^{D_i}_I$ corresponding to the presence of many detectors. Furthermore, let

$$\rho^Q_I = |\psi^Q\rangle\langle\psi^Q| = \sum_i |c_i|^2 \hat{\xi}^Q(i) + \sum_{i\neq j} c_i c_j^* |i\rangle\langle j|, \tag{15}$$

where $|i\rangle = |v_i\varphi_i\rangle$ and $\hat{\xi}^Q(i) = |i\rangle\langle i|$. Consequently, we obtain

$$\hat{\Xi}^{\text{tot}}(t) = \sum_i |c_i|^2 \hat{\Xi}^{ii}(t) + \sum_{i\neq j} c_i c_j^* \hat{\Xi}^{ij}(t), \tag{16}$$

† In the literature[11] the step of spectral decomposition is often confused with the measuring process because of the formal similarity between (4) and (14). We point out emphatically that there is no measurement without detection.

in which

$$\hat{\Xi}^{ii}(t) \xrightarrow[t \to \infty]{} \prod_{j \neq i} \omega_j \cdot \left\{ \omega_i \cdot e^{-iH_0 t/\hbar} \hat{S} \left[|i\rangle\langle i| \otimes \hat{\rho}_I^{D_i} \otimes \prod_{j \neq i} \hat{\rho}_I^{D_j} \right] \hat{S}^+ \, e^{iH_0 t/\hbar} \right\} \tag{17a}$$

$$\hat{\Xi}^{ij}(t) \xrightarrow[t \to \infty]{} \prod_{k = i,j} \omega_k \cdot \left\{ \omega_i \cdot \omega_j \cdot e^{-iH_0 t/\hbar} \hat{S} \left[|i\rangle\langle j| \otimes \hat{\rho}_I^{D_i} \otimes \hat{\rho}_I^{D_j} \right. \right.$$

$$\left. \left. \otimes \prod_{k \neq i,j} \hat{\rho}_I^{D_k} \right] \hat{S}^+ \, e^{iH_0 t/\hbar} \right\}. \tag{17b}$$

At this point, we should mention an important role of the S-matrix. It has been repeatedly noted that one collision with a local system (but not with the whole apparatus system) is sufficient to cause the reduction of wave packet within the detection efficiency given by S-matrix. One of the important properties of system A is that its local system has a finite size much larger than an atomic dimension. The theory of nuclear reactions tells us that the S-matrix for collision of a particle with a finite-size system can be decomposed as follows:

$$\hat{S} = e^{i\hat{\delta}} \frac{1 + i\hat{K}}{1 - i\hat{K}} e^{i\hat{\delta}} \tag{18}$$

where $\hat{\delta}$ is diagonal and $\hat{K}$ off-diagonal in the channel representation. The finite-size effect is reflected in diagonal elements of $\hat{\delta}$ equal to $-\bar{p}l/2\hbar$, where $\bar{p}$ is the effective momentum of the particle and l the linear size of the target system. In nuclear reactions, $\bar{p}$ is equal to the incident particle momentum and l is of the order of the size of compound nucleus. Detailed discussions on the structure of the S-matrix will be given later through a few solvable models of detectors.

The S-matrix in the detection process described by (16) and (17) brings interactions only between the two states connected by the dashed lines, thereby producing the phase shift factors: $\exp[-i\bar{p}_i l_i/\hbar]$ for the pair $(|i\rangle, \hat{\rho}_I^{D_i})$, $\exp[i\bar{p}_j l_j/\hbar]$ for the pair $(\langle j|, \hat{\rho}_I^{D_j})$ and so on. The averaging procedures in $\omega_i \omega_j$ give us the integrals

$$\left[\int dl_i W(l_i) e^{-i\bar{p}_i l_i/\hbar} e^{i\bar{p}_i l_i/\hbar} \cdots \right] \left[\int dl_j W(l_j) \cdots \right] \tag{19a}$$

in $\hat{\Xi}^{ii}(t)$, and

$$\left[\int dl_i W(l_i) e^{-i\bar{p}_i l_i/\hbar} \cdots \right] \left[\int dl_j W(l_j) e^{i\bar{p}_j l_j/\hbar} \cdots \right] \tag{19b}$$

in $\hat{\Xi}^{ij}(t)$. Note that the phase shift factors cancel each other out in (19a) but are kept in (19b). The remaining integrands ... are smooth functions of l_i or l_j, so that both integrals in (19b) vanish for (7), i.e. $\bar{p}_i \gg \Delta p_i = h/\Delta L_i$ and $\bar{p}_j \gg \Delta p_j = h/\Delta L_j$, because of the Riemann–Lebesgue lemma, while (19a) never vanishes due to the lack of phase factors. Thus we are led to the asymptotic behavior of the detection process:

$$\hat{\Xi}^{\text{tot}}(t) \xrightarrow[t \to \infty]{} \sum_i |c_i|^2 \hat{\xi}_F^Q(i) \otimes \hat{\sigma}_F^{D_i} \otimes \prod_{j \neq i} \hat{\sigma}_I^{D_j}, \tag{20}$$

where $\hat{\xi}_F^Q(i)$ is the final free statistical operator of system Q with momentum $\mathbf{q}_i$, and $\hat{\sigma}_F^{D_i}$ represents the final statistical operator of detector D_i. Note that $\omega_i \cdot \hat{S}[|i\rangle\langle i| \otimes \hat{\rho}_I^{D_i}]\hat{S}^+ = \hat{\xi}_F^Q(i) \otimes \hat{\sigma}_F^{D_i}$. Eq. (20) has the same form as (3) which we wanted to have as the final goal of the theory of measurement. Note that $\hat{\sigma}_{F(i)}^A = \hat{\sigma}_F^{D_i} \otimes \prod_{j \neq i} \hat{\sigma}_I^{D_j}$.

Application of the above procedure to the negative-result-measurement case is straightforward. Consider an experiment of the Stern–Gerlach type in which we have two states $i = a, b$. In this case (20) becomes

$$\hat{\Xi}^{\text{tot}}(t) \xrightarrow[t \to \infty]{} |c_a|^2 \hat{\xi}_F^Q(a) \otimes \hat{\sigma}_F^{D_a} \otimes \hat{\sigma}_I^{D_b} + |c_b|^2 \hat{\xi}_F^Q(b) \otimes \hat{\sigma}_I^{D_a} \otimes \hat{\sigma}_F^{D_b}. \tag{21}$$

In the negative-result-measurement case, we have to remove detector D_b. We have also to remove ω_b and $\hat{\rho}_I^{D_b}$ in (17a) and (17b), and the l_b-integral in (19a) and (19b) while retaining (16) with $i = a, b$. Even in this case there still remains the l_a-integral of the form,

$$\left[\int dl_a \, W(l_a) \, e^{-i\bar{p}_a l_a/h} \cdots \right] \tag{22}$$

in $\hat{\Xi}^{ab}(t)$, which vanishes for $\bar{p}_a \gg h/\Delta L_a$ because of the Riemann–Lebesgue lemma. For the same reason, $\hat{\Xi}^{ba}(t)$ also vanishes. Thus, we obtain the asymptotic behavior,

$$\hat{\Xi}^{\text{tot}}(t) \xrightarrow[t \to \infty]{} |c_a|^2 \hat{\xi}_F^Q(a) \otimes \hat{\sigma}_F^{D_a} + |c_b|^2 \hat{\xi}_I^Q(b) \otimes \hat{\sigma}_I^{D_a} \tag{23}$$

which certainly expresses the reduction of wave packet in the negative-result-measurement case.

If we remove the averaging procedure ω or put $W(l) = \delta(l - l_0)$, then we can no longer obtain the reduction of wave packet because the Riemann–Lebesgue lemma does not work. This means that we go back to the case of the Wigner theorem formulated within the framework of one Hilbert space. Even keeping the averaging procedure, we still have the phase correlations and then cannot obtain the reduction of wave packet, if the condition (7) is not satisfied. In such a case, the macroscopic system is not adequate as a measuring apparatus, even though some kind of interesting macroscopic quantum effects may take place. Generally speaking, (7) gives us a measure of the detection efficiency of the apparatus.

VI. SOLUBLE MODELS

We demonstrate occurrence of reduction of wave packet by four solvable models in the following.

VI.1. Perfect Mirror Model[6]

A perfect mirror can be used to measure momentum of a microscopic particle with very high energy (for example, $\gtrsim 10^{19}$ eV $\simeq 1$ Joule) such as a cosmic ray proton. A small wagon having a perfect mirror on the outside of the rear wall is put on a horizontal railway. The particle colliding on the mirror has energy enough to move the wagon by a macroscopic distance.

We shall treat the particle-mirror system within the framework of quantum mechanics. The "large" statistical operator for the inner state of the mirror can be written, similarly to (11), as

$$\hat{\sigma} = \int dl \, W(l - L)\hat{\rho}(l)$$

where l is the length of the mirror-wagon. L is the center of distribution of l, the classical thickness of the mirror. The exact S-matrix element for the collision of a particle with the mirror in one "small" Hilbert space is easily given by

$$\langle \, p', P' |\hat{S}| p, P \rangle = \exp[-ipl/\hbar], \qquad (24)$$

where $p(p')$ and $P(P')$ are momenta of the particle and the wagon before (after) the collision. The phase factor on the right-hand side of (24) gives rise upon integration over the continuous parameter l, to the reduction of wave packet in the measurement of the particle's momentum.

VI.2. Collision of an Energetic Particle with a Harmonic Lattice of Hard Atoms

Recently Nakajima[12] has pointed out that the collision of an energetic particle with a harmonic lattice of hard atoms provides a simple model of the averaging procedure. He has shown that the non-diagonal part of the statistical operator after the collision is proportional to a Debye–Waller-like factor. In this theory the reduction of wave packet results from taking the average over the phonon degrees of freedom not observed in the measurement. This implies that the rigid body in VI.1 can be replaced by a more realistic one, and that we can construct the averaging procedure, i.e. the "continuous"-superselection-rule space on a physical basis.

VI.3. The Dirac Comb Model

A component of the detecting apparatus in one Hilbert space of (5) is assumed to be expressed by N δ-function potentials spaced with equal distance, i.e. the Dirac comb potentials spaced with equal distances, i.e. the Dirac comb

$$V(x) = \Omega a \sum_{n=0}^{N-1} \delta(x - x_n), \qquad x_0 = 0, \quad x_{n+1} = x_n + a, \qquad (25)$$

where Ω may be positive or negative. Scattering of a non-relativistic particle by this potential was worked out in detail by Kiang.[13] The S-matrix element connecting the incoming wave from the left and the outgoing wave to the right is given by

$$S = \frac{z_+ - z_-}{z_+^N (z_+ - b^*) - z_-^N (z_- - b^*)} e^{2i\delta} \qquad (26)$$

$$\delta = -\tfrac{1}{2} kl, \qquad (27)$$

where $k = (2mE/\hbar)^{1/2}$, $l = Na$, m and E are the mass and the energy of the incoming particle, respectively, and

$$z_{\pm} = c \pm (c^2 - 1)^{1/2}, \qquad c = \text{Re } b, \quad b = \left(1 + i\frac{m\Omega a}{\hbar k}\right) e^{ika}.$$

The phase $-kl$ in (26) and the structure of the statistical operator given by (11) cause the reduction of wave packet as described in a general form in the preceding sections.

VI.4. One-dimensional Model of Photographic Emulsion

We first consider a component of the state of emulsion in one Hilbert space. Before the interaction we have AgBr molecules at N sites with an equal spacing, and during the interaction some of them dissociate into Ag atoms. These two states may be designated as the up and down states, respectively. We assume the following Hamiltonian,

$$\hat{H} = \hat{H}_0 + \hat{H}', \tag{28}$$

$$\hat{H}_0 = c\hat{p}, \tag{29}$$

$$\hat{H}' = \sum_{n=1}^{N} V(\hat{x} - na)\hat{\sigma}_1^{(n)} \tag{30}$$

where $\hat{x}$ and $\hat{p}$ are the coordinate and the momentum of the particle, respectively, and $\hat{\sigma}_i^{(n)}$ is the "spin" operator in the spatial direction i ($i = 1, 2, 3$) at the nth site. We further assume

$$V(x - na) = -V_0\theta(na - x)\theta(x - (n - 1)a), \tag{31}$$

with $V_0 > 0$ and evaluate the time evolution operator in the interaction representation to obtain

$$\hat{U}(t) \xrightarrow[t \to \infty]{} \exp\left[-\frac{i}{\hbar}\bar{p}l\hat{\Sigma}_1^N\right], \tag{32}$$

where

$$\bar{p} = V_0/c \tag{33}$$

is the effective momentum of the particle in the emulsion, and

$$\hat{\Sigma}_1^N \equiv \frac{1}{N}\sum_{n=1}^{N}\hat{\sigma}_1(n) \tag{34}$$

is the average spin operator. The maximum expectation value of $\hat{\Sigma}_1^N$ is in a component Hilbert space. Appearance of the factor $\bar{p}$ in the phase of the time evolution operator for the many-Hilbert-space formalism gives rise to the reduction of wave packet also in this case.

In the limit $N \to \infty$, the operators of the type given by (34),

$$\hat{\Sigma}_i^\infty \quad (i = 1, 2, 3) \tag{35}$$

commute with one another. If the limit existed, they would belong to the center of a von Neumann algebra and describe classical observables in a global equilibrium state of the system.

The probability for the transition, after the time evolution (32), from AgBr molecule to the Ag atom at each site is given by

$$P = \sin^2(V_0 a/hc) \tag{36}$$

where P gives the efficiency of the emulsion used in our model.

VII. CONCLUDING REMARKS

In the preceding sections we have seen that our theory of measurement certainly breaks through the Wigner theorem and derives the reduction of wave packet even for the negative-result-measurement case. It is very important that derivation of wave packet can be achieved on the following two bases: The first is the macroscopic nature of a local system of apparatus represented in a continuous direct sum of a large number of Hilbert spaces and the second is the phase shift proportional to the finite size of the local system. Both give the vanishing integral of the form $[\int dl \, W(l) \exp(\pm i\bar{p}/h) \cdots]$ to the cross-correlation parts of the asymptotic total statistical operator, thereby enabling us to derive the reduction of wave packet. Especially, it should be remarked that the first has been formulated at an infinite particle number limit of a local system which interacts with the object system, and that some kind of preparation such as the condition (7) is also necessary to the apparatus system.

References

1. See, e.g., B. d'Espagnat, *Conceptual Foundation of Quantum Mechanics* (W. A. Benjamin, Reading, Mass., 1976), 2nd ed.; *Selected Papers on the Theory of Measurement in Quantum Mechanics*, ed. M. M. Yanase, M. Namiki and S. Machida (Phys. Soc. Japan, Tokyo, 1978).
2. See, e.g., J. von Neumann, *Mathematische Grundlagen der Quantenmechanik* (Springer, Berlin, 1932); E. P. Wigner, *Am. J. Phys.* **31**, 6 (1963).
3. See, e.g., H. S. Green, *Nuovo Cim.* **9**, 880 (1958); A. Daneri, A. Loinger and G. M. Prosperi, *Nucl. Phys.* **33**, 297 (1962).
4. See Wigner's paper in Ref. 2 and A. Fine, *Phys. Rev. D2*, 2783 (1970); A. Shimony, *Phys. Rev. D9*, 2321 (1974).
5. M. Renninger, *Z. Phys.* **158**, 417 (1960).
6. S. Machida and M. Namiki, *Prog. Theor. Phys.* **63**, 1457 (1980); ibid. **63**, 1833 (1980); M. Namiki, *J. Japan Ass. Phil. Sci.* (in Japanese) **15**, 45 (1981).
7. H. Araki, *Prog. Theor. Phys.* **64**, 719 (1980).
8. S. Watanabe, private communication.

9. J. M. Jauch, E. P. Wigner and M. M. Yanase, *Nuovo Cim.* **48B**, 144 (1967).
10. For a critical review of the controversy, see Machida and Namiki's papers in Ref. 6 and in *Proceedings of the International Symposium on Foundations of Quantum Mechanics—in the Light of New Technology*, ed. S. Kamefuchi *et al.* (Phys. Soc. Japan, Tokyo, 1984).
11. See, e.g., Refs. 2 and 9. The present authors have discussed this confusion. See Ref. 6.
12. S. Nakajima, a comment on Namiki's talk in Ref. 10 and a forthcoming paper.
13. D. Kiang, *Am. J. Phys.* **42**, 785 (1974).

THE QUANTUM MECHANICS OF VISION

John A. Schumacher

Department of Philosphy, Rensselaer Polytechnic Institute, Troy, NY 12181

What do the experimental tests of Bell's Theorem imply about the visual process? In particular, do they constitute the ground on which we can seriously question the long-held belief that the visual experience itself is a kind of "inner show", the last of a series of events that begins in "an outer world"? To address these questions we need to compare the order of description of the visual experience to the quantum mechanical order of description. An extended thought experiment indicates that the visual process requires an order of description that is not based on the family of local, realistic theories.

I. INTRODUCTION

The recent experiments[1] based on Bell's Theorem have stimulated a lively discussion about the foundations of quantum mechanics. As yet, however, the discussion has not been extended to areas beyond physics proper. What can we say about how the human nervous system works? The aim of this paper is to address this question, with a special focus on vision. Along the way we will also extend the discussion about the foundations of quantum mechanics.

First, we need to understand this discussion in its own terms. These terms will be represented here by the work of Bohr and Bohm, set against the background of the work of Newton. We need to provide this historical background because even the great interpreters of quantum mechanics did not entirely abandon the classical point of view. Hence, among other things, both Bohr and Bohm concentrated on the role of artifactual instruments in quantum mechanics. Herein we will turn our attention to the role of the human body, beginning with the work of the eyes regarded as instruments in themselves.

Based on selected research in sensory physiology, neurophysiology, and psychology, we will then speculate about how quantum mechanics could apply to the rest of the work of the nervous system in vision. We will also ask whether or not the quality of a visual experience requires this sort of physical understanding. The answer will help us to understand how the classical point of view is eliminated in physics.

II. THE CLASSICAL POINT OF VIEW: NEWTON

What are the events relevant to a point of view if that point of view is a visual experience? To begin with, said Newton,[2] "all natural bodies are variously qualified to reflect one sort of light in greater plenty than another". The reflected light must carry the information required to resolve a visual experience. Here Newton was confined by his understanding of light and natural bodies:

> The ends of the capillamenta of the optic nerve, which front or face the retina, being such refracting surfaces, when the rays impinge upon them, they must there excite these vibrations, which vibrations (like those of sound in a trumpet) will run along the aqueous pores or crystalline pith of the capillamenta, through the optic nerve. (Ref. 2, pp. 97–98)

A certain kind of vibration carries the required information. Elsewhere Newton expanded upon this idea:

> If when we look but with one eye it be asked why objects appear thus and thus situated one to another, the answer would be because they really are so situated among themselves and make their pictures in the retina so situated one to another as they are; and those pictures transmit motional pictures into the sensorium in the same situation. (Ref. 2, p. 102)

When we use two eyes in "their natural posture", moreover, the respective motional pictures "come together and become coincident".

What next? "By the situation of those motional pictures the soul judges of the situation of things without," said Newton (Ref. 2, p. 102). We can illustrate the relevant series of events as follows:

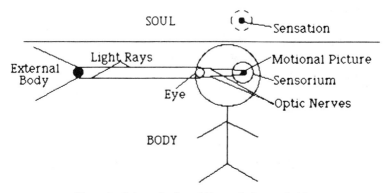

Figure 1   Schematization of Newton's theory of vision.

In the case of color, however, it is hard to imagine how the motional pictures provide the information required by the soul. The appropriate vibrations must again run through the optic nerve "into the sensorium (which light itself cannot do), and there affect the sense with various colors according to their bigness and mixture" (Ref. 2, p. 98). Certainly Newton was more confident about the case of the primary quality of spatial situation than about the case of the secondary quality of color. A cornerstone of the very distinction between primary and secondary qualities was the ease of imagining how motional pictures could be used in the former case.

One of the few classical philosophers-scientists to distrust this distinction, Berkeley still believed that, were there external bodies, colors could not be inherent in them:

> In case colors were real properties or affections inherent in external bodies, they could admit of no alteration, without some change wrought in the very bodies themselves; but is it not evident from what has been said, that upon the use of microscopes, upon a change happening in the humors of the eye, or a variation of distance, without any manner of real alteration in the thing itself, the colors of any object are either changed, or totally disappear? Nay, all other characteristics remaining the same, change but the situation of some objects, and they shall present different colors to the eye. (Ref. 3, p. 21)

Notice the classical separability of the observing instrument and the observed object: they are, as Bohm put it,[4] "autonomously existent". The most interesting example for us herein is certainly the microscope. One way to change the situation of an external body is to put it in the target of a microscope. From a classical point of view, the body will not undergo any "real alteration". Another way to change the situation of an external body is to put it, so to speak, in the target of an eye. Again, the body will not undergo any "real alteration". In both cases we must be able to assign definite properties to the body in isolation from the observing instrument as an autonomous existent. To reconsider the work of an eye, let us therefore turn to the Heisenberg thought experiment about microscopes.

III. THE HEISENBERG THOUGHT EXPERIMENT: BOHR AND BOHM

Let us briefly consider Bohm's account[4] of Heisenberg's thought experiment. Formulated prior to experimental tests based on Bell's Theorem, this account still anticipates their results in all the necessary ways.

In classical terms we suppose that a particle is at rest in the target of a microscope. The microscope is designed to focus another particle deflected by the target particle so that the deflected particle enters a photosensitive emulsion, leaving what we can observe as a track in the emulsion. Given the track and an adequate description of how the microscope works, we make inferences about the properties of the target particle; we come to know *both* its position and the momentum imparted to it at the time of deflection. All the events leading up to the track in the emulsion are, as we say, time-like separated in such a way that the past of the track can be uniquely determined. Hence, the description of the experimental conditions drops out of the description of the final result.

Now, let us suppose that the microscope is an electron microscope, and that the deflected particle is an electron. What must we say in quantum mechanical terms? On Bohm's account,[4] Heisenberg "evidently brought in the four primarily significant features of the quantum theory":

1. He describes the link electron *both* as a wave (while it is passing from object O through the lens to the image P) *and* as a particle (when it arrives at the point P and then leaves a track T).

2. The transfer of momentum to the "observed atom" at O has to be treated as discrete and indivisible.

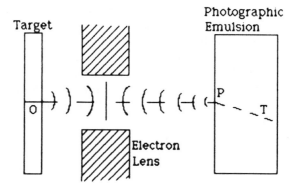

Figure 2   Bohm's schematization of the experiment.

3. Between O and P the most detailed possible description of the link electron is in terms of a wave function that determines only a statistical distribution of potentialities whose actualization depends on the experimental conditions (e.g. the presence of sensitive atoms in the emulsion, which can reveal the electron).

4. The actual results (the spot P, the track T, and the properties of the atom O) are correlated in the way indicated by Einstein, Podolsky, and Rosen, which cannot be explained in terms of the propagation of signals as chains of causal influence. In a more detailed mathematical treatment according to the quantum theory, the "wave function" of the "observed object" cannot be specified apart from a specification of the "wave function" of the "link electron", which in turn requires a description of the overall experimental conditions. (Shimony[5] coined the epithet "entangled potentialities" to refer to these correlations, which involve "spatially separated particles", though not necessarily anything like action at a distance or even at speeds faster than that of light.)

Together these four features of the quantum mechanical description imply that the precision of the interences we make about the position and the momentum of the target particle is limited: $\Delta x \Delta p \geqslant h$, where h is Planck's constant.

Can we still believe that something like a classical particle exists in the target, and in turn that the quantum mechanical description of the particle is simply incomplete? Theorists such as Einstein, Podolsky, and Rosen[6] believed that a hidden variables theory could salvage the world of the special theory of relativity and its time-like causality, as opposed to the space-like correlations involved in the quantum theory. But Bell's Theorem[7] shows that no such hidden variables theory "can reproduce all of the statistical predictions by quantum mechanics". Though this result covers only "idealised situations", Bell's Theorem "can be extended to cover actual systems". Clauser and Shimony[8] asserted "with reasonable confidence that the experimental evidence to date"—and there is even more evidence today[1]—supports the quantum theory over members of "the family of local, realistic theories".

After presenting their evidence, Clauser and Shimony claimed that "Bohr's position remains as one of the few feasible options concerning the foundations of quantum mechanics". Each measurement of the link electron, as Bohr put it,[9] affects "the very conditions which define the possible types of predictions regarding the future behavior of the system": that one track obtains as opposed to another deprives us of the opportunity to make certain predictions about the target particle. Position and momentum are "complementary physical quantities" in such a way that any experimental procedure capable of providing an "unambiguous definition" of one such quantity excludes all experimental procedures capable of providing a definition of the other, and vice versa. Bohm offered the following interpretation:

> This means that the description of the experimental conditions does not drop out as a mere intermediary link of inference, but remains inseparable from the description of what is called the observed object. The "quantum" context thus calls for a new kind of description that does not imply the separability of the "observed object" and "observing instrument". Instead, the form of the experimental conditions and the meaning of the experimental results have now to be one whole, in which analysis into autonomously existent elements is not relevant. (Ref. 4, p. 133)

We can no longer give unambiguous definitions to the observing instrument and the observed object in isolation from each other as autonomous existents. We *must draw* the line separating the observing instrument from the observed object. The necessity of drawing this line, is, as Bohr put it,[9] "the principal distinction between classical and quantum mechanical description of physical phenomena".

Bohr also believed that the necessity of drawing this line "has its roots in the indispensable use of classical concepts in the interpretation of all proper measurements, even though the classical theories do not suffice in accounting for the new types of regularities with which we are concerned in atomic physics"; we use quantum mechanics "to predict the result obtained by a given experimental arrangement described in totally classical terms". What exactly does it mean to adopt a classical point of view with new limits to the meaning of such terms as "position" and "momentum"? According to Clauser and Shimony,[8] these limits undermine "the whole realistic viewpoint": "the term 'reality' can be used unambiguously in microphysics only when the experimental arrangement is specified". Notice nevertheless the curious lack of any specification regarding the person who arranges the experiment.

In the same spirit, Bohm did not extend "the whole" of the Heisenberg thought experiment beyond the track in the emulsion. To discriminate between the observing instrument and the observed object, Bohm regarded the description of the track in the emulsion as the condition that resolves the relevant wave function. He spoke of an order of distinctions "relevated (and recorded) by our instruments", in this case by the emulsion; hence the track in the emulsion is "directly observable". Though Bohm did not agree with Bohr that the results of experiments could be described in totally classical terms—Bohm's own version of the lack of separability in the quantum context aimed more to found a new family of realistic theories than to abandon "the whole realistic viewpoint"—Bohm still assumed the classical point of view: he did not extend "the whole" to the observer in the experiment. The track in the emulsion may be said to be directly observable only if we take a classical point of view of it. Or again, as Bohm put it, "such a track is evidently to be regarded as no more than an *aspect* appearing in immediate perception"—"an inner show".[4,10] Let us now turn to a consideration of this position.

IV. THE QUANTUM MECHANICS OF LOOKING AT EMULSIONS

First we must note what the track in the emulsion *is*. The relevant order of movement cannot be that of a classical particle (see Ref. 4, pp. 154–155). Must we not treat the properties of the particles of the emulsion in the target of an eye as an order of quantum mechanical description analogous to that of the properties of the particle in the target of Heisenberg's microscope? The answer requires another thought experiment, following Heisenberg's very closely so as to assure that the reasoning is the same: the particles of the emulsion are the observed object, an eye is the observing instrument, and we make inferences about the properties of the observed object. We assume that the particles of the emulsion are initially at rest, and that a light wave of known energy is directed at them in such a way that the wave is diffracted to the eye (through a vacuum, let us say), which in turn focuses the wave in such a way that it leaves a track in the retina of the eye. Here the retina plays the role of an emulsion.

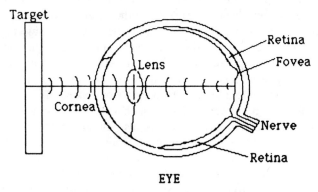

Figure 3   Schematization of the experiment.

Given the above figure the comparison to the original thought experiment is evident: *the eye is a Heisenberg microscope.*

We have come to know a great deal about how the eye works, indeed, to know so much that the assumptions at work in this additional thought experiment are not really assumptions at all. Let us concentrate on the retina's role as a photochemical emulsion:

> The absorption of quanta by the visual pigment of receptor cells leads to at least one chemical event that is indispensable for triggering a light sensation. In retinas that contain rhodopsin, when a quantum is absorbed a molecule of the latter shifts to prelumirhodopsin, i.e., from cis-form to trans-form. This first chemical event is followed by many others, but as yet there is no proof that the latter are indispensable to vision.... The absorption of a quantum of any wavelength must give rise to the same sequence of events in a given rod receptor, for the spectral efficiency curve of scotopic vision [the threshold of which is moonlight], after correction for light losses in the ocular media, coincides with the absorption curve of rhodopsin. (Ref. 11, p. 30)

A shift from the cis-form to the trans-form of rhodopsin constitutes, let us say, a track in the retina. Though this track is not as accessible as the track in the emulsion in Heisenberg's original thought experiment, we still know that, were we to observe it, we could make inferences on that basis about the interaction at the target, namely, the original emulsion itself. To paraphrase Bohm, the wave function of the particles of the emulsion cannot be specified apart from the wave function of the link photons, which in turn cannot be specified apart from the overall experimental conditions, including the track in the retina. By correlating the entry into the retina with the properties of the particles of the emulsion, we are, again, discriminating between the observing instrument—the eye—and the observed object—the particles of the emulsion.

So, what keeps us from taking the eye to be the observing instrument in *both* thought experiments? We now know that the track in the retina is correlated to the properties of the particles of the emulsion in exactly the same way as the track in the emulsion is correlated to the properties of the target particle. But what *is* the track in the emulsion—or perhaps more carefully, what can we *say* that the track in the emulsion is—but the properties that the particles of the emulsion can be said to have when they are "directly observed" by the eyes? Hence, why not suppose that "the whole" of the original thought experiment can be extended to the eyes of the observer? We imagine—and this is the additional thought experiment, remember—another observer who uses the track in the first observer's eyes to make inferences about the properties of the particles of the emulsion, and in turn about the properties of the target particle. Note that the additional thought experiment does not make any assumptions about what happens in the vision of the first observer after the track in the retinas. We assume only that the track in the retinas can be used as a track in an emulsion must be used in any version of the original thought experiment.

Though in extending this experiment we are concerned with an order of error that can be neglected—and here I am thinking about the difference between what the first observer concludes on the basis of direct observation of the emulsion, and what the second observer concludes on the basis of the track in the retinas of the first observer—we are also concerned with *another order of description*, the one that both Bohr and Bohm were willing to apply only to the experimental arrangement. Whether or not we need to extend "the whole" of the original thought experiment all the way to the retinas of the first observer, the quantum mechanical order of description is required by a crucial portion of the very process of vision that is supposed to be capable of description in totally classical terms. Stretching from any target into the retinas of an observer, at least up to "the first chemical event", the relevant order of movement is "a whole" that cannot be analyzed into "autonomously existent elements".

We are presented with two alternatives: (1) we may extend "the whole" through the nervous system of an observer to eliminate the separability of an event and its observation, thereby undermining the classical point of view; or (2) we may still draw a line to separate the classical from the quantum mechanical domains of description. Whereas formerly we drew the line at the track in the emulsion, now we may draw it at the track in the eyes of an observer, at the verge of nervous system activity.

We cannot avoid considering these two alternatives, it should be noted, since we can easily ask what the track in the first observer's eyes *is*, and create a third thought experiment about it. Indeed, from what we have already said about "the first chemical event", we must evidently consider drawing the line in (2) even deeper in the nervous system. Technically speaking, we would continue to express the physical situation in terms of wave functions even through the first chemical event, which itself is understood at the quantum mechanical level of description: did a molecule of rhodopsin absorb "a quantum" or not? We have been pretending so far that we *need* to let the description of the track in the retina serve as the means for drawing the line to separate the observing instrument from the observed object, and this kind of pretense is found throughout the practice of quantum mechanics: how do we *know* where to draw the line? To the extent that it is, as Bohr said,[9] "a free choice", our additional thought experiment exposes the pretense, forcing us to confront the dilemma represented by (1) and (2).

We can understand this dilemma even better against the background of more facts about scotopic vision. Studies have been conducted to compare the frequency of an observer's response near the threshold levels of stimulation in scotopic vision. Flashes of various energy levels at a specific wavelength are correlated to the frequency of their perception by different people, and from this comparison inferences are made about the lowest number of quanta that must be absorbed by the rods to yield a threshold response. The resulting figures match previously established figures for the number of quanta required at the cornea level for a threshold response. According to Baumgardt,[11] "we are thus led to the conclusion that in these experiments the uncertainty of seeing is predominantly due to fluctuations in the actual numbers of quanta absorbed by the retinal rods".

If the quantum mechanical level of organization of the physical environment reaches into an observer's body, at least as far as the retinas of the eyes, why doesn't the body *use* this reach? What should we expect if indeed the body does us it? Note again that the uncertainty of seeing in scotopic vision is predominantly due to fluctuations in the actual numbers of quanta absorbed by the retinal rods, and that the first chemical event is followed by many others, but as yet there is no proof that the latter are indispensable to vision. Better yet, we can understand all of these events as *aspects of one whole*, not analyzable into autonomously existent elements.

This understanding of scotopic vision tips the scales toward (1) rather than (2). If the relevant nervous system activity after the first chemical event were essentially classical, it would be hard to imagine quite how the activity could make use of the whole that reaches to the first chemical event; the retina could play no role except that of a kind of emulsion. Yet the facts of scotopic vision imply a certain wholeness that reaches to the vision itself. It is interesting to suppose that this wholeness arises because the nervous system itself *works* at the quantum mechanical level of organization. What does this mean?

Notice that the description of the track in the emulsion of the original thought experiment allows us to proceed to the stage of making an inference about the properties of the observed object; the basis of the inference is the "actualization" of one potentiality among others represented by the relevant wave function. To make use of the quantum mechanical level of organization that reaches to the first chemical event, the relevant nervous system activity must constitute the very feat in question here, the "actualization". The nervous system activity is not just another aspect of "the whole"; it resolves "the whole", as if the track in the additional thought experiment were to extend beyond the retina through the entire nervous system activity relevant to vision. In Bohm's terms, the nervous system is an instrument that relevates (and records) an order of distinctions, that of visual observation. Bohm himself said much the same thing,[4] though he failed, as we discussed in the previous section, to draw the conclusion that Heisenberg's reasoning must be applied in this case as well.

In our discussion of Bohr's position, we noted that experimental procedures capable of providing unambiguous definitions of certain pairs of physical conditions—either position and momentum, or time and energy—are mutually exclusive in such a way that a procedure for position or time excludes a procedure for momentum or energy, respectively. We therefore reach the limits of classical description. In every experiment we must discriminate between the observing instrument and the observed object: the discrimination is no longer merely a given, but must be chosen, so that descriptions of experimental results—still expressed in classical terms, according to Bohr—include the discrimination and in turn the whole experimental situation. Our additional thought experiment extends Bohr's position to the supposed classical observation itself. We regard it too as a kind of experiment that requires discrimination between the observing instrument and the observed object in such a way that the whole observational situation can no longer drop out of the description of the observed object. Herein we are imagining the human body to be one among other instruments, quite a bit more sophisticated, but not essentially different in any way that allows it to be free from inclusion in the descriptions of the results of experiments it performs.

We therefore reach the limit, not of descriptions from a classical point of view, but of the classical point of view itself. Even if "the whole" reaches only to our skin—and here I am thinking of the skin as the place of the first chemical event of perception, as in the case of the eyes—we can no longer separate our bodies from the world we observe, even in those situations in which we typically assume a classical point of view. We begin to draw the line that Bohr made so much of with every move to observe the world, even before we reach our artifactual instruments: only by reference to our bodies can we give an unambiguous description of the world. Indeed, should (1) hold as we have suggested, any urge to continue to talk in terms of "an inner show" separable from the events leading up to it must also be quieted. Whatever show takes place, it cannot be thought of as "inner" as opposed to "outer". "Inner" and "outer" are precisely the classical terms that our discussion tends to undermine, reducing their supposed referents to aspects of "one whole". Or again, in Newton's terms, the nervous system activity relevant to vision is not ultimately present to the soul; the soul's work becomes an aspect of "the whole" at the quantum mechanical level of organization of the nervous system and the physical environment. We speculate that this whole *is* experience of the world, in which we find *both* us and the world.

V. THE SIGNIFICANCE OF THE QUANTUM MECHANICS OF VISION

As Gibson was fond of saying,[12] the contemporary psychology of vision is still seduced by the classical point of view. His epithet for the resulting theory is "the little man in the brain theory". Again a progression of events leads to "constructing a phenomenal environment out of spots of differing brightness and color", based on a "correspondence of intensity to brightness and of wavelength to color". Unable to imagine how such a process could succeed, Gibson even went so far as to conclude that transmission along the optic nerve is not essential to vision:

> It is not necessary to assume that *anything whatever* is transmitted along the optic nerve in the activity of perception. We can think of vision as a perceptual system, the brain being simply part of the system. The eye is also part of the system, since retinal inputs lead to ocular adjustments and then to altered retinal inputs, and so on. The process is circular, not a one-way transmission. The eye-head-brain-body system registers the invariants in the structure of ambient light. The eye is not a camera that forms and delivers an image, nor is the retina simply a keyboard that can be struck by fingers of light. (Ref. 12, p. 61)

But neither does the eye-head-brain-body system simply register the invariants in the structure of ambient light. This way of putting the matter will return us to the separability of this system and the ambient light. Gibson did not think that physics could help him, but then again he was thinking of classical physics. On our approach, quantum mechanics not only supports Gibson's conclusion, but also extends it: the system in question should be the eye-head-brain-body-physical-environment system, in which light plays the role we discussed in the previous section. Gibson's remarks about a circular process then take their place with references to "the whole" of this larger system. The invariants of the structure of ambient light are no more independent from the smaller system than the elements of the smaller system are independent from each other.

There are three phenomena to which we can apply this conclusion: (1) macular sparing, (2) the occluding edge, and (3) the resolution of ambiguous figures. In these applications we are extending the discussion of the previous section, though we are not actually making an argument that the speculations in that discussion are correct. We are instead suggesting how to proceed further along the lines of that discussion; we are opening ourselves to the ways in which that discussion could be supported by future research in the relevant sciences.

(1) The transmissions along the optic nerve eventually reach the lateral geniculate body (LGB) and then the visual cortex. Specific places on the retina excite particular cells in both the LGB and the visual

cortex, though the sides get switched along the way, that is, the left visual field ends up correlated to the right visual cortex, and vice versa. The macular zone at the center of the retina also excites specific cells of the visual cortex—in the rear tip of the lobe—on the opposite side of the half visual field. Again, each half of the macular zone excites cells in *only one* half of the visual cortex. Hence, we should expect that seeing the central field is correlated with the specific cells of the visual cortex to which the macular zone is correlated. Yet whenever either half of the visual cortex is destroyed, there typically arises the phenomenon of macular sparing: a spared zone, exactly that of the central field, exists in the associated visual field. (Why is the central field so special? Thinking of it from the side of the observer, we realize that the central field is *the point of view itself*.) Pietsch concluded,[13] "There is not an exclusive center in the brain for seeing the central field." This sort of flexibility fits well with Gibson's position, and may well indicate another feature of vision that requires the quantum mechanical level of organization of the nervous system. Note again the lack of an exclusive center for *seeing* the central field: the localization in space-time of all the events leading up to seeing the central field does not imply that the seeing itself is similarly localized. The preceding events can be so localized only from a classical point of view, and on our speculation the seeing cannot be so localized: it is an aspect of "the whole" that "actualizes".

(2) Here is Gibson on the recent discovery of the occluding edge:

> The crucial experiment was performed by Kaplan, and involved kinetic, not static, displays of information. Each display was a motion picture shot of a random texture filling the screen, with a progressive deletion (or accretion) of the optical structure on one side of a contour and preservation of the structure on the other side. Photographs of a randomly textured paper were taken frame by frame, and successive frames were modified by careful paper cutting. No contour was ever visible on any single frame, but progressive decrements (or increments) of the texture were produced on one side of the invisible line by cutting off thin slices of paper in succession. This particular kind of decrementing or incrementing of structure had not previously been achieved in a visual display. (Ref. 12, p. 189)

The result is that all observers see one surface going behind or coming from behind another surface, a process of covering or uncovering, respectively. The surface going out of sight is never seen as going out of existence, nor is the surface coming into sight ever seen as coming into existence. One surface is seen *behind* another *at an occluding edge*. Gibson offered the following conclusion:

The surface that was being covered was seen to persist after being concealed, and the surface that was being uncovered was seen to pre-exist before being revealed. The hidden surface could not be described as remembered in one case or expected in the other [instead of *seen* in both]. A better description would be that it was perceived retrospectively and prospectively. It is certainly reasonable to describe perception as extending into the past and the future, but note that to do so violates the accepted doctrine that perception is *confined* to the present. (Ref. 12, p. 190)

That is, it violates the accepted doctrine not only that perception is "an inner show", constructed from input confined to a certain moment of reception, but also that such input is confined to events simply localized in space-time. Even Gibson spoke of the occluding edge as if it were simply localized in space-time, and in turn either simply covering or not covering another surface behind it. To paraphrase Heisenberg, it is "meaningless to speak of the place of the edge with a definite velocity". It has such a place *only if the display is no longer working*; hence, the kinetic display and "the invisible line" have a significance that Gibson could not explain. Once we extend the relevant perceptual system to the physical environment, we can understand how the two surfaces are connected by the light that interacts with them. They are connected to each other as well as to the eye-head-brain-body system; they are all aspects of "one whole" that represents the "actualization" of "seeing one surface behind another", even when in classical terms the localization of the occluding edge would deprive us of information about the hidden surface.

(3) Consider the figure shown in Fig. 4. Either square side of the figure may be seen as the front side of a rectangular solid. Notice that, from a classical point of view, the light that reaches the retinas of an observer's eyes is just as ambiguous as the figure itself. Somehow the nervous system is supposed to resolve the figure, but only in "an inner show" separable from the interaction of the light with the figure. Hence, the

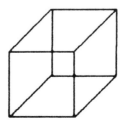

Figure 4   An ambiguous figure.

figure itself cannot be involved in the resolution; or rather, its properties are supposed to be independent of the resolution. What keeps us from supposing, on the contrary, that the figure is as involved in the resolution as are the two surfaces in (2)? (Doesn't the figure always seem to rearrange itself when we switch between the teo rectangular solids?) Notice that the ambiguity here involves a mutually exclusive pair of definite figures. Can there be a connection between our ability to resolve such ambiguities and our ability to resolve the ambiguity represented by the celebrated complementarity of position and momentum, or of time and energy? In both cases, we may speculate, we are dealing with an eye-head-brain-body-physical-environment system that is "one whole": the physical environment must be involved in its own resolution.

It is not speculation, however, that at least through the event of the absorption of light by the retinas, we are dealing with "one whole", the analysis of which into merely time-like separated events is not possible. The time-like separation of the relevant events must come along *with* "the whole" in question. Here the order of movement of light, an order of separation at the limit of movement, must have a deeper order, an order of wholeness, not analyzable into autonomously existent elements. If in the special theory of relativity we can be said to penetrate inside Newtonian body—and here I am thinking of Einstein's example of the box that contained a distribution of radiant electromagnetic energy[10]—then in the quantum theory we can be said to penetrate inside Einsteinian light, though in the latter case we no longer refer to an order of inner movement. Light cannot be said to have inner movement, but presumably can be said to have inner correlations or relationships, exactly those that constitute the lack of separability implied by the quantum theory. Herein we have speculated that visual experiences themselves must enter into the *same* relationships.

So, finally, let us consider the quality of a visual experience: we pay attention to the visual object *at a distance at once*. Certainly this is how we talk in everyday terms. Though we do switch from one object of attention to another, it makes no sense to say that we catch attention travelling to its object; attention is always already caught. (Or again, to put the matter so as to strike an anlogy with the quantum context, the at-a-distance-at-once quality of a visual experience does not itself entail any action at a distance.) In the case of the resolution of the ambiguous figure in (3) above, first one rectangular solid, then the other, *stands apart at once, as if it were resolved in "an outer show"*.

The at-a-distance-at-once quality of a visual experience may now be compared to the quality of "the whole" of the relevant physical process, the eye-head-brain-body-physical-environment system: *they have the same form*. The physical process has the same space-like quality that we typically assign to the visual experience itself. Hence, our speculation that the relevant whole *is* experience of the world amounts to connecting "inner" and "outer" exactly as they are connected in everyday terms. Our common sense that we pay attention *to the physical environment* does indeed have a physical foundation: the object of attention *is* the physical environment, which hereby participates in its own resolution. Quantum mechanics is the science that turns common sense into the truth.

References

1. A. Aspect, J. Dalibard and G. Roger, *Phys. Rev. Lett.* **49**, 1804 (1982).
2. H. S. Thayer (ed.), *Newton's Philosphy of Nature: Selections of His Writings* (Hafner Press, New York, 1974).
3. G. Berkeley, *Three Dialogues between Hylas and Philonous*, ed. R. M. Adams (Hackett, Indianapolis, 1982).
4. D. Bohm, *Wholeness and the Implicate Order* (Routledge and Keegan Paul, London, 1980), pp. 128–155.
5. A. Shimony, "Hidden Variables Theories, Bell's Theorem and Locality" presented at the SUNY-Albany Symposium on Fundamental Questions in Quantum Mechanics (April 12–14, 1984).
6. A. Einstein, B. Podolsky and N. Rosen, *Phys. Rev.* **47**, 777 (1935).
7. J. S. Bell, *Physics* **1**, 195 (1964).
8. J. F. Clauser and A. Shimony, *Rep. Progr. Phys.* **41**, 1881 (1978).
9. N. Bohr, *Phys. Rev.* **48**, 696 (1935), reprinted in *Physical Reality*, ed. S. Toulmin (Harper and Row, New York, 1970), pp. 138–140.
10. D. Bohm, *The Special Theory of Relativity* (Benjamin, New York, 1965).
11. E. Baumgardt, "Threshold Quantal Problems" in *Handbook of Sensory Physiology*, VII/4, 29 (1972).
12. J. J. Gibson, *The Ecological Approach to Visual Perception* (Houghton and Mifflin, Boston, 1979).
13. P. Pietsch, *Shufflebrain: The Quest for the Hologramic Mind* (Houghton and Mifflin, Boston, 1981), p. 21.

MEASUREMENT THEORY AND A QUANTUM MECHANICAL AUTOMATON

David Z. Albert

Department of Physics and Astronomy, University of South Carolina, Columbia, SC 29208

A gedankenexperiment involving a Quantum-Mechanical automaton is described. The experiment sheds a curious new light on an old question in the theory of measurement.

A recent study[1] of a certain hypothetical quantum-mechanical automaton has raised, in a new and particularly dramatic way, a very old and menacing question about quantum theory. I should like to consider that question here; it turns out that a gedankenexperiment involving a quantum-mechanical automaton sheds a very novel light on it.

The study I mentioned above established that such an automaton, if it were capable of measuring certain of its own physical properties and certain properties of the external world (ordinary physical properties like electron spins, for example) would behave in a very remarkable way. The empirical description which such an automaton would produce of itself would be unlike anything imagined in the conventional theory of measurement.

Those considerations assumed, of course, that measuring-instruments can be considered as quantum-mechanical systems (that is what one usually assumes, say, in the many-worlds interpretation of quantum mechanics); and (consequently) that the measuring-process is in principle reversible. The question is (and this is the old question I should like to discuss here): is that a reasonable assumption? Is it even *imaginable* that a measurement can be a reversible, quantum-mechanical process? The conventional answer is no. What I have to say here, on the other hand, will suggest that the answer (to the second question, at least) is yes.

A quantum-mechanical automaton was imagined in Ref. 1 which can "measure" certain physical properties of the world (the z-spin of an electron, say) in the following sense: the automaton can switch on interactions which produce *correlations* between a certain element of its memory-bank and the physical property (the z-spin) in question. It was shown that if such an automaton were capable as well of measuring a certain observable of the composite system consisting of that electron and the memory-element wherein the result of the first measurement is stored (an observable that *does not* commute with the z-spin of the electron), then *that* "measurement" might also be carried out and recorded elsewhere in the memory *without* disturbing the correlation established by the first "measurement". Such an automaton could in such a way come to "know", to be able to "predict", accurately and simultaneously, the values of two *non*-commuting observables!

Whether these "measurements" (these quantum-mechanical, reversible procedures which produce correlations), or any purely quantum-mechanical processes whatever, can really be *measurements* is what I should like to consider here.

Now, if *any* such procedure can really be a measurement, then doubtless many of them can, so it will be sufficient for our purposes to focus on one of them, a particularly simple one: the "measurement" (mentioned above) of the z-spin of a spin-$\frac{1}{2}$ particle. Suppose (to simplify matters still further) that the relevant memory-element within the automaton is a spin-$\frac{1}{2}$ particle too, and that the result of the z-spin "measurement" on the first particle is recorded in the z-spin of the second. We shall call the first particle (the "measured" system) S, and will call the second one (the memory unit) M. We shall need to suppose that if M and S are brought close together a spin-dependent interaction arises between them (that, of course, will almost invariably be the case: if M and S have magnetic moments, their spins will couple directly; if M and S have electric charge, the spin of each will couple to the orbital angular momentum of the other as well).

The outer body of the automaton is a macroscopic box (such as is described in Fig. 1). When the automaton is instructed to measure and record the value of σ_z on S, a door opens, and S is sucked into the

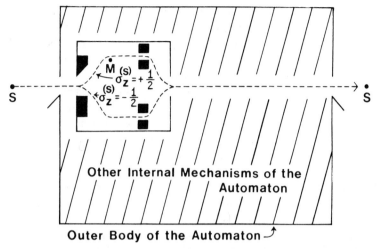

Figure 1   Schematic plan for the measurement of $\sigma_z^{(S)}$. The solid areas represent Stern–Gerlach magnets.

automaton and passed through a Stern–Gerlach magnet[2] which divides its wavefunction into $\sigma_z = +\frac{1}{2}$ and $\sigma_z = -\frac{1}{2}$ beams. Then, one of those beams (the $\sigma_z = +\frac{1}{2}$ beam, say) is made to pass close enough to M to produce an interaction, and everything is arranged so that the net effect of this interaction is to flip the z-spin of M and to leave the $\sigma_z = +\frac{1}{2}$ beam of S unaffected.[3] Then the two S-beams are re-united (by means of another S–G magnet) and S exits the automaton by a back door, and now the first measurement is complete: $\sigma_z^{(S)}$ is unaffected, and the z-spin of M ($\sigma_z^{(M)}$) is the negative of its original value if and only if $\sigma_z^{(S)} = +\frac{1}{2}$.

Notice (since, in what follows, much will be made of this) that the automaton that measured $\sigma_z^{(S)}$ and that remembers, now, what $\sigma_z^{(S)}$ is, is perhaps macroscopic and very complicated, perhaps it speaks English, perhaps it can walk; but its memory of $\sigma_z^{(S)}$ is stored in a purely *microscopic*, purely quantum-mechanical degree of freedom, and is subject to purely quantum-mechanical laws. Consequently, if S is prepared initially in the state

$$|\sigma_y^{(S)} = +\tfrac{1}{2}\rangle = 2^{-1/2}[|\sigma_z^{(S)} = +\tfrac{1}{2}\rangle + |\sigma_z^{(S)} = -\tfrac{1}{2}\rangle] \qquad (1)$$

and if M is prepared initially in the state $|\sigma_z^{(M)} = -\frac{1}{2}\rangle$, then the composite state, when the measurement is done, will be:

$$2^{-1/2}[|\sigma_z^{(M)} = +\tfrac{1}{2}\rangle\,|\sigma_z^{(S)} = +\tfrac{1}{2}\rangle + |\sigma_z^{(M)} = -\tfrac{1}{2}\rangle\,|\sigma_z^{(S)} = -\tfrac{1}{2}\rangle]. \qquad (2)$$

Now that the physical situation is well defined, we can address our original question. We have been calling the evolution from (1) to (2) a "measurement" of $\sigma_z^{(S)}$, and something is uncomfortable about that. Certainly we should *like* to be able to avoid speaking of superpositions of "measuring that $\sigma_z^{(S)} = +\frac{1}{2}$" and "measuring that $\sigma_z^{(S)} = -\frac{1}{2}$" (wherein it seems that the measurement has some result, but no *particular* one!). Indeed, it was in order to avoid speaking of such things that we began to suppose that the measuring-process cannot be described quantum-mechanically; that, at some point, an un-quantum-mechanical, irreversible "act of amplification" must enter the picture.

Now, the exact point at which that irreversible act is supposed to enter the picture is unclear, but certainly the area in which it could possibly enter can be bounded: nothing of that sort can happen to an isolated microscopic system (we know that from experiment) and something of that sort *must* happen (if it is to happen at all) in time to preclude superpositions (like the one described above) of the knowledge of macroscopic observers.

I am about to describe a thought experiment that suggests, for a certain particular case, that these two boundaries lie at precisely the same point, with no space whatever between them wherein anything irreversible might occur.

Imagine that (in the year 3000, say) a z-spin measuring-device such as was just described is surgically implanted into the brain of Mr. Smith, and is connected to his neurons in a very particular way. We might think of Smith as an automaton of the kind described above, which "measures" the z-spins of spin-$\frac{1}{2}$ particles by the method described, and which is in all other respects like a man.

Now the following sequence of events takes place:

(A) A spin-$\frac{1}{2}$ particle S is prepared, say, in the state $|\sigma_z^{(S)} = +\frac{1}{2}\rangle$, and is presented to Smith.

(B) A small door opens on the side of Smith's head and admits S, and S is passed through the z-spin measuring-device, and then S emerges from Smith's head by another door.

(C) Smith claims (albeit that M has as yet interacted *only* with S; that $\sigma_z^{(S)}$ is as yet recorded *nowhere* in Smith's brain other than in $\sigma_z^{(M)}$) that he "knows" the value of $\sigma_z^{(S)}$; he testifies, moreover, that his awareness of $\sigma_z^{(S)}$, is a *psychological* experience, a *psychological* awareness, no less direct and no less genuine than any other.[4]

(D) If, thereafter, Smith is requested to predict the result of a subsequent re-measurement of $\sigma_z^{(S)}$, he invariably predicts (and in the process, for the first time, he couples M to the speech center of his brain), correctly, that $\sigma_z^{(S)} = +\frac{1}{2}$.

Smith's claims in (C) look odd, but some reflection will persuade the reader that they are very hard to contest. He is, after all, in a position to assert, correctly, that $\sigma_z^{(S)} = +\frac{1}{2}$; and presumably no one is better qualified than Smith himself to judge the "genuineness" of his own psychological experience!

Well, suppose that Smith's claims *are* correct. Then (as Smith is undoubtedly a macroscopic observer) steps (B) and (C) constitute a measuring-process. On the other hand, M (Smith's memory about S) and S constitute a *microscopic*, and (as of step (C)) dynamically isolated system; so this measuring-process (albeit that it *is*, genuinely, a measuring process) can entail nothing irreversible and nothing un-quantum-mechanical; indeed, it entails nothing even in the nature of an amplification.

References and Notes

1. D. Z. Albert, *Phys. Lett.* **98A**, 6 (1983).
2. S–G magnets are, of course, not very efficient at resolving *charged* particles into spin-beams. But there is no need here for very much efficiency; we can wait.
3. That may take some doing (external magnetic fields, for example, will be required to undo the effects of M on S), but the reader can easily persuade himself that such arrangements are not impossible and not even *impractical*.
4. This may seem absurd. How can it be (it might be asked) that $\sigma_z^{(M)}$ forms a part of Smith's present awareness, without presently interacting with the rest of Smith's brain? There is a mistake in such questions. If we suppose that Smith can presently be aware of the information recorded in some subsystem X of his brain only if X is presently interacting with other such systems, then the same reasoning can be applied to this *group* of systems, and to the next group, and so on *ad infinitum*. The proper description of the content of present awareness is a *dispositional* one: Smith is presently thinking of the information stored in X if, say, on being asked, now, "What are you thinking of?", his speech center would (in his present physical state) couple to X.

THE NEUTRON INTERFEROMETER AND THE QUANTUM MECHANICAL SUPERPOSITION PRINCIPLE

Daniel M. Greenberger

City College of the City University of New York, NY 10031

The neutron interferometer allows one to construct a neutron wave packet of macroscopic size, split it into two beams separated by centimeters, and coherently recombine them. This amazing device has allowed a number of hitherto impossible experiments to be performed which test in a simple way the conclusions of quantum theory. The emphasis in this paper is on a series of beautiful experiments utilizing the coherence of spatially separated spin states, which verify the superposition principle in a very direct way. These experiments also illustrate some of the limitations imposed by quantum measurement theory. Objections have been raised against the standard interpretation of the theory in analyzing these experiments. I discuss a number of these objections and show that they are incorrect.

I. REMARKS ON SUPERPOSITION

I would like to describe for you some really beautiful experiments performed with neutrons by Rauch and his colleagues in Vienna, with an emphasis on the insight they give into the superposition principle. I do not mean to imply in my remarks that one can give an iron-clad proof of superposition. Personally, I do not believe physics gives such proofs. What it does do is provide "natural" explanations for certain phenomena, in terms of which it is very easy to describe what one sees, and to design new experiments.

Alternative explanations tend to be "strained" and can interpret data only after the fact. During revolutionary periods, all explanations tend to be strained and it is the unease caused by this that produces the drive to develop new theoretical syntheses.

Today quantum theory produces a very natural explanation of many phenomena, and the superposition principle is a very fundamental ingredient in the theory. What tension there is, is caused by the fact that quantum measurement theory is then grafted onto quantum theory proper, in a somewhat ad-hoc manner. And even though the theory works well in practice, the description of what happens when a measurement is actually made bothers many people. This has led to a continuing search for alternative descriptions, which to date have not produced any really convincing theories.

I mention this because the experiments I will describe have a very natural and clean explanation in terms of standard quantum theory. But this explanation has been attacked on the grounds that it is actually inconsistent with the tenets of quantum measurement theory, and that therefore these experiments disprove the standard theory. But in my opinion these criticisms are ill-founded. They misconstrue the nature of the measuring process in quantum theory.

I will analyze some of these objections because they can be used to shed light on the nature of the kind of information one can actually obtain in a quantum measurement. But let me reiterate that the standard theory is quite capable of interpreting these experiments, and that these objections are wrong. I do not believe that attacks along the lines I shall describe pose any threat to the standard theory.

First, does the neutron truly have wavelike properties? Let me show you the results of a lovely experiment performed by Zeilinger et al.[1] at the ILL at Grenoble. It shows the pattern (see Fig. 1) made by collimated, monoenergetic neutrons emerging from a single slit and double slit interference device. This pattern fits perfectly with the calculations of scalar diffraction theory. The appearance of alternate maxima and minima in the two-slit pattern argues very convincingly that the neutrons possess a phase as well as an amplitude, and that the pattern on the screen is the sum of coherent contributions from each slit. Shull[2] has pointed out that neutrons have been shown to possess wave properties over an energy range from 10^{-7} eV to several hundred MeV, covering over 15 orders of magnitude!

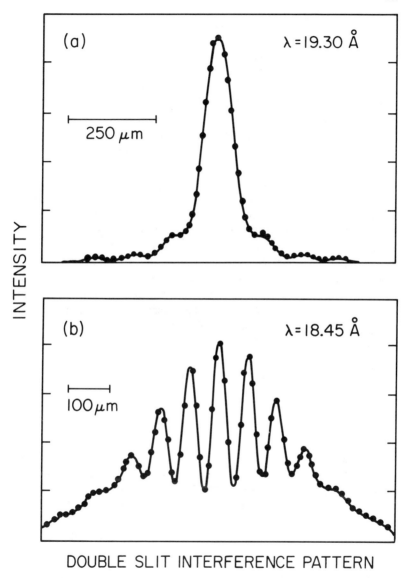

Figure 1   Neutron diffraction patterns. (a) The pattern produced by a neutron beam after passing through a single slit. The solid line is the pattern calculated from scalar diffraction theory. (b) The pattern produced by passage through a double slit. See Ref. 1.

II. THE NEUTRON SPIN

Besides intensity, which is the property of "being there", or localizability, and which is tested in the above experiment, the neutron has another property, its spin, which is highly non-classical and yet which is subject to the superposition principle. Let me just state a few simple properties of the spin, and then describe a simple but beautiful gedankenexperiment which has a simple explanation in terms of superposition, and yet which illustrates some complications introduced by quantum measurement theory. Then I will show how the Rauch group actually implemented this experiment with the neutron interferometer.

The neutron has spin-$\frac{1}{2}$. It has two states ($s_z = \pm \hbar/2$)—spin up and spin down—which we can denote by $|+\rangle$ and $|-\rangle$. If now one makes a measurement on an arbitrary neutron state, in order to determine the z-component of spin, one will find that it is either in the state $|+\rangle$ or in the state $|-\rangle$, never in between. If the initial state is $|\psi\rangle = a|+\rangle + b|-\rangle$, then the measurement will reveal a probability for finding spin up of $p_+ = |a|^2$, and a probability for spin down of $p_- = |b|^2$, where $|a|^2 + |b|^2 = 1$. But although $|\psi\rangle$ may have been in a superposition of up and down, the result of the measurement is always up or down. All these statements are also true if the measurement is made along any other direction.

Now if the spin happens to lie along some direction in the x–y plane, at an angle φ from the x-axis (see Fig. 2), then the wave function will be

$$|\psi_n\rangle = 2^{-1/2}(|+\rangle + e^{i\varphi}|-\rangle).$$

In other words, a measurement in the z-direction will give equal probability for finding the particle up or down. The alignment in the x–y plane is determined by the relative phase of the two components.

If one could translate these statements into an experiment, it would be a clear example of superposition. For example, if the particle were lined up along the $(+x)$-axis, a measurement along x would yield 100% chance of finding the particle along $+x$ and 0% chance of finding it along $-x$. Yet a measurement along the z-axis would give 50% for finding it up or down.

Thus while one measurement reveals that half the particles are along $+z$ and half are along $-z$, a different measurement would reveal that they are all along $+x$, a direction orthogonal to both $+z$ and $-z$. So the

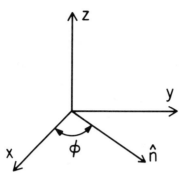

Figure 2   Spin wave function. For a neutron polarized along the direction $\hat{n}$ in the x–y plane at an angle φ to the x-axis the wave function will be $|\psi\rangle = (|+\rangle + e^{i\varphi}|-\rangle)/\sqrt{2}$.

state of the particle contains information that cannot be revealed by a simple intensity measurement.

The simple gedankenexperiment I mentioned is the following. A beam of neutrons, all polarized spin up, impinges upon a double slit. The neutrons passing through slit a are unaffected, and so preserve their polarization (see Fig. 3). However, just behind slit b there is a flip coil, so that all neutrons passing through slit b are flipped, and they emerge spin down. Subsequently, at a screen, the neutron beams are coherently recombined.

A detector runs along the screen in the z-direction and measures the neutron intensity. Because of the different distance to the detector from each of the two slits, the neutron wave function at the detector will be proportional to

$$|\psi\rangle = 2^{-1/2}(|+\rangle + e^{i\varphi(z)}|-\rangle).$$

If the detector is set to only measure polarization along the $+z$-axis, it will only see those neutrons in the state $|+\rangle$, which are those which passed through the slit a, and the intensity pattern will be that labelled by I_a in Fig. 3. If it is set to only detect spin down, it will detect the intensity pattern I_b. In neither case will there be an interference pattern. But if the detector is set to detect polarization along the $+x$-axis, the intensity will go as $\cos^2 \varphi(z)/2$, and as φ changes the intensity will vary between 0 and 1, giving a 2-slit interference pattern (pattern I_{a+b}).

This result is consistent with the conclusion of measurement theory that one cannot simultaneously see an interference pattern and determine which slit the neutron passed through. In the case where the

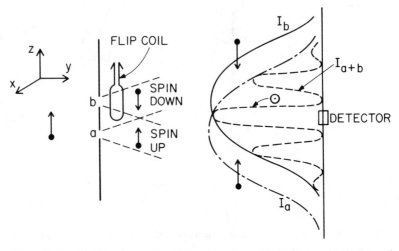

Figure 3   Double-slit spin experiment. A neutron beam polarized spin up along the z-axis is sent through a double slit. Immediately behind slit b is a flip coil, so all neutrons emerging from slit b are spin down. A detector moving along a screen will see no interference if it can detect only neutrons along the $+z$ (pattern I_a) or $-z$ (pattern I_b) direction. However, if it can detect neutrons along the x-axis it will observe interference (pattern I_{a+b}).

detector saw only the state $|+\rangle$ or $|-\rangle$, one knew through which slit the particles passed, and there was no interference. In the case where particles were detected polarized along the x-direction, there were contributions from both slits, and thus an interference pattern was possible. Another beautiful facet to this experiment is that in the last case, the particles are detected polarized along the x-direction, but the particles passing through each of the two slits separately are each polarized perpendicularly to this direction. So clearly superposition is at work, since neither beam alone has the property in question.

A further interesting feature of this experiment is that if there were no spin involved, the 2-slit interference pattern would have appeared when one measured the intensity of the beam. In the spin case it appears in a polarization measurement, and not in the total intensity, thus showing the coherence of the total wave function, i.e. the spin and space parts together.

III. THE NEUTRON INTERFEROMETER

We need only a few properties of the neutron interferometer to understand its role in the experiments to be described. For more information see Werner.[3] The typical neutron interferometer is composed of three successive slabs (called "ears") of a single perfect crystal, as in Fig. 4(a). The reason one needs a perfect crystal is so that the atoms in the second and third ear are perfectly aligned with those in the first ear. The beam enters the first ear and is Bragg scattered (see the top view, Fig. 4(b)) off the planes perpendicular to the entrance plane of the

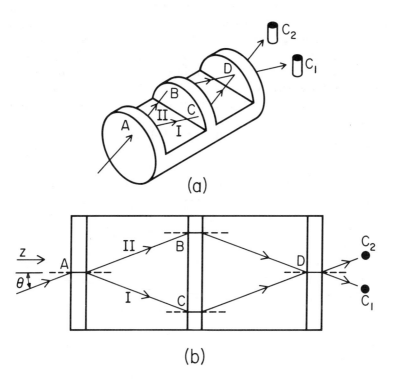

(a)

(b)

Figure 4   The neutron interferometer. (a) Three "ears" are cut from the same perfect crystal, ensuring coherence over the entire crystal (about 10 cm). The incident beam is split at point A into two beams I and II. These are redirected at B and C and recombined in the last ear at D. The relative phase at D determines the counting rates at the detectors C_1 and C_2. (b) The top view of the interferometer. The beam is split by Bragg scattering off the atomic planes perpendicular to the surface of the ear (dashed lines).

ear. At the far end of the first ear two beams emerge, one at the incident angle, and the other at the Bragg scattered angle, each at θ from the normal to the ear.

At the second ear, these beams are rescattered, and they are recombined at the third ear. The beams entering each of the detectors I and II then consist of coherently recombined beams from each of the paths ABD and ACD. If the relative path length is such that, say all of the intensity goes into counter I, then if one of the two beams is altered by one-half of a wavelength, all the intensity will appear at counter II. Thus by monitoring the intensity of one of the counters, changes in path length of about 1% of a wavelength can be detected.

A typical Bragg angle θ might be 20–30°, the crystal is up to 10 cm in length, the atomic place spacing, a, is about 10^{-8} cm, and the energy of the neutrons is typically about .02 eV, or thermal energy, with a wavelength correlated to a, $\sim 10^{-8}$ cm, by the Bragg formula. In a typical reactor, about one coherent neutron per second will pass through the instrument, with a velocity $v \sim 10^5$ cm/sec. So it stays in the interferometer about 10^{-4} sec, and thus only one neutron at a time is in the device.

The width of the Bragg scattering "window" is about $\delta\theta \sim \delta k/k \sim 10^{-6}$ rad, and since $k = 2\pi/\lambda \sim 10^8$, then $\delta k \sim 10^2$. Since $\delta k \cdot \delta z \sim 1$ then $\delta z \sim .01$ cm $\sim .1$ mm. This is the theoretical width of the wave packet, of macroscopic dimensions. And since the beam geometry limits the beam area to about 1 cm², the wave packet is about the size and shape of a small postage stamp. This packet is split in two parts by the interferometer and so one has these two macroscopic wave packets representing the neutron, and coherently separated by a distance of several cm. The interferometer is therefore a really fantastic device for showing quantum interference effects over macroscopic distances (see Greenberger[4]).

The gedankenexperiment above was actually performed by Rauch's group[5] by sending a polarized beam in the z-direction into the interferometer (see Fig 5). At point E a flip coil rotated the polarization of one beam to the $-z$-direction and the beam was detected at counter I. Before the detector was an analyzer system which had the effect of making the detector rotatable, so that polarization in any direction could be detected.

The result was that when the polarization was detected in the $+z$-direction there was no interference pattern, and similarly for $-z$. This was because in either case one could tell by which path the beam had

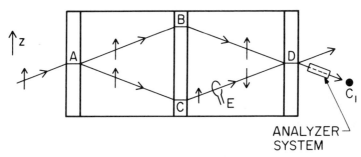

Figure 5   Spin superposition experiment. A highly idealized diagram of the experiments performed by Rauch's group.[5,6] A beam polarized along $+z$ was sent into the interferometer. A flip coil at E coherently flipped the spin of the neutrons in this path. The analyzer had the effect of rotating the detector so that it could detect spin in any direction. When the detector was pointing along $\pm z$ there was no interference pattern (as one could tell which beam the neutron was in). However, when the detector was pointed along x (out of the plane of the paper) interference was detected. This experiment was done in both a DC and an AC version.

arrived. However, if the polarization was detected in the x-direction (out of the plane of the paper) then there was an interference pattern—for exactly the same reason as described in our gedankenexperiment. This beautiful experiment was actually the first time ever that a spin-$\frac{1}{2}$ particle had been coherently split and then recombined. (The first time in a real laboratory. Theorists do it daily with Pauli matrices!) It is both a wonderful confirmation of the superposition principle, and also of some of the statements of the limits on simultaneous measurability implied by measurement theory.

IV. DC AND AC VERSIONS OF THE EXPERIMENT

The experiment just described was in fact performed in two versions, a DC version and an AC one. The physics involved is quite different in each case and this has led to the criticism of quantum theory referred to earlier. In the DC version of the experiment, the flip coil consisted of a static magnetic field aligned in a different direction to that of the overall uniform magnetic field on the apparatus, which was used to define the direction of polarization of the beam. The beam then precessed around the field in the coil to assume a new direction.

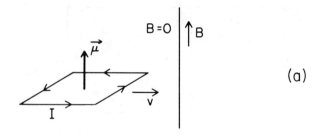

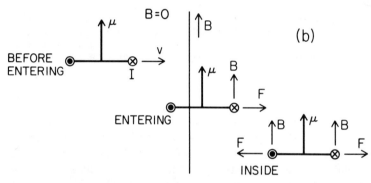

Figure 6 Current loop entering a static magnetic field. (a) The magnetic moment is parallel to the field B. The current loop has not yet entered the field. (b) Side view: Before entering the field there is no force on the current loop. As the loop enters the field there is a force on the forward part which pulls it into the field and accelerates it. Once inside the field there is a compensating backward force. Inside the field the loop has greater kinetic energy and lower potential energy than outside.

To understand the energetics in this case it is easiest to look at a classical analog. Imagine a coil with a current I, moving with speed v, and entering a region of uniform magnetic field B (see Fig. 6). After the front half of the coil enters the field, there is a force on it. This accelerates the coil. Once the second half of the coil enters, there is a compensating backward force, and no further acceleration. So it is only while the current loop is crossing the spatially inhomogeneous boundary region that there is a force on the loop.

One can express this analytically by saying there is a potential energy present, $V = -\mu \cdot B$, and a force, $F = (\mu \cdot \nabla)B$. This force changes the momentum of the loop and therefore, in the quantum analogy, the wavelength. Since the field is static, however, overall energy is conserved. The loss in potential energy is made up for by the increase in kinetic energy caused by the force. If the wave function goes as $\psi \sim \exp[(i/\hbar)(\mathbf{p} \cdot \mathbf{x} - Et)]$ then p changes, λ changes, and the wave fronts shift, causing a change in the interference pattern after the two beams recombine. In this experiment the B field was about $20\,g$, so the distance moved by the neutron in flipping was about $d = vT$, $T = 2\pi/\omega$, $\omega = \mu B/\hbar$. So for a neutron, for which $\mu \sim 10^{-11}$ eV/g, $v \sim 10^5$ cm/sec, $\hbar \sim 10^{-15}$ eV$\cdot$sec, we have $d \sim 1$ cm.

In the second version of the experiment,[6] an r–f flip coil was used to rotate the neutron, as in a Rabi resonance experiment. In this case $B = f(t)$. So now it is not a spatial inhomogeneity, but a time dependence, which is causing the spin flip. In this case if B is uniform then p will not change, nor λ, but the overall energy will not be conserved. Thus in this case the neutron is actually absorbing energy from the external field. It has been put that, since the neutron is absorbing a single photon from the external field, an actual measurement is being made on the neutron, in the quantum mechanical sense, which should destroy its coherence. This argument will be discussed later—but it is incorrect. The neutron remains coherent.

The way in which this experiment was actually done, was to have another coil in the analyzing system, just before the detector in Fig. 5. This coil rotated the neutron again, and was equivalent to rotating the detector. The signal in this coil was then beat against that in the original flip coil, so that the relative phase between the flip coil and the neutrons hitting the detector could be determined. The result again showed that, during a cycle, when the neutrons were polarized in the $\pm z$-direction there was no interference, while when they were polarized along the x-axis, there was interference. Thus this experiment confirms in a dynamic situation the enormous power of the superposition principle, just as the static result did in the previous experiment.

V. THE MEASURING PROCESS FOR FLIPPING NEUTRONS

The objection has been raised[7] that in a time dependent experiment, when a neutron absorbs a photon from the external magnetic field, a measurement has been made, and as a result the neutron must become incoherent with its counterpart in the other beam. Also, according to this argument, it has been determined in which beam the neutron resides, namely the beam in which the photon was present. However, since the experiment clearly shows that an interference pattern still exists, the argument goes, this shows that one has obtained both a knowledge of which beam the neutron was in, and the phase information inherent in the interference process. To know both of these facts simultaneously violates the precepts of quantum theory, but on the other hand is perfectly consistent with a "realistic" theory, in which one has knowledge of the trajectories of individual particles.

But this argument is incorrect. One does not have any knowledge of which beam the neutron was in after this experiment. The interference pattern is not caused by the absorption of the photon, but by the beating of the two amplitudes, one for the photon having been absorbed, and the other for the photon having not been absorbed. It is also incorrect to think of the absorption process as having determined which beam the neutron was in. We pointed out that the width of the "Bragg window" (the barrier separating Brillioun zones in the crystal, which determines over what range of values of k Bragg scattering occurs) is about equivalent to $\delta\theta \sim 10^{-6}$. Therefore $\delta k/k \sim 10^{-6}$, and so $\delta k \sim 10^2$ for neutrons passing through the interferometer (since $k \sim 10^8/cm$).

In Rauch's experiment the angular frequency of the flip coil, ω_f, was about $3.5 \times 10^5/sec$, and so $\delta k/k_f \sim 10^7$. Therefore when one photon is absorbed from the beam it does not absorb nearly enough momentum to knock it out of the beam (by a factor of 10^7), and thus the beam remains coherent.

If instead, one sets up an experiment to monitor the energy, rather than momentum, of the recoiling neutron, one sees that the energy of an absorbed photon is $\hbar\omega_f \sim 4 \times 10^{-10}$ eV. In terms of energy, the width of the Bragg window is given by $\delta E/E \sim 2\delta k/k$ (since $E \sim k^2$), or $\delta E \sim 2 \times 10^{-6}E \sim 4 \times 10^{-8}$ eV.

So in terms of energy the photon energy is about 100 times too small to remove the neutron from the beam, and again coherence is preserved. In fact if it were otherwise, there could be no coherent phase effects in classical physics, as every individual absorption event would disrupt the system. Thus it is untrue that individual absorption events from the external field constitute measurements of the neutron position in the quantum mechanical sense, or that they disrupt the coherence of the beam.

A variation of this argument, also incorrect, is the claim that one can simultaneously determine both the phase of the photon field and the number of photons absorbed. In this connection it is pointed out that the often quoted relation, $\Delta N \cdot \Delta \varphi \sim 1$, where N represents the number of quanta (photons) and φ the phase, is not right, and that one should use a periodic function such as $\cos \varphi$ in evaluating uncertainties. But in fact in many experiments the expression $\Delta N \cdot \Delta \varphi$ can be given a firm meaning. If one has a way of keeping continuous track of φ over many periods, then one can obtain valid results. For example in our experiment the neutron is moving down the interferometer, and $x = vt$, so that $\delta x = v \delta t$.

As the neutron moves through the flip coil, one may keep track of the phase as it precesses, since $\varphi/2\pi = t/T_f$, where T_f is the flip period, and so $\varphi = \omega_f t$, and $\delta \varphi = \omega_f \delta t$, or $\delta \varphi = \omega_f \delta x/v$. Therefore one can monitor the phase of the neutron in this experiment, by observing its position.

If one wanted to measure φ accurately so that $\delta \varphi \ll 2\pi$, one could do so, but not if one absorbed only one photon from the beam. For the total momentum one absorbs from the beam, $p_{ph} = \hbar k = N \hbar k_f$, is a measure of the number of photons absorbed. But then $N = k/k_f$, and $\delta N = \delta k/k_f$. Therefore

$$\delta N \cdot \delta \varphi = \left(\frac{\delta k}{k_f}\right)\left(\frac{\omega_f \delta x}{v}\right) = \left(\frac{c}{v}\right)\delta k \cdot \delta x,$$

and since $\delta k \cdot \delta x > 1$, we have in this experiment that $\delta N \cdot \delta \varphi > 10^5$.

Again, one could do much better by monitoring energy rather than momentum, but of course one still cannot beat the uncertainty principle. In this case $E_{ph} = N \hbar \omega_f = \hbar \omega$, and so $\delta N = \delta \omega/\omega_f$. Then

$$\delta N \cdot \delta \varphi = \frac{\delta \omega}{\omega_f} \frac{\omega_f}{v} \delta x$$

$$= \delta \omega \frac{\delta x}{v} = \delta \omega \delta t > 1$$

and we see again that we cannot simultaneously know both how many photons have been absorbed and the phase. In this case knowledge of phase and photon number is determined by measurements of more conventional quantities like momentum and position, and so these relationships take on a concrete meaning. So even though it is true that one cay say that if the neutron has flipped, it must have absorbed one photon, nonetheless, if one sets about to determine its phase, one cannot measure that it has absorbed one photon. The standard quantum mechanical interpretation is perfectly adequate to describe these experiments.

VI. CONCLUSION

The neutron interferometer is an amazing instrument which allows one to set up a coherently split neutron wave packet, of macroscopic size, and separated spatially by several centimeters. With this device one can actually perform a number of experiments which would have previously been considered to be gedankenexperiments. To date, all of these experiments have verified the predictions of quantum theory in great detail.

I have discussed the first experiments to fully verify the superposition principle for spin, which incidentally also extend the concept of the two-slit optical experiment. We have also pointed out that some objections which have been raised to the interpretation of these experiments are based on an incorrect conception of what corresponds to a quantum measurement.

Acknowledgments

This work was supported in part by grants from the National Science Foundation and the PSC-CUNY Research Foundation. It was also supported in part by the NSF and OED through the MIT Neutron Diffraction Laboratory, and I owe a great debt of gratitude to its director, Prof. Cliff Shull. The paper was written during a stay at the Institute for Theoretical Physics at the University of California at Santa Barbara, and special thanks are owed to Profs. Vinay Ambegaokar, Tony Leggett, and Walter Kohn, and to the Institute staff.

References

1. A. Zeilinger, R. Gaehler, C. G. Shull and W. Treimer, *Symposium on Neutron Scattering, Argonne National Laboratory* (Amer. Inst. Phys., 1981).
2. C. G. Shull, *Fiftieth Anniversary of the Neutron* (Cambridge University Press, Cambridge, 1983).
3. S. A. Werner, *Physics Today*, Dec. 1980, p. 24.
4. D. M. Greenberger, Rev. Mod. Phys. **55**, 875 (1983).
5. J. G. Summhammer, H. Badurek, H. Rauch and U. Kischko, *Phys. Lett. A***90**, 110 (1982).
6. G. Badurek, H. Rauch and J. Summhammer, *Phys. Rev. Lett.* **51**, 1015 (1983).
7. C. Dewdney, Ph. Gueret, A. Kypryanidis and J. P. Vigier, *Phys. Lett. A***102**, 291 (1984), and various preprints.

MACROSCOPIC QUANTUM COHERENCE IN SUPERCONDUCTING INTERFERENCE DEVICES

Sudip Chakravarty

Department of Physics, State University of New York at Stony Brook, Stony Brook, NY 11794

This paper reviews the considerable progress that has been made in the subject of macroscopic quantum coherence in superconducting quantum interference devices (SQUID) in the past three years. The focus of attention is the quantum fluctuations of a macroscopic variable, the total magnetic flux through the superconducting loop.

I. INTRODUCTION

In this paper I shall try to explore the question as to what extent the behavior of a superconducting quantum interference device (SQUID)[1] can be affected by the quantum fluctuations of the macroscopic variable, the total magnetic flux through the loop. It is hoped that some of these ideas can be tested with the present-day micro-fabrication and low temperature cryogenic techniques; in fact, currently, a number of experiments are in progress. The question of general interest is as follows:[2] Is it possible that the state of a complex macroscopic system (the SQUID in the present case) is a linear superposition of two distinct macroscopic states (the distinct flux states of the SQUID in the present discussion)? Having stated the problem in its utmost generality, I shall

131

retreat into the fascinating world of superconductors, and shall have very little to say about the quantum measurement theory.

II. A SIMPLIFIED THEORETICAL PICTURE OF A SQUID

It is worth our while to recapitulate the arguments due to Bloch.[3] First, consider a superconducting ring, i.e. one without a weak-link (see Fig. 1(a)); the external magnetic field vanishes (Meissner effect) in the region occupied by the ring so that its flux through any closed curve around the ring has the same value. Similarly, if we take into account the field contributed by the particles, the same considerations as above would

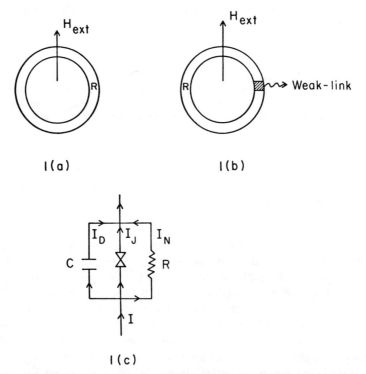

Figure 1 (a) A continuous superconducting ring. (b) A rf-SQUID; a superconducting ring interpreted by a weak link. (c) An electrical circuit description of the current flowing in a rf-SQUID.

apply to the total flux. Now comes the crucial assumption which is appropriate for most practical situations, but, as we shall see, not for our purpose. One assumes that while the particles obey the laws of quantum statistical mechanics, one can ignore, with entirely negligible errors, the quantum and statistical fluctuations of the total vector potential $\mathbf{A}(\mathbf{r})$ (induced plus external). If this is so, the total $\mathbf{A}(\mathbf{r})$ can be taken to be uniquely determined, and the free energy F can be calculated from the Hamiltonian

$$H = H[\mathbf{p}_j - e_j \mathbf{A}(\mathbf{r}_j)/c, \mathbf{r}_j], \qquad H\psi = E\psi.$$

The Schrödinger equation is to be solved subject to the condition that the wave function is single valued in all particle coordinates $\mathbf{r}_j$. Since there is no magnetic field in the interior of the ring, one can write $\mathbf{A} = \nabla\chi$. A gauge transformation then leads to

$$H_0\psi_0 = E\psi_0,$$

where H_0 does not contain $\mathbf{A}$ any more. However, since ψ is single-valued and since the line integral $\int \mathbf{A} \cdot d\mathbf{s} = \phi$, the wave function ψ_0 is multiplied by $\exp(-ie_j\phi/hc)$ when the particle j is brought around the ring. Therefore, since charge is quantized in units of e, the phase factor repeats itself whenever ϕ changes by hc/e. Remembering that the boundary condition and the Schrödinger equation *together* determine the energy eigenvalues, it follows that the energy eigenvalues are also periodic functions of $\phi/(hc/e)$. This implies that the free energy F is an even (by time reversal invariance) periodic function of $\phi/(hc/e)$. Quite generally we can write

$$F = \sum_{n=0}^{\infty} F_n \cos(2\pi n\alpha)$$

where $\alpha = \phi/(hc/e)$. This argument is obviously incapable of determining what these coefficients F_n are. In particular, it may be that all the F_n's for $n \geqslant 1$ are zero. This is actually what happens in a normal metal. One ought to pause here and admire the argument; the implied assumptions, except one, are so general that there is no room for any modification: (a) the total vector potential has neither statistical nor quantum fluctuations; (b) charge is quantized in units of e; (c) the invariances under time reversal and gauge transformations hold; (d) the wave function is single-valued and continuous. It is only the first assumption that in principle can be modified and we shall later see how. We should also note that the above result holds irrespective of any specific properties of the

ring and thus remains valid even if a weak link is added to the system
(Fig. 1(b)). The concept of free energy of course refers to thermal
equilibrium at a fixed value of the flux. However, if we assume that ϕ
varies so slowly in time that the system is in equilibrium at each instant,
we may write (V the voltage and I the current),

$$\frac{dF}{dt} = IV,$$

and hence from Faraday's law,

$$I = -c\frac{dF}{d\phi},$$

where the current I is now given by,

$$I = \sum_{n=1}^{\infty} I_n \sin(2\pi n\alpha),$$

and $I_n = 2\pi ne F_n/h$. We still do not know whether the I_n's are non-zero or
not. At this point we invoke other considerations; from the concept of
off-diagonal long range order introduced earlier by Yang,[4] Bloch
concludes that the lowest non-vanishing term is the one with $n = 2$.
Bloch then shows that if there exists a weak-link which presents a barrier
to the electrons in the ring, I_n for higher n become progressively smaller,
i.e. $I_n \sim \theta^n$, where θ is the quantum mechanical barrier transmission
factor. We then arrive at the celebrated expression for the supercurrent
which was first derived by Josephson:[5]

$$I = I_2 \sin(2\pi\phi/\phi_0),$$

where $\phi_0 = hc/2e$ (note $2e$ here). We have taken this particular route in
deriving the result in order to make the assumptions clearer; so that we
can see how the present subject of macroscopic quantum coherence
abandons one of the assumptions listed above.

The total free energy of the system is the sum of the free energy of the
electrons and the *energy* stored in their accompanying field (consistent
with the assumptions above, we are ignoring the fluctuations of the field;
hence we need not distinguish between the energy and the free energy of
the field). The energy stored in the circulating current is $\frac{1}{2}LI^2$, where L is
the self-inductance of the ring. We can also express this energy as
$(\phi - \phi_{ext})^2/2Lc^2$; ϕ_{ext} is the flux due to the external field through the ring
which we take to be an experimentally controlled c-number. I shall now

change to MKS units for later convenience and write for the total free energy F_{tot}:

$$u(\phi) \equiv F_{tot} = \frac{(\phi - \phi_{ext})^2}{2L} - \frac{I_c \phi_0}{2\pi} \cos\left(\frac{2\pi\phi}{\phi_0}\right),$$

$$\phi_0 = h/2e = 2.07 \times 10^{-15} \text{ Webers}$$

where I have set $I_2 \equiv I_c$; I_c is known as the critical current of the junction. If we now recognize that any time dependence of the total flux will induce a voltage across the capacitor (the weak link now acts as the capacitor), the dynamics of the flux will be given by,

$$C\ddot{\phi} = -\frac{\partial}{\partial\phi} u(\phi),$$

because the stored capacitive energy is $\frac{1}{2}C\dot{\phi}^2$. There is one further consideration that we must not forget and that is dissipation. A phenomenological approach is to add to the above equation a term $\frac{1}{R}\dot{\phi}$ due to the presence of dissipative normal current, where R is the resistance of the weak link. Thus we have

$$C\ddot{\phi} + \frac{1}{R}\dot{\phi} = -\frac{\partial}{\partial\phi} u(\phi).$$

This is an extremely successful phenomenological equation on which all common experimental considerations of the operation of a SQUID is based, and which goes under the name of RSJ (Resistively Shunted Junction) equation. Figure 1(c) shows the engineering equivalent of this equation; the ideal Josephson element is in parallel with a capacitor transmitting a displacement current ($C\ddot{\phi}$), and a resistor which carries a normal current ($\dot{\phi}/R$). The Josephson supercurrent is, as above, $I_c \sin(2\pi\phi/\phi_0)$.

In order to understand the operation of a SQUID, it is helpful to look at $u(\phi)$ in more detail. This quantity, according to the equation above, can be interpreted as the potential energy of a Newtonian particle whose coordinate is ϕ. When $\phi_{ext} = 0$, the particle sits at the stable minimum $\phi = 0$ of $u(\phi)$ (Fig. 2(a)) and hence the supercurrent flowing through the ring is zero. If we now turn on ϕ_{ext}, the particle will be trapped in a metastable minimum as shown in Fig. 2(b) (we are ignoring fluctuations), and a supercurrent $I_c \sin(2\pi\phi/\phi_0)$ will flow through the ring. If we keep on increasing ϕ_{ext} we would ultimately reach a situation in which the

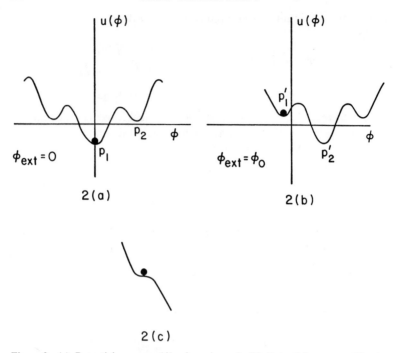

Figure 2 (a) Potential energy $u(\phi)$ when $\phi_{ext} = 0$. (b) Potential energy $u(\phi)$ when $\phi_{ext} = \phi_0$. (c) Classical break-point: If ϕ_{ext} is increased any further the particle will roll down the hill.

metastable minimum would disappear, as shown in Fig. 2(c), and the particle will roll down the hill, ultimately settling down at the stable minimum with no supercurrent flowing through the ring any more. The mechanics described here is precisely the basis of operation of a SQUID.

III. THERMAL ACTIVATION

The time is now ripe to re-examine one of the "standard" assumptions mentioned earlier, namely that the accompanying electromagnetic field of the particles has no thermal fluctuations. This would be true if the barriers in $u(\phi)$ were strictly macroscopic but they are only of the order of $I_c\phi_0$, which for rings under present experimental conditions is about 10–15 K. One must note that $I_c\phi_0$ is the total energy barrier, *not* the energy barrier per electron. Therefore the particle does not have to wait until the metastable minimum turns into a point of inflection, but can be thermally activated over the barrier. Thus the whole SQUID can change its macroscopic state, characterized by the total flux enclosed, due to thermal fluctuations. One may now ask if this can be experimentally observed. The answer is yes, and it was unambiguously observed almost ten years ago by Jackel et al.[6] A related experiment on a current biased Josephson junction was done by Fulton and Dunkelberger.[7] These experiments have been recently repeated by Schwartz et al.[8] with experimental parameters more relevant to the present problem, i.e. the problem of quantum tunneling which I shall discuss below. The experimental results unambiguously follow the Kramer's expression for the activation rate which was first analyzed in the present context by Kurkijärvi.[9] The activation rate over a barrier of height Δu at a temperature $1/\beta$ is given by,

$$\frac{1}{\tau} = \frac{\omega e^{-\beta \Delta u}}{2\pi}.$$

The prefactor ω is an "attempt frequency" and depends on dissipation; i.e. the resistance of the weak link in the present case. The experiments are not yet accurate enough to detect subtle variations of the prefactor ω with R.

IV. QUANTUM FLUCTUATIONS AND DISSIPATION

Armed with the assurance that the macroscopic state of the SQUID can change due to thermal fluctuations, the theory of which is well understood and in agreement with experiments, we now turn to the discussion of quantum fluctuations. Furthermore, the above discussion was designed to motivate the fact that the macroscopic state of the SQUID can be well characterized by a single collective coordinate, the total flux through the loop. We see no reason why this should be untrue in a quantum mechanical treatment. Let us now allow the electromagnetic field accompanying the particles to have quantum fluctuations as well. Is this realistic? People do not often realize how remarkably realistic this is. To substantiate this view consider the number of photons contained in this field; the stored energy in the field is of the order of $\phi_0^2/2L$ ($\phi - \phi_{\text{ext}} \sim \phi_0$ as we shall see later); if we now divide this energy by $\hbar\omega_0$, where ω_0 is a typical small oscillation frequency around the bottom of the wells in $u(\phi)$, we get, for a relevant set of experimental parameters, a number anywhere between 1 and 100. It is clear that a set of experimentally accessible realistic parameters exists for which we are far from the classical limit. What is remarkable is that the description in terms of the collective variable ϕ is still meaningful provided we introduce explicitly the dissipative degrees of freedom of the environment which couple to ϕ; clearly the basic phenomenon of superconductivity, i.e. the coherent nature of the macroscopic state, plays an important role here.

Being a macroscopic variable, the total flux ϕ is strongly coupled to the environment and no meaningful discussion can be carried out without explicitly introducing the dissipative degrees of freedom. This is a point that has been emphasized by Leggett on numerous occasions, and has been expounded at great length by Caldeira and Leggett.[10] In the related problem of a current biased Josephson junction a first principles microscopic formulation has been given by Ambegaokar et al.[11] Here I shall follow Caldeira and Leggett who have essayed to give a rather general formulation of the problem. They have argued that provided any one degree of freedom of the environment is weakly perturbed, the environment can be considered as a collection of independent harmonic oscillators coupled to the system. They have also argued, quite reasonably, that the coupling of the system to the environment is linear in ϕ in the present problem. We therefore arrive at the Lagrangian

$$\mathscr{L} = \tfrac{1}{2}C\dot{\phi}^2 - u(\phi) + \sum_\alpha \tfrac{1}{2}m_\alpha(\dot{x}_\alpha^2 - \omega_\alpha^2 x_\alpha^2) - \phi \sum_\alpha f_\alpha x_\alpha .$$

One also needs a counter term;[10] I shall assume that it is always present. Here x_α represents the degrees of freedom of the environment. If one now demands that the classical equation of motion for ϕ should reduce to the phenomenological (observable) RSJ equation mentioned earlier, one must demand that the spectral density $J(\omega)$ of the environment be given by

$$J(\omega) = \frac{\pi}{2}\sum_\alpha \frac{f_\alpha^2}{m_\alpha \omega_\alpha}\delta(\omega - \omega_\alpha)$$

$$= \frac{1}{R}\omega, \qquad \omega < \omega_c$$

$$= 0, \qquad \omega \geq \omega_c$$

where ω_c is a microscopic (generally fast) frequency scale. While none of the important results will change if we take a different form of $J(\omega)$ which behaves linearly as $\omega \to 0$ and vanishes as $\omega \gg \omega_c$, it is absolutely crucial to note that it is only the linear behavior of $J(\omega)$ as $\omega \to 0$ which gives the characteristic linear dissipation of the RSJ model, as one can readily verify by writing down the classical equation of motion for ϕ and eliminating the x_α variables. The phenomenological RSJ equation has thus produced a strong constraint on the spectral density of the environment. It is now evident that dissipation in the present framework is nothing but the transfer of energy from the system to the infinite number of degrees of freedom of the environment. It is assumed that once the energy is transferred out it does not return on a physical time scale. Given the total Lagrangian it is trivial to quantize the complete system using the Feynman path integral formalism. The quantum mechanical nature of ϕ is nothing but the manifestation of the quantum mechanical fluctuations of the electromagnetic field which we have argued earlier cannot be treated classically in the present context.

V. MACROSCOPIC QUANTUM TUNNELING

Consider now the situation shown in Fig. 3(a), i.e. the flux in the SQUID is experimentally trapped in a metastable state. One would like to know the rate at which it could quantum mechanically tunnel out of this state. With not much loss one can simplify the problem as shown in Fig. 3(b). Furthermore, considering specifically the case of a SQUID, one can approximate $u(\phi)$ by a cubic potential, i.e.

$$u(\phi) = A\phi^2 - B\phi^3.$$

This approximation is valid as long as tunneling predominantly occurs close to the classical break point, which is defined to be the point at which the metastable local minimum turns into a point of inflection. The tunneling rate at zero temperature was calculated by Caldeira and Leggett[10] and can be summarized in the formula:

$$\frac{1}{\tau} = A(\eta) e^{-B(\eta)/\hbar}.$$

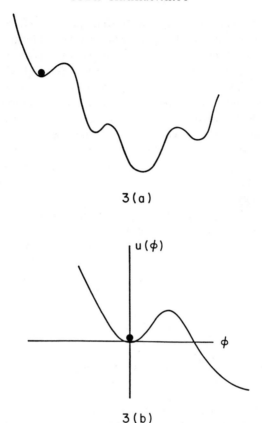

3(a)

3(b)

Figure 3   (a) A schematic picture of the situation in which the particle is trapped in a metastable well. (b) The cubic potential well.

They obtained the following analytic results:

$$B(\eta)/(\tfrac{1}{2}C\omega_0(\Delta\phi)^2) \equiv b(\eta) = b(0) + 1.86\eta + O(\eta^2), \qquad \eta \to 0$$

$$= \frac{8\pi}{9}\eta + \frac{2\pi}{9}\frac{1}{\eta} + O\!\left(\frac{1}{\eta^2}\right), \qquad \eta \to \infty$$

where $\eta = (2\omega_0 RC)^{-1}$ and ω_0 is the small oscillation frequency at the bottom of the metastable well. For intermediate values of η, they gave some good variational bounds. In a recent paper the present author in collaboration with L.-D. Chang[12] has obtained very accurate numerical results for all relevant values of η. We now know $A(\eta)$ with an accuracy better than 2% and $b(\eta)$ better than 0.1%. All of these calculations employ the so-called instanton method which was invented by Langer[13] and was later popularized by Callan and Coleman.[14]

The conclusion that we can draw is that while dissipation reduces the tunneling rate, it does not reduce it to the point of unobservability. For a readily accessible set of parameters, $R \sim 10^2$ to $10^4\,\Omega$, $I_c \sim 10^{-6}$ to $10^{-8}\,A$, $C \sim 10^{-13}$ to $10^{-14}\,F$, and $L \sim 5 \times 10^{-10}\,H$, quantum tunneling should be observed (1 to 10^6 per second) at sufficiently low temperatures. In fact, two recent experiments[15,16] claim to have done so. At present a considerable amount of work is in progress which is aimed at calculating the rate at finite temperatures, particularly the crossover temperature between the thermally activated regime and the quantum regime. David Waxman[17] has recently made some progress in this direction.

VI. MACROSCOPIC QUANTUM COHERENCE

We now look at a somewhat different problem. When $\phi_{ext} = \phi_0/2$ and $2\pi L I_c/\phi_0 > 1$, $U(\tilde{\phi})$ consists of two symmetrical wells shown in Fig. 4. Here we have defined $\tilde{\phi} = \phi - \phi_{ext}$. Classically the zero temperature state would correspond to a situation where the particle sits at the bottom of either one of the wells for ever. Which particular well it sits in will depend on how the system was prepared, since the energies are the same for them both. The two different minima correspond to two

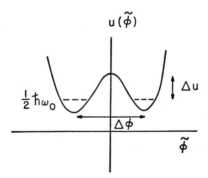

Figure 4 $U(\tilde{\phi})$ when $\phi_{ext} = \phi_0/2$: The double well.

different senses of the rotation of the supercurrent $I_c \sin(2\pi\tilde{\phi}/\phi_0)$, which are of equal magnitudes. If we ignore dissipation altogether, it is easy to calculate the quantum mechanical tunnel splitting. The tunnel splitting $2\Delta_0$ can be written as

$$2\Delta_0 = 4\sqrt{3}\, h\omega_0 \left(\frac{s_0^{1/2}}{2\pi h}\right) e^{-s_0/h}$$

where

$$\frac{s_0}{h} = \frac{16}{3}\left(\frac{\Delta u}{h\omega_0}\right).$$

Here Δu is the barrier height and ω_0 the small oscillation frequency. If this were all there was to it one could conclude that the quantum mechanical ground state is a linear superposition of two distinct macroscopic states, and the average supercurrent will be zero due to the quantum superposition principle.

The above discussion is highly simplified and must not be trusted without a serious investigation of the effect of dissipation. To do this, we again consider the Lagrangian given earlier, but now the potential energy $u(\tilde{\phi})$ represents a symmetric double well. Instead of the general form $u(\phi)$, we use a truncated basis: one state per well. For experimentally realizable situations this can be shown to be reasonable. The mapping of the double well coupled to the environment to an effective two-level system requires some careful consideration of the parameters of the effective problem.[18] We therefore consider the following Hamiltonian (σ's are the usual Pauli matrices)

$$H = -\tfrac{1}{2}h\Delta_0\sigma_x + \sum_\alpha \frac{1}{2}\left(\frac{p_\alpha^2}{m_\alpha} + m_\alpha\omega_\alpha^2 x_\alpha^2\right) + \tfrac{1}{2}\Delta\phi\sigma_z \sum_\alpha f_\alpha x_\alpha$$

where the spectral density of the environment, $J(\omega)$, is given by

$$J(\omega) = \frac{\pi}{2}\sum_\alpha \frac{f_\alpha^2}{m_\alpha\omega_\alpha}\, \delta(\omega - \omega_\alpha).$$

But we have argued previously that $J(\omega)$ must be of the form $\dfrac{1}{R}\omega$ for $\omega \ll \omega_c$, and vanish when $\omega \gg \omega_c$. Furthermore, for our problem $\Delta_0/\omega_c \ll 1$; $h\Delta_0$ is the tunnel splitting discussed earlier (renormalized by a factor discussed in Ref. 18) which is of the order of mK or less, and ω_c is of the order of the small oscillation frequency (see Ref. 18) for an explicit expression) which is of the order of a few K (or more precisely $h\omega_c \sim$ a few degrees). Let us also introduce a parameter called α by,

$$\alpha \equiv \frac{(\Delta\phi)^2}{2\pi\hbar R} = \frac{x^2}{4}\frac{(h/e^2)}{R},$$

where the separation $\Delta\phi$ between the two minima is some fraction x of the flux quantum ϕ_0.

The problem is the following: Given that for $t < 0$ the system is known to be localized in the state corresponding to $\sigma_z = +1$, what is the value $p(t) \equiv \langle \sigma_z(t) \rangle$ for $t > 0$, and in particular how far is the characteristic oscillatory behavior $\cos \Delta_0 t$ of the uncoupled system preserved for finite α? The answer is rather striking and has been recently discussed by the present author and A. J. Leggett.[19] Let me summarize the conclusions. Let us define a parameter Δ_r by

$$\Delta_r \equiv \begin{cases} \Delta_0 \left(\dfrac{\Delta_0}{\omega_c}\right)^{\alpha/(1-\alpha)}, & \alpha < 1 \\ 0, & \alpha > 1 \end{cases}.$$

We can then show that (to lowest order in Δ_r/ω_c, $kT/\hbar\omega_c$):

1. For all $\alpha kT \gg \hbar\Delta_r$, relaxation of the spin is incoherent and is given by

$$\frac{1}{\tau} = \left(\frac{\Delta_0^2}{\omega_c}\right)\left(\frac{\sqrt{\pi}}{2}\right)\left(\frac{\Gamma(\alpha)}{\Gamma(\alpha + \frac{1}{2})}\right)\left(\frac{\pi kT}{\hbar\omega_c}\right)^{2\alpha-1},$$

i.e.

$$P(t) = e^{-t/\tau}.$$

2. For $T = 0, \frac{1}{2} < \alpha < 1$, we also find an incoherent relaxation but at a rate $\sim \Delta_r$; there is also a power law background.[20]

3. For $T = 0, 0 < \alpha < \frac{1}{2}$, we find damped oscillations with frequency $\sim \Delta_r$ and Q-factor $\frac{1}{2}\cot\left[\left(\dfrac{\pi}{2}\right)\alpha/(1-\alpha)\right]$ plus a power law background. More precisely

$$P(t) = \frac{A(\alpha)}{1-\alpha}\cos\left\{\left[\Delta_{\text{eff}}\cos\left(\frac{\pi}{2}\frac{\alpha}{1-\alpha}\right)t\right]\exp\left\{-\left[\Delta_{\text{eff}}\sin\left(\frac{\pi}{2}\frac{\alpha}{1-\alpha}\right)\right]t\right\}\right.$$

$$+ P_{\text{inc}}(t) + \Delta P(t)$$

where

$$\Delta_{\text{eff}}t = \{[1 + q(\alpha)](\cos \pi \alpha)[\Gamma(1 - 2\alpha)]\}^{1/2(1-\alpha)}\Delta_r t.$$

$q(\alpha)$ tends to 0 (and $A(\alpha)$ to 1) both to first order in α and for $\alpha \to \frac{1}{2}$. We have a strong though not rigorous argument to indicate that $q(\alpha)$ is at most a few percent for all α. $P_{inc}(t)$ is a "cut" contribution which is negative for all t and at large t gives a power-law decay,

$$P_{inc}(t) \approx -\frac{1}{\pi} \sin(2\pi\alpha)\Gamma(2 - 2\alpha)(\Delta_{eff}t)^{-2(1-\alpha)}.$$

Although for sufficiently long times this term will dominate the behavior (except for $\alpha = 0$ and $\alpha = \frac{1}{2}$), from the perspective of the "macroscopic quantum coherence" problem the important point is that at relatively short times the coherent oscillations occur for $\alpha < \frac{1}{2}$ with a Q-factor which is given by $\frac{1}{2}\cot\left[\left(\frac{\pi}{2}\right)\alpha/(1-\alpha)\right]$.

4. The crossover from the thermal to the zero temperature behavior takes place at $\alpha kT \sim \hbar\Delta_r$. This complicated crossover phenomenon has been studied to some extent by A. K. Garg.[20]

Conclusions derived in this calculation are consistent with the earlier work of the present author[21] and Bray and Moore[22] who found that the ground state was two-fold degenerate for $\alpha > 1$ but not for $\alpha < 1$. Consideration of the experimental situation leads one to believe that unless $\alpha < 0.1$ and the temperature is down to a few mK or less, the possibility of observing coherent oscillation is grim. However, the above restrictions are not entirely out of sight given the present-day technology.

VII. PHOTOINDUCED MACROSCOPIC QUANTUM TUNNELING

By adjusting ϕ_{ext} one can also produce an asymmetric double well. It was shown by the present author in collaboration with S. Kivelson[18] that a time-varying external magnetic field through the SQUID loop can cause transitions between the flux states. Once again, given $J(\omega) \sim \omega/R$ for $\omega \ll \omega_c$, the transition rate shows some remarkable power law behavior at the threshold. A transition rate of the order of 10^2–10^3/sec does not seem to be difficult to achieve. In fact this may be one of the cleanest experiments to perform in this subject, since the transition threshold can be experimentally controlled, thereby allowing us to determine whether the flux changes taking place are due to spurious noise or are due to the theoretical mechanism described.

VIII. CONCLUSION

I hope that I have been able to convince you that at very low temperatures superconducting quantum interference devices can be used to ask many interesting questions about quantum mechanics.

Acknowledgments

I would like to thank L.-D. Chang and D. Waxman for many interesting discussions. This work was supported by a grant from the National Science Foundation (DMR-83-01510). The author would also like to thank the Alfred P. Sloan Foundation for a fellowship.

References

1. See for example, M. Tinkham, *Introduction to Superconductivity* (McGraw-Hill, New York, 1975).
2. A. J. Leggett, *Prog. Theor. Phys. Suppl.* **69**, 80 (1980).
3. F. Bloch, *Phys. Rev.* **B2**, 109 (1970).
4. C. N. Yang, *Rev. Mod. Phys.* **34**, 694 (1962).
5. For a text book style discussion of the Josephson effect see for example Ref. 1.
6. L. D. Jackel, W. W. Webb, J. E. Lukens and S. S. Pei, *Phys. Rev.* **B9**, 115 (1974).
7. T. A. Fulton and L. N. Dunkelberger, *Phys. Rev.* **B9**, 4760 (1974).
8. D. Schwartz, J. E. Lukens, C. N. Archie and B. Sen (private communication).
9. J. Kurkijärvi, *Phys. Rev.* **B6**, 832 (1972).
10. A. O. Caldeira and A. J. Leggett, *Ann. Phys. (N.Y.)* **149**, 374 (1983).
11. V. Ambegaokar, U. Eckern and G. Schön, *Phys. Rev. Lett.* **48**, 1745 (1982); see also V. Ambegaokar in *Proceedings of the NATO-ASI Summer School Lectures on Percolation, Localization, and Superconductivity* (Pergamon, New York, 1984).
12. L.-D. Chang and S. Chakravarty, *Phys. Rev.* **B29**, 130 (1984); also erratum, to be published.
13. J. S. Langer, *Ann. Phys.* **41**, 108 (1967).
14. C. G. Callan and S. Coleman, *Phys. Rev.* **D16**, 1762 (1977).
15. R. F. Voss and R. A. Webb, *Phys. Rev. Lett.* **47**, 265 (1981).
16. L. D. Jackel *et al.*, *Phys. Rev. Lett.* **47**, 697 (1981).
17. D. Waxman (private communication).
18. S. Chakravarty and S. Kivelson, *Phys. Rev. Lett.* **50**, 1811 (1983).
19. S. Chakravarty and A. J. Leggett, *Phys. Rev. Lett.* **52**, 5 (1984).
20. A. K. Garg (private communication).
21. S. Chakravarty, *Phys. Rev. Lett.* **49**, 681 (1982).
22. A. J. Bray and M. A. Moore, *Phys. Rev. Lett.* **49**, 1546 (1982).

FLUX PERIODIC EFFECTS IN SMALL ONE-DIMENSIONAL NORMAL METAL RINGS

M. Büttiker

IBM T. J. Watson Research Center, Yorktown Heights, NY 10598

Small one-dimensional rings of normal metal, driven by an external magnetic flux, act like superconducting rings with a Josephson junction except that 2e is replaced by e. The resistance of such a ring connected to current leads is periodic in the flux with the single charge flux quantum. We point out a connection between the flux periodic effects in the closed ring and the oscillation in the resistance of the ring connected to current leads.

I. RESISTANCE AND IRREVERSIBILITY

In 1957, Landauer succeeded in relating the electric resistance of a barrier in a one-dimensional metal to the quantum mechanical transmission probability $T(E)$ of carriers through the barrier.[1] Landauer found that the zero temperature resistance in the absence of inelastic scattering is,

$$R_{el} = \frac{\hbar\pi}{e^2} \frac{1 - T(E_F)}{T(E_F)}, \tag{1}$$

where E_F is the Fermi energy. Subsequently, Landauer pointed out that this same relationship could be applied to the transmission coefficient of the whole conductor.[2] The potential of this result has only recently been appreciated. Eq. (1) has proven to be most fruitful in the investigation of

147

highly disordered systems,[3,4] where the Boltzmann–Bloch theory is inapplicable.

To derive Eq. (1), Landauer considered an obstacle between two reservoirs (Fig. 1) and studied the current flow and density gradients associated with the wave functions at the Fermi level.[1,5] Four wave functions have to be considered; in addition to the two wave functions describing waves of unit amplitude incident on the barrier, there are the complex conjugate (time reversed) wave functions which describe waves of intensity T and R incident from both reservoirs to produce a reflected wave of unit amplitude. These later two wave functions which require *coherent* incident streams from both reservoirs have to be eliminated in order to obtain the positive resistance of Eq. (1). This elimination is achieved by assuming that the reservoirs cause phase randomization. The processes "carrier enters reservoir" and "carrier leaves reservoir" are not reversible. It is this irreversible property of the reservoirs which gives rise to the positive resistance given by Eq. (1). Note, also, that there is no Joule heat produced at the location of the barrier (the sample). In the situation depicted by Fig. 1, energy relaxation processes occur only in the reservoirs. This situation is analogous to that found in the calculation of residual resistance through elastic scattering by point defects. The elastic scattering at the point defect determines the resistance, but the actual energy dissipation arises in inelastic events elsewhere. For further discussions, see Ref. 5.

Thus the following question arises: What happens if we eliminate the reservoirs?

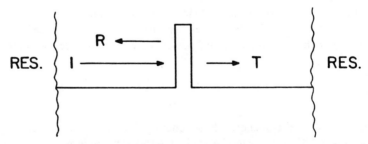

Figure 1   Wave incident from left produces transmitted and reflected waves. The complex conjugate wave function has two incident streams, with a well-defined phase relationship, and a single wave of unit amplitude leaving the obstacle.

II. THE RING WITHOUT RESERVOIRS

To eliminate the reservoirs we close the specimen on itself,[6] making it into a closed loop. To study transport, we apply a (changing) magnetic flux Φ. The motion of electrons in the potential $V(x)$ around the ring and subject to the flux Φ is determined by the Schrödinger Equation with a (time dependent) vector potential. As pointed out in Ref. 6 electrons circuiting in the ring behave exactly like electrons traversing a periodic structure with potential $V(x + L) = V(x)$, where L is the circumference of the ring. Indeed, the Schrödinger Equation of the ring can be mapped onto the Schrödinger Equation of the periodic lattice. A Bloch state of the periodic structure $\psi_n(k, x) = e^{ikx}u_n(k, x)$, $u_n(k, x + L) = u_n(k, x)$ corresponds to an eigenstate $u_n(k_0\Phi/\Phi_0, x)$ of the ring. To obtain the eigenfunctions of the ring, we have to replace the lattice momentum k, $-k_0/2 \leqslant k \leqslant k_0/2$, where $k_0 = 2\pi/L$ is the width of the Brillouin zone, by $k_0\Phi/\Phi_0$, where $\Phi_0 = hc/e$ is the single charge flux quantum. Correspondingly, the energy bands $E_n(k)$ of the lattice (see Fig. 2) yield the one-electron energy spectrum of the ring,[6]

$$E_n(k_0(\Phi + \Phi_0)/\Phi_0) = E_n(k_0\Phi/\Phi_0). \tag{2}$$

The eigenstates of the ring can thus be represented by a ladder of Bloch states with a k selected by the flux Φ. All states on the ladder up to the Fermi energy are filled with one electron (two, if the spin is taken into account).

The mapping discussed above, now allows us to apply the usual solid-state schemes to calculate the transport properties of the ring. Consider a *time-independent* applied flux Φ. The current,

$$I = e\frac{1}{L}\sum_n v_n(k) = -c\sum_n \partial E_n/\partial\Phi,$$

where n labels all occupied states up to the Fermi energy, is non-zero if the flux is not a multiple of $\Phi_0/2$. The velocities in successive bands alternate, but higher lying bands, typically, have larger magnitudes of currents. Thus, typically, the sign of the current will be determined by the highest lying occupied band. Thus, we obtain a *persistent* current which is a periodic function of the applied flux with period Φ_0. Consider a flux which increases *linearly* with time, $d\Phi/dt = cV$, where V is the voltage induced in the ring. According to Bloch an electric field F leads to a change in the crystal momentum given by $dk/dt = -eF/\hbar = -eV/\hbar L$.

M. BÜTTIKER

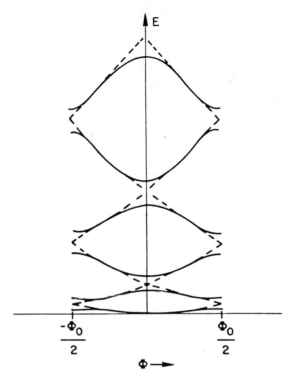

Figure 2 One-electron states of the closed ring as a function of flux. The case without elastic scattering is represented by the dashed lines. Elastic scattering opens gaps in the center and at the boundary of the Brillouin zone.

Thus the whole ladder of states in Fig. 2 is pushed at constant speed through the Brillouin zone. The gaps in Fig. 2 will prevent the electrons from making transitions into higher lying unoccupied bands (Zener tunneling). Therefore, the current oscillates with a frequency $\omega = eV/\hbar$, where $T = 2\pi/\omega$ is the time it takes the ladder to traverse the Brillouin zone. Thus we have a *Josephson* frequency, with a single electronic charge, instead of that of a pair.[6]

The Josephson-like effects discussed above are thus a consequence of the elimination of the reservoirs. Despite the elastic scattering due to the nonuniform potential $V(x)$, the wave functions in the ring are *coherent* and thus sensitive to the flux through the loop.

Up to now, we have completely neglected inelastic scattering. Such scattering will give rise to excitations of electrons into higher lying unoccupied bands into states with the same k-value. These excitations have a small probability as long as kT is small compared with the level spacing. The level spacing is of the order of E_F/N, where N is the number of atoms in the ring. For $N \sim 10^4$ the level spacing is of the order of 10^{-3} eV which is equal to kT for a few degrees Kelvin. Thus for small enough rings and low enough temperatures inelastic scattering is modest and does not destroy the effects described above.

We have now stressed the similarities between such a one-dimensional ring and a superconducting ring[7] with a Josephson junction. We should, however, not overlook the fact that there are also profound differences. In the superconducting loop all electrons condense into the same wave function and all pairs contribute, therefore, with equal velocity to the current. Thus the amplitude[8] of the current $I(\Phi)$ is of order N, whereas $I(\Phi)$ in the normal metal ring is only of order 1. In a superconducting loop, excitations above the ground state (breaking pairs), give rise to a relative change in current of order $1/N$, whereas, in the normal metal ring, excitation of an electron into a higher lying unoccupied band gives rise to a current which typically is larger than the current of the ground state and has the opposite sign.

III. THE RING BETWEEN RESERVOIRS

Gefen, Imry and Azbel[9] have studied the current transport through a one-dimensional ring connected to current leads (Fig. 3). The current leads must eventually be connected to reservoirs, and we have, therefore, again the situation considered in Section I. The only difference is that our obstacle now is a ring instead of a simple barrier or sequence of barriers. To obtain the resistance of this ring, we apply the Landauer formula, Eq. (1). We need to calculate the probability $T(E, \Phi)$ for an electron incident from the left current lead for transmission to the right current lead. The calculation proceeds in the following way;[9] the potential $V(x)$ along a branch of the ring can formally be replaced by a single scatterer (squares in Fig. 3) which gives rise to the same scattering-matrix as the potential $V(x)$. The 2×2 scattering-matrix relates the amplitudes of the two incoming waves to the amplitudes of the two waves leaving the branch. The scattering-matrices are specified by the transmission amplitude t and reflection amplitude r from the left and by the transmission amplitude t' and reflection amplitude r' from the right. Time-reversal

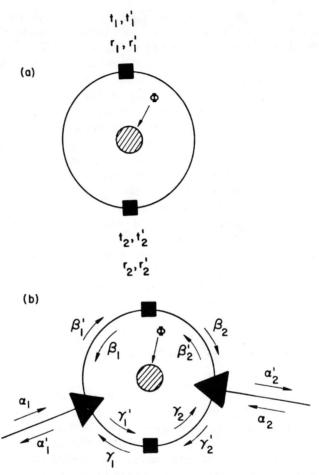

Figure 3   (a) Closed ring with two elastic scatterers. (b) Ring connected to current leads with the same elastic scatterers as in Fig. 3(a).

invariance and current conservation require $t = t'$ and $t/t^* = -r/r'^*$. The junctions of the current leads to the ring are described by 3×3 scattering-matrices (triangles in Fig. 3(b)) which relate the three incoming waves (one from the current lead, and one from each of the two branches of the ring) to the three outgoing waves. The applied flux Φ leads to additional phase changes θ_1 and θ_2, (both taken in a counterclockwise sense), along the two branches of the ring. θ_1 and θ_2

depend on the length of the two branches or, if the ring is circular, on the position of the current leads. But, in any case, the total phase change is $\theta_1 + \theta_2 = 2\pi\Phi/\Phi_0$ with the single charge flux quantum. The principle result of such a calculation is that the trnasmission probability $T(E, \Phi)$ and hence the resistance of the ring is a periodic function of the flux,[9]

$$R_{el}(\Phi + \Phi_0) = R_{el}(\Phi), \tag{3}$$

with a period $\Phi_0 = hc/e$.

Again we would like to stress that the Aharonov–Bohm interferences which give rise to the periodic variation in the resistance occur despite elastic scattering. In fact, the mean elastic scattering length can even be short compared with the circumference of the ring. On the other hand, we must, as for the closed ring, require that inelastic scattering is modest, i.e. the mean inelastic scattering length has to be larger than the circumference of the ring.

IV. RESONANCES IN THE TRANSMISSION PROBABILITY OF THE RING

In this section we want to establish a connection[10] between the flux periodic effects in the closed ring (Fig. 3(a)) and the flux periodic effects in the resistance of the connected ring (Fig. 3(b)). We have shown that sharp peaks in the transmission probability are of the Breit–Wigner form,

$$T(E, \Phi) = T_{res} \frac{\Gamma_n^2(\Phi)}{(E - E_n(\Phi) - \Delta E_n(\Phi))^2 + \Gamma_n^2(\Phi)}. \tag{4}$$

Here, $T_{res} \leqslant 1$, is the value of the transmission probability at resonance. $E_n(\Phi)$ is the energy of an electronic state of the closed ring, Eq. (2), and ΔE_n is a small shift away from this energy. Γ_n is the width of the resonance. In order to obtain sharp resonances, the width of the resonance has to be small compared with the gaps in Fig. 2. Eq. (4) predicts that the position of the peaks in the transmission probability is a function of the flux through the loop. The energy range over which the position of the peak varies as the flux is increased by a flux quantum is equal to the width of the band $E_n(k)$. The resistance of the ring is, using Eq. (1), determined by the transmission probability at the Fermi energy, $T(E_F, \Phi)$. If the Fermi energy lies close to a band $E_n(k)$ or lies in the band, the resistance of the loop will vary from a huge nonresonant value to a very small value at resonance. Strong oscillations in the resistance are thus connected to the eigenstates of the closed loop.

We have identified two mechanisms which give rise to sharp resonances and we will now discuss these in some detail.

The quantum mechanical coupling of the current leads to the ring is variable.[10] We have invoked a 3×3 scattering-matrix (triangles in Fig. 3(b)) which has the property that an electron approaching the ring from the current lead is reflected back into the lead with probability $1 - 2\varepsilon$. The parameter ε plays the role of a coupling strength. For $\varepsilon = \frac{1}{2}$, the reflection probability is zero and the leads are strongly coupled to the ring. For $\varepsilon = 0$ the reflection probability is one and the ring and leads are completely decoupled. In the later case, the electronic states of the ring are obviously the eigenstates discussed in Section II. If we now switch on the coupling, an electron initially in a ring state will remain a long time $\hbar/\Gamma_n$ in the ring as long as the coupling to the leads is poor. We have shown that for a perfectly symmetric ring with no elastic scattering in the branches, $\Gamma_n \propto \log(1 - 2\varepsilon)$ for small ε. The width of the resonances is thus directly related to the reflection probability of the junction. The perfectly symmetric ring without elastic scattering is specified by transmission amplitudes $t_1 = t_2 = e^{i\phi_s}$, where ϕ_s is the phase increment along a branch of the ring and by vanishing reflection amplitudes $r_1 = r_2 = r_1' = r_2' = 0$. The spectrum of the closed symmetric and uniform ring is determined by[10]

$$\cos^2 \phi_s = \cos^2 \pi\Phi/\Phi_0. \tag{5}$$

The solutions of Eq. (5) are $\phi_s = \pi\Phi/\Phi_0$ or since $k = k_0\Phi/\Phi_0$, we have $\phi_s = kL/2$. This is the free electron spectrum indicated in Fig. 2 by dashed lines. The transmission probability of this ring coupled with strength $\varepsilon = 1/16$ to the current leads is shown in Fig. 4. The position of the peaks in Fig. 4 are to very high accuracy given by Eq. (5).

The second mechanism which gives rise to sharp peaks in the transmission probability is strong elastic scattering. We demonstrate this by studying the simple example of a symmetric ring with elastic scattering. We put $t_1 = t_2 = T_s^{1/2} e^{i\phi_s}$ for the transmission amplitude of the scatterers and $r_1 = r_2 = r_1' = r_2' = e^{-i\pi/2} R_s^{1/2} e^{i\phi_s}$, $R_s = 1 - T_s$, for the reflection amplitudes. The eigenvalue spectrum of the closed ring is determined by[10]

$$\cos^2 \phi_s = T_s \cos^2 \pi\Phi/\Phi_0. \tag{6}$$

For $T_s = 1$, Eq. (6) is identical with Eq. (5). For $T_s < 1$ the spectrum

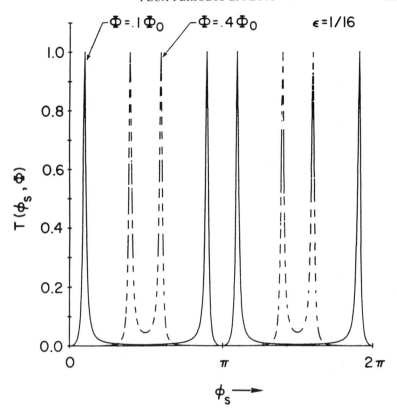

Figure 4   Transmission probability of a symmetric ring with no elastic scattering on its branches in presence of a flux $\Phi = .1\Phi_0$, and $\Phi = .4\Phi_0$ for a coupling strength $\varepsilon = 1/16$. ϕ_s is the phase of the transmission amplitudes, $t_1 = t_2 = e^{i\phi_s}$ of the scatterers.

obtained from Eq. (6) exhibits gaps at $\Phi = n\Phi_0$, i.e. in the center of the Brillouin zone. Since the ring with two equal scatterers corresponds to a periodic structure with identical "atoms" per unit cell, no gaps appear at the Brillouin zone boundary. For $T_s \ll 1$, the case of interest here, the solutions of Eq. (6) are $\phi_s = (2n + 1)\pi/2 \pm T_s^{1/2} \cos(\pi\Phi/\Phi_0)$. In this limit the bands are narrow and the range of allowed phases is limited to a narrow range around $(2n + 1)\pi/2$. Fig. 5 shows the transmission probability of this ring in the strong coupling limit $\varepsilon = \frac{1}{2}$, for $T_s = .25$. Peaks appear only for the eigenstates of the closed ring which have nodes at the junctions and thus couple poorly to the leads. The width of the poles is, in this case, given by $\Gamma_n \propto \log(1 - T_s \sin^2 \pi\Phi/\Phi_0)$ which is small

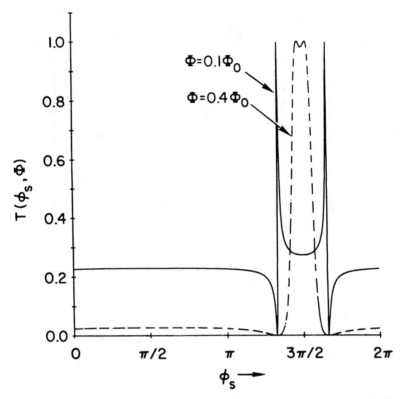

Figure 5   Transmission probability of a symmetric ring with equal elastic scattering in both branches in presence of a flux $\Phi = .1\Phi_0$ and $\Phi = .4\Phi_0$. ϕ_s is the phase of the transmission amplitude of the scatterers $t_1 = t_2 = T_s^{1/2} e^{i\phi_s}$. $T_s = .25$.

for small T_s. The peaks in Fig. 5 are again determined to high accuracy by the eigenvalue equation of the closed ring, Eq. (6).

To summarize: In the case of poor coupling between current leads and the ring, *or* in the case of strong elastic scattering, the ring connected to current leads mirrors the properties of the closed ring. Such a connection seems absent in the case of strong coupling *and* weak elastic scattering. In this latter case, the resistance of the ring between reservoirs is closely related to the Aharonov–Bohm interferences in a vacuum experiment.

V. THEORY AND EXPERIMENT

Al'tshuler, Aronov and Spivak[11] have presented a calculation of the resistance in small normal metal cylinders and small rings, assuming small *elastic* scattering. The result of their perturbation analysis is a resistance which is periodic with period $\Phi_0/2$. Compared with out exact calculations, which yield a period Φ_0, this is just a second harmonic effect. However, at present, the relation between the two approaches is not clear to us. Sharvin and Sharvin[12] have measured the resistance of hollow Mg and Li cylinders and have obtained results which clearly confirm the predictions of Al'tshuler *et al.* Similar experiments on hollow cylinders have since then been performed by Ladan and Maurer[13] and Gijs, Van Haesendonck and Bruynseraede,[14] both yielding a resistance periodic in $\Phi_0/2$.

Lithographic technics have been used by Blonder[15] to produce GaAs rings and by Umbach, Washburn, Laibowitz and Webb[16] to produce Au rings. These experiments yield a resistance for the rings which is a highly structured and complex function of the flux through the ring but seem more compatible with a period Φ_0 given by Eq. (3). To make contact with these experiments, it is necessary to extend our one-dimensional discussion to a real three-dimensional ring with its much higher density of levels.

Note added in proof. The resistance oscillations for a small normal (metal) ring, with a flux period $\Phi_0 = hc/e$, have now been seen in several experiments in a striking and clear fashion.[17]

References

1. R. Landauer, *IBM J. Res. Dev.* 1, 223 (1957).
2. R. Landauer, *Phil. Mag.* 21, 863 (1970).
3. P. Erdös and R. C. Herndon, *Adv. Phys.* 31, 65 (1982).
4. Y. Imry, in *Proceedings of the 1983 NATO Advanced Study Institute on Percolation, Localization and Superconductivity*, ed. A. Goldman and S. Wolf (Plenum, New York, 1984), p. 189.
5. R. Landauer, *Z. Phys. B* 21, 247 (1975); in *Proceedings of the International Conference on Localization, Interaction and Transport Phenomena in Impure Metals*, ed. G. Bergmann, Y. Bruynseraede and B. Kramer (Springer-Verlag, Berlin, 1985), p. 38.
6. M. Büttiker, Y. Imry and R. Landauer, *Phys. Lett. A* 96, 365 (1983).
7. F. Bloch, *Phys. Rev. Lett.* 21, 1241 (1968).
8. N. Byers and C. N. Yang, *Phys. Rev. Lett.* 7, 46 (1961).
9. Y. Gefen, Y. Imry and M. Ya. Azbel, *Phys. Rev. Lett.* 52, 129 (1984).
10. M. Büttiker, Y. Imry and M. Ya. Azbel, *Phys. Rev. A* 30, 1982 (1984).
11. B. L. Al'tshuler, A. G. Aronov and B. Z. Spivak, *JETP Lett.* 33, 94 (1981).

12. D. Yu. Sharvin and Yu. V. Sharvin, *JETP Lett.* **33**, 499 (1981); *JETP* **34**, 273 (1981).
13. F. Ladan and J. Maurer, *C. R. Acad. Sci. Ser. B (France)* **297**, 227 (1983).
14. M. Gijs, C. Van Haesendonck and Y. Bruynseraede, European Physical Society Meeting, March, 1984.
15. G. Blonder, *Bull. Am. Phys. Soc.* **29**, 535 (1984).
16. C. Umbach, *Bull. Am. Phys. Soc.* **29**, 535 (1984).
17. R. A. Webb, S. Washburn, C. P. Umbach and R. B. Laibowitz, *Phys. Rev. Lett.* **54**, 2696 (1985); V. Chandrasekhar, M. J. Rooks, S. Wind and D. E. Prober, *Phys. Rev. Lett.* **55**, 1610 (1985); S. Datta, M. R. Melloch, S. Bandyopadhyay, R. Noren, M. Vaziri, M. Miller and R. Reifenberger, *Phys. Rev. Lett.* **55**, 2344 (1985).

OPTICAL INTERFEROMETER DATA IN SUPPORT OF LOCAL THEORIES

J. D. Franson

Applied Physics Laboratory, The Johns Hopkins University, Laurel, MD 20707

All local theories require that optical interference effects become negligible at single-photon intensities for sufficiently large separation of the optical paths through an interferometer. The quantum theory, on the contrary, predicts that optical interference effects will be independent of the intensity, path length, or path separation, provided that the difference in path lengths remains small. Measurements made using a Jamin interferometer 25 m in length show a significant reduction in the visibility of the interference pattern at single-photon intensities, consistent with the requirements of any local theory. No significant reduction in the visibility was observed at high intensities and large distances, nor at any intensity and short distances. These results are in apparent disagreement with recent experiments by Aspect et al.

All local theories require[1] that there be no single-photon interference effects in an interferometer having sufficiently large separation between the two optical paths. The quantum theory, on the contrary, predicts that optical interference effects will be independent of the intensity, path length, or path separation, provided that the difference in path lengths remains negligible. Preliminary measurements[2] using a Jamin interferometer 25 m in length have shown a significant reduction in the visibility of the interference pattern at single-photon intensities. In contrast, no significant reduction in the visibility was observed at high intensities and large distances, nor at any intensity and short distances. These results are consistent with the requirements of all local theories

159

and are in apparent disagreement with the predictions of the quantum theory. A recent experiment by Aspect *et al.*[3] has been widely accepted[4] as providing conclusive evidence against all local theories. It will be shown here, however, that the interpretation of that experiment contains a hidden assumption, and that the results of the experiment are actually consistent with a plausible local theory.[5] This paper is intended to provide an overview of the work which has been completed and other work which is still in progress. Experimental details and a more rigorous theoretical discussion can be found in other publications.[1,2,5]

Perhaps the most straightforward way to describe the basis for this experiment is to make an analogy with the better-known experiments based upon Bell's theorem.[6] In those experiments, two photons known to have the same polarization travel in opposite directions toward two widely-separated polarizers. The polarization of the two photons is initially uncertain, and this uncertainty is reflected in the quantum-mechanical field describing each of the photons. However, the instant the polarization of one photon has been measured, the polarization of the other photon is determined as well. In the quantum theory, such a measurement produces an instantaneous change (partial reduction) in the field describing the second photon, regardless of its distance from the first. Einstein and others[7] have objected to the non-local nature of this measurement process, and have argued, instead, that the polarizations of both photons must have been determined all along. Bell[6] was later able to show that the quantum theory and any local theory predict different results for experimentally measurable correlations in the polarizations of two such photons.

The experiment to be described here is based, instead, on the properties of single photons passing through a very large interferometer.[1] The situation is actually analogous to that of Bell's theorem, however, as can be understood by referring to Fig. 1. In the quantum theory, a single photon passing through a large interferometer will be described by a field which propagates along both paths of the interferometer. Interference effects will thus occur regardless of the length or separation of the two paths, provided that the difference in path lengths is less than the coherence length of the light source; it will be convenient to assume that the two optical paths are of equal length. However, if the photon should happen to be absorbed at some position x, then the quantum theory requires that its field be set to zero instantaneously at all other positions x', regardless of the distance between x and x'; this is necessary to prevent a subsequent absorption of the same photon. The non-local nature of this process is analogous to the instantaneous change in the photon field discussed previously in connection with the EPR paradox and Bell's theorem. It might be argued that the quantum-mechanical field associated with a photon has a purely probabilistic interpretation, and that it is thus perfectly natural to expect such probabilities to be changed when measurements provide additional information. It is worth noting in that regard, however, that the field associated with the photon must be capable of interacting with both mirrors and of measuring their relative position in order to produce an interference pattern of the correct phase. In any event, the process is non-local.

In a local theory, the situation is quite different, as is represented schematically in Fig. 1. It is an experimental fact that a photon can be absorbed in a finite amount of time by an absorber of finite dimensions, such as an atom. As a result, a photon must be localized within a region whose dimensions are on the order of the speed of light multiplied by the maximum time interval required to definitely absorb such a photon.[1] A single photon in a local theory would thus have to choose between one path or the other through an interferometer whose dimensions were much larger than the characteristic dimensions of the photon itself; single-photon interference effects could not occur under such conditions. A more rigorous discussion of these restrictions has been published elsewhere.[1]

Roughly speaking, the time required to absorb a photon must be comparable with the time required to emit a similar photon, provided that the conditions are such that time-reversal invariance would be expected to apply. This leads[1] to a characteristic dimension l for a single photon which is on the order of

$$l \sim 3\lambda^2/4\pi^2 r_c \qquad (1)$$

where λ is the wavelength and r_c is the classical radius of the electron. It must be emphasized, however, that there are many situations in which the time-reversal argument does not apply. In particular, the time-reversal argument and Eq. (1) do not apply to the light emitted by a laser or any other coherent or correlated source. In addition, the usual coherence length of a light source often represents the statistical properties of a large number of photons, rather than the properties of any one, and is not at all equivalent to the characteristic length of Eq. (1). Because of the possible existence of coherent or correlated effects of the type discussed by Dicke[8] and Senitzke,[9] the ideal light source for an

LOCAL THEORY:

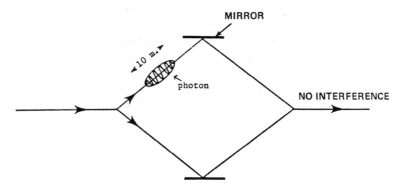

QUANTUM THEORY:

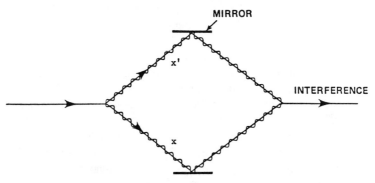

Figure 1   A single photon passing through a large interferometer. In any local theory, a single photon must be confined to one optical path or the other if the distance between the two paths is sufficiently large. In the quantum theory, the field associated with a single photon need not be so confined, and the detection of the photon at position x may require an instantaneous change in its field at a distant position x'.

experiment of this type is one in which there is at most one excited atom at any given time.

The experiment of interest thus consists of determining whether or not interference effects exist in the limit of low source intensity in an interferometer whose dimensions are much larger than the characteristic length of Eq. (1). At optical wavelengths, l is on the order of 10 m. No suitable experiments had previously been performed.[1] The apparatus used in this experiment is outlined in Fig. 2, and consisted primarily of a

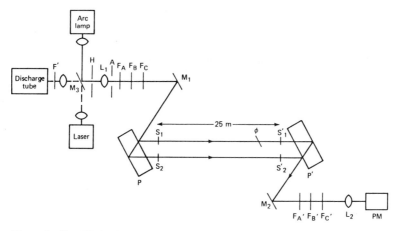

Figure 2 Simplified schematic of the experimental apparatus. The light from a Kr discharge tube, arc lamp, or He–Ne laser could be focused onto a pinhole H by means of three lenses and an adjustable mirror M_3. An achromatic lens L_1 and aperture A produced a collimated beam. Natural density filters mounted on filter wheels $F_A, F_B, F_C, F_{A'}, F_{B'}$, and $F_{C'}$ could be inserted into the beam preceding or following the interferometer; F' was a narrow-band spectral filter. Glass plates P and P' formed the Jamin interferometer, while mirrors M_1 and M_2 were used to aim the beam. S_1, S_2, S_1', and S_2' were shutters which could be inserted into the two beams. The superimposed beams were focused onto the photo-multiplier tube PM by means of lens L_2. The relative phase was varied with fused silica plate ϕ. An evacuated pipe with optical windows (not shown) occupied the space between the shutters.

Jamin interferometer 25 m in length and a thin plate ϕ used to shift the phase of one optical path with respect to the other. The experimental technique has been described in detail elsewhere,[2] and included the normalization of the measured visibilities of the interference pattern to that obtained at the highest available intensity. The normalized visibility V_N obtained in this way had the advantage of cancelling out any variations in the intrinsic visibility of the interferometer itself.

Measurements were first performed using a He/Ne laser and an interferometer length of 25 m, the results of which are shown in Fig. 3. As discussed previously, the time-reversal argument does not apply to a laser and one would expect the interference to persist at all intensities, since the photons from such a source need not be localized even in a local theory.[10] It can be seen that the visibility remained constant to better than 10% as the laser intensity was reduced by eleven orders of magnitude by the insertion of neutral density filters between the laser and the interferometer. The lowest intensity corresponded to a photon

J. D. FRANSON

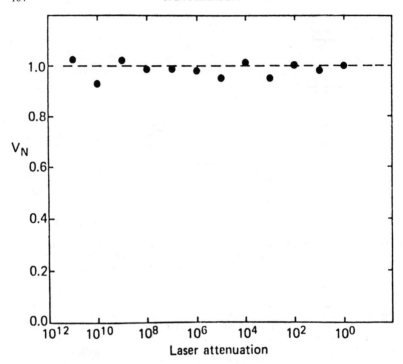

Figure 3   The normalized visibility, V_n, measured using a He–Ne laser and an interferometer length of 25 m, as a function of the laser attenuation. The largest attenuation (10^{11}) corresponded to a counting rate of 0.09 photons/sec.

counting rate of 0.1/sec. Fig. 3 simply provides an indication of the accuracy of the experimental technique.

The solid points in Fig. 4 show the first set of measurements of the normalized visibility V_N of the Kr 587.1 nm spectral line, as a function of the photon counting rate R and for an interferometer length of 25 m. A significant reduction in the visibility was observed, and is roughly consistent with what would be expected on the basis of any local theory; a simple model[11] predicts a reduction of approximately 51% at this distance and wavelength. Perhaps the most obvious alternative interpretation of these results is that the spectral quality of the light source was degraded in some way at the lowest intensities. This possibility was ruled out by repeating the measurements using the same source and interferometer at an interferometer length of 0.7 m, the results of which are shown by the open points of Fig. 4. No significant decrease

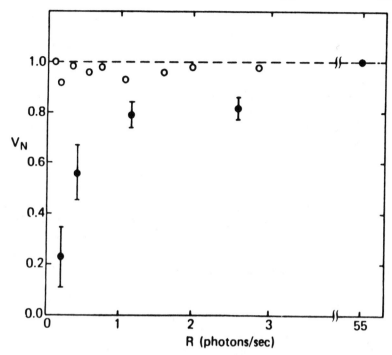

Figure 4   The normalized visibility, V_n, measured using a Kr discharge, as a function of the photon counting rate, R. The solid points correspond to an interferometer length of 25 m, while the open points correspond to an interferometer length of 0.7 m.

in the visibility was observed at the shorter distance. The measurements were repeated several times, alternating back and forth between the short and long distances, and with various combinations of silvered and unsilvered interferometer plates. Similar results were obtained in all cases, and are summarized elsewhere.[2] Several possible sources of systematic error have been considered, but no error source consistent with the nature of the experimental results has been suggested.

The original apparatus which produced the data of Figs. 3 and 4 was necessarily constructed with limited funding. The interferometer plates in particular were not of the highest quality, and the intrinsic visibility of the interferometer was relatively low (0.7). In addition, the data was recorded manually, which made it impractical to perform systematic studies, such as continuously varying the length of the interferometer. For these reasons, a new apparatus is under construction and is now near completion. The new apparatus will utilize custom-made optical components of the highest available quality, and will have a variable path length up to 45 m. The difference in path lengths will be maintained at less than one part in 10^9 by periodic alignment to the white-light fringes. The new interferometer will be automatically aligned and operated by a computer-automated control system. The computer will be able to position 22 motorized micrometers equipped with optical encoders to a resolution of 0.1 μm. The new apparatus will also allow parametric studies of the dependence of the visibility on the wavelength and other properties of the source.

No discussion of these experimental results would be complete without some comment on the recent experiment[3] of Aspect et al., which is in apparent agreement with the quantum theory. Their experiment included the use of optical switches to rule out the possibility of an exchange of information between the two photon detectors, a possibility which had caused earlier experiments based upon Bell's theorem to be inconclusive. However, the interpretation of the experiment by Aspect et al. can be seen to contain a hidden assumption.

In the Aspect experiment, the accidental coincidence rate was comparable with the true coincidence rate, the accidental rate being due almost entirely to photons emitted by two different atoms in the light source. The accidental rate was therefore determined at a large coincidence-circuit delay time and subtracted from the total counting rate at zero delay time. The validity of this procedure is dependent upon the implicit assumption that photons emitted by two different atoms are statistically independent; otherwise, the accidental rate may be a function of the delay time. In the high-intensity source used by Aspect et al., an atomic beam was excited by two lasers incident from opposite directions and focused into a small volume. The absorption of two laser photons and the subsequent emission of two other photons can be viewed as a four-photon Raman scattering process. The conditions necessary for coherent, stimulated emission of the scattered photons appear to be met, in which case the pairs of photons would have highly correlated polarizations.

A Monte Carlo simulation of the experiment by Aspect et al. which included possible correlations between photons emitted by different atoms was performed; the details of the simulation have been submitted for publication elsewhere,[5] and only the basic results will be shown here. Fig. 5 shows the coincident photon counting rate $R(\theta)$, as measured by Aspect et al., as a function of the angle θ between the two distant

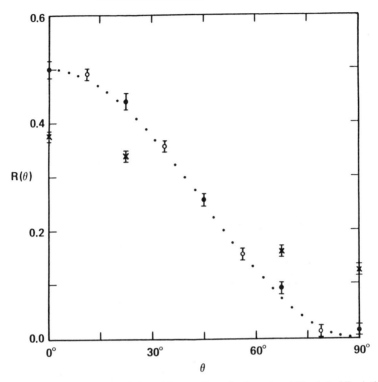

Figure 5   Monte Carlo simulation of the experiment by Aspect *et al*. The dotted line is the analytic prediction of the quantum theory, while the open points are the results of the Monte Carlo simulation based on the quantum theory, both in the limit of low source intensity. The *x*'s represent the results of the simulation based on the local theory in the limit of low source intensity. The solid points represent the results of the simulation based on the local theory for the actual source intensity used in the experiment.

polarizers. The dotted line represents the analytic prediction of the quantum theory in the limit of low source intensity, where such correlations are negligible. The open points represent the results of the Monte Carlo simulation for the same case, and are in reasonable agreement with the analytic results. The *x*'s represent the Monte Carlo results for a local theory in which the probability of the passage of a photon through a polarizer is simply proportional to $\cos^2(\Delta\theta)$, where $\Delta\theta$ is the angle between the polarization of the photon and the axis of the polarizer; these points also correspond to the limit of low source intensity where the correlations between photons emitted by different atoms have negligible effect. Finally, the solid points in Fig. 5 correspond to the Monte Carlo simulation for the local theory at the actual source intensity quoted by Aspect *et al.* There are no free parameters in this calculation, but the agreement with the dotted curve is excellent nevertheless. The plausibility of the correlations discussed here is perhaps debatable, but there would seem to be no doubt that there exists at least one local theory which is consistent with the experimental results.

In summary, a significant reduction in the low-intensity visibility of the 587.1 nm Kr line has been observed at large path separations. These results are consistent with the requirements of any local theory and are in apparent disagreement with the predictions of the quantum theory. If substantiated, these results would resolve the long-standing dispute between the Bohr and Einstein schools of thought: a photon could then be viewed as a soliton of finite dimensions, behaving like a particle and preserving locality over large distances, while propagating as a wave and obeying quantum mechanics on a smaller scale.

Note added in proof. Aspect has obtained more recent data at lower intensities which reportedly cannot be explained by the local theory described here.

Acknowledgments

The author would like to acknowledge K. A. Potocki, who is collaborating on the experimental measurements. This work was supported by U.S. Navy contract #N00024-83-C-5301, and by the APL Development Fund.

References

1. J. D. Franson, *Bull. Am. Phys. Soc.* **26**, 531 (1981); *Phys. Rev.* **D26**, 787 (1982).
2. J. D. Franson and K. A. Potocki, *Bull. Am. Phys. Soc.* **28**, 26 (1983); and submitted to *Phys. Rev. A.*
3. A. Aspect, J. Dalibard and G. Roger, *Phys. Rev. Lett.* **49**, 1804 (1982).
4. A. L. Robinson, *Science* **219**, 40 (1983); F. Rohrlich, *Science* **221**, 1251 (1983).
5. J. D. Franson, submitted to *Phys. Rev. Lett.*
6. J. S. Bell, *Physics* **1**, 195 (1964); for a review, see J. F. Clauser and A. Shimony, *Rep. Prog. Phys.* **41**, 1881 (1978).
7. A. Einstein, B. Podolsky and N. Rosen, *Phys. Rev.* **47**, 777 (1935).
8. R. H. Dicke, *Phys. Rev.* **93**, 99 (1954).
9. I. R. Senitzky, *Phys. Rev.* **121**, 171 (1961).
10. The possibility cannot be ruled out that single photons from a laser beam filtered to a low intensity may have to travel a relatively large time and distance before being localized into solitons of characteristic dimensions given by Eq. (1), even in a local theory.
11. A photon was assumed to propagate primarily down one of the two paths, with its intensity decreasing exponentially along the other path with a decay length of 9.3 m.

EXPERIMENTAL CONFIRMATION OF THE AHARONOV–BOHM EFFECT BY ELECTRON HOLOGRAPHY

Akira Tonomura

Central Research Laboratory, Hitachi Ltd., Kokubunji, Tokyo 185, Japan

Recent arguments against the existence of the Aharonov–Bohm effect have been investigated experimentally. In order to avoid any possible influence due to fringing fields from the finite whiskers or solenoids used in previous experiments, we employed ferromagnets of toroidal geometry, which had often been proposed but never realized. A phase difference was detected between two electron beams passing inside and outside the toroids. The measured values agreed with the theoretical ones, which reinforces the validity of the AB effect. The amount of leakage flux from the toroids was confirmed to be negligible by electron holographic measurement. Further experiments were carried out to test for possible quantization of the magnetic flux in a toroidal magnet and the effect of an electron beam penetration into the magnet. All the results support the existence of the AB effect.

I. AHARONOV–BOHM EFFECT

In 1959, Aharonov and Bohm proposed paradoxical experiments in a paper entitled "Significance of electromagnetic potentials in quantum mechanics".[1] Although the proposed experiments concerned both electrostatic and vector potentials, only the latter will be discussed here.

When an electric current is applied to an infinite solenoid as shown in Fig. 1, a magnetic field is produced only within that solenoid. If two

169

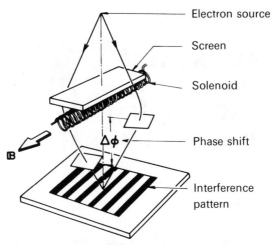

Figure 1   Schematic of Aharonov–Bohm effect. Two electron beams pass on both sides of an infinitely long solenoid, and are brought together by a biprism. Even though there are no magnetic fields along the electron paths, a phase difference is produced, proportional to the magnetic flux inside the solenoid.

electron beams starting from the same point are made to pass on both sides of it, and then to overlap, a phase difference, $\Delta\phi$, is produced between two electron beams. This phase difference is proportional to the magnetic flux inside the solenoid, as

$$\Delta\phi = -\frac{e}{\hbar}\oint \mathbf{A}\cdot d\mathbf{s}, \tag{1}$$

with the integral performed along the closed path determined by the two beams. This relation indicates that a phase difference proportional to the enclosed magnetic flux ($\int \mathbf{B}\cdot d\mathbf{S} = \oint \mathbf{A}\cdot d\mathbf{s}$), is produced irrespective of the existence of magnetic fields along the electron paths.

It is inconceivable in classical physics that electrons outside "know of" the magnetic flux inside. However, this paradoxical phenomenon can be explained in terms of vector potentials: Outside the solenoid there is no $\mathbf{B}$-field, but there is an $\mathbf{A}$-field, and outside electrons have direct interaction with vector potentials.

Experiments on this AB effect were carried out by several authors[2] soon after it was proposed. However, it did not become the object of general attention until Wu and Yang accepted the AB effect as experimental evidence of the gauge principle.[3] In gauge theory, vector potentials are extended to gauge fields, which are considered to be fundamental physical quantities.

Although the significance of the AB effect has long been discussed, quite recently even its existence was strongly denied by Bocchieri and Loinger.[4] They asserted that the AB effect is of purely mathematical origin. Experiments in the past[2] were also questioned from the standpoint that electrons were affected by inevitable leakage flux from finite whiskers or solenoids used in these experiments. Although these assertions have since then been disputed theoretically,[5] the controversy has not fully abated.

Lyuboshits et al. proposed an interference experiment free from leakage fields using a toroidal solenoid.[6] Kuper recommended employment of a superconducting toroid[7] in which fluxons are trapped. Greenberger et al. used a toroidal ferromagnet to test for the existence of AB effect with a neutron beam.[8]

II. EXPERIMENT ON THE AB EFFECT

The author's group carried out an interference experiment[9] using a toroidal ferromagnet with the help of electron holography. However, before going into the details of the experiment, let us discuss electron holography a bit.

II.1 Electron Holography

Electron holography[10] is, in a word, two-step photography. A hologram of an object is formed in an electron microscope; then this image is reproduced with a laser beam. By such imagery, electron wavefronts are transformed into light ones. Thus, the microscopic world, which can only be observed with the aid of an electron beam with a $1/100$ Å wavelength, is displayed on an optical bench. The liabilities inherent to electron microscopes have become possible with the help of versatile optical techniques. One such technique is interference electron microscopy, which makes observable the phase distribution of an electron beam. Even though it makes use of light, the micrographs are considered to be interference micrographs obtained with electrons.

Although electron holography was invented prior to 1948,[10] it only recently reached a practical stage using a coherent electron beam:[11] A field emission electron beam has made it possible to produce 3000 interference fringes. This is in contrast to 300, at most, with a conventional thermionic beam.

II.2. Experimental Method

In our experiment, toroidal ferromagnetis were used instead of whiskers or solenoids to remove ambiguities concerning leakage-field effects. Magnetic flux rotates inside toroids and forms a complete magnetic circuit. Since there are no leakage fields, we cannot even tell from the outside whether they are magnets or not. Sample sizes had to be as small as a few μm for two electron beams passing through the free space outside and inside the toroid to interfere with each other. The toroidal ferromagnets were prepared by photo-lithography and put on a carbon thin film.

The amount of leakage was measured from holographic interference electron micrographs. An example is shown in Fig. 2, together with a Lorentz micrograph. It was verified that contour lines in the interference micrographs follow magnetic lines of force as viewed from the direction

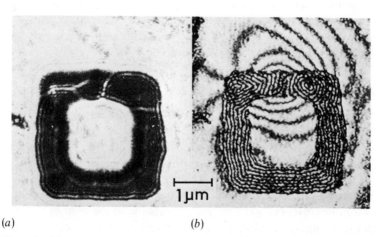

(a)　　　　　　　　　　　　　　(b)

Figure 2　Observation of leakage flux from a toroidal magnet. (a) Lorentz micrograph. (b) Interference micrograph (two-times phase amplification). It is difficult to estimate the leakage flux from a toroid in Lorentz micrograph (a), which provides the highest-resolution domain structure information. In interference micrograph (b), however, the amount of leakage flux can be measured by counting the contour fringes. A magnetic flux of $h/2e$ flows between two adjacent contour lines.

of the electron beam, and that the magnetic flux between two adjacent contour lines is a constant value of h/e.[12] Therefore, the amount of leakage flux can be quantitatively determined. In this contour map (b), almost all the flux is from leakage. In contrast, no leakage fields show up in the Lorentz micrograph (a). Only samples having leakage flux of less than $h/2e$ were employed for the experiment.

II.3. Experimental Results

The phase distribution of a 100 kV electron beam passing through a toroid was measured as shown in Fig. 3. These photographs were obtained optically as interference micrographs in the optical reconstruction stage of electron holography. In interferogram (b), the phase distribution is observed as deviation from the reference system of regular fringes. Therefore, the interference fringes should be along a straight line inside and outside the toroid, if there is no AB effect. However, this photograph reveals that a phase difference really exists

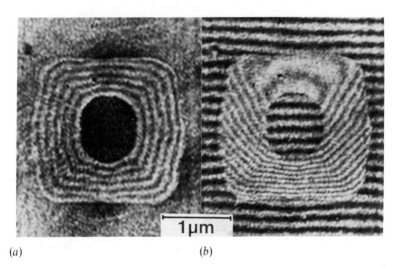

(a)   (b)

Figure 3   Interference micrographs of a toroidal magnet. (a) Contour map. (b) Interferogram. Using a tiny toroidal magnet, the phase distribution of an electron beam passed through the magnet is observed as an interferogram (b). The leakage flux is estimated to be $h/2e$ from contour map (a). If there is no AB effect, any fringe in photograph (b) will lie along a straight line in the two regions inside and outside the toroid. This picture clearly indicates the existence of a phase difference between the two regions, even though there are no magnetic fields.

between the two electron beams that have passed through the field-free regions inside and outside the toroid. In addition, the phase difference of 5.5 λ agrees with the theoretical value within an error of 20%. From the contour map (a), obtained optically from the same hologram, it can be confirmed that leakage fields are insignificantly small.

Two kinds of problems were presented against our experimental results. One was the problem of flux quantization in a toroidal ferromagnet, which was pointed out by Costa de Beauregard.[13] If the magnetic flux in a perfect toroidal magnet is quantized in h/e units, the fringe shift must be integral, and consequently the AB effect could never be observed. This possibility could not be determined using square toroids because of disturbance of the leakage flux. In order to reduce the leakage flux to $h/10e$, new circular toroids were produced as samples. The resulting interferogram is shown in Fig. 4.[14] Since this micrograph is phase-amplified by a factor of two, the magnetic flux is quantized neither in h/e nor $h/2e$ units.

The other problem was presented by Bocchieri et al.[15] They claimed that electrons partly penetrated the toroidal magnet, and consequently the observed effect was not due to the AB effect but to the Lorentz force effect. We responded to this assertion by carrying out an experiment to

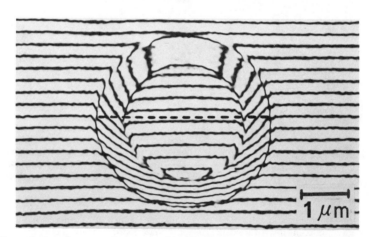

Figure 4   Interferogram of a magnet with a circular-toroid geometry. In order to test the predicted flux quantization, a tiny magnet with an extremely small leakage flux ($h/10e$) was produced. Interferogram of the toroid indicates that the magnetic flux inside a toroidal ferromagnet is quantized neither h/e nor $h/2e$ units. This provides further definite evidence for the existence of the AB effect.

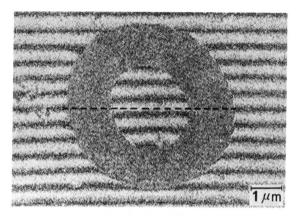

Figure 5   Interferogram of a toroidal magnet shielded against electron penetration. A toroidal magnet was covered with gold film thick enough to prevent electron penetration, and the existence of a phase difference was checked to dispel the claim that the observed phase difference in Fig. 3(b) is not due to the AB effect, but to the Lorentz force effect. The resultant interference fringes inside and outside the toroid are not in line with each other.

test whether the phase difference might not depend on the degree of electron penetration into the magnet.[16] It is impossible to realize a situation in which the tail of the electron wavefunction does not penetrate into the magnet.

Toroidal ferromagnets were covered with gold film thick enough to prevent electron penetration. This film was about 0.4 μm thick, which is 40 times as thick as the magnet. In the interferogram shown in Fig. 5, there is no read-out of information from inside the toroid. However, the fringes inside and outside the toroid are not in line with each other.

III. CONCLUSIONS

There has been much argument about the significance of the Aharonov–Bohm effect since its prediction.[17] Recently even the existence of the AB effect has been questioned, and a hot dispute has been revived from both theoretical and experimental aspects.

Coincidentally, we have just developed a coherent field emission electron microscope, and also a method for measuring quantitatively microscopic magnetic fields using electron holography.

Experiments have been carried out to test for the existence of the AB effect with special reference to the questioned points: leakage field effect, the possibility of flux quantization and electron penetration effect into the magnetic field regions.

All the experimental results support the existence of the AB effect and remove the questioned ambiguities.

References

1. Y. Aharonov and D. Bohm, *Phys. Rev.* **115**, 485 (1959).
2. R. G. Chambers, *Phys. Rev. Lett.* **5**, 3 (1960); H. A. Fowler *et al.*, *J. Appl. Phys.* **32**, 1153 (1961); H. Boersch *et al.*, *Z. Physik* **165**, 79 (1961); G. Möllenstedt and W. Bayh, *Phys. Bl.* **18**, 299 (1962).
3. T. T. Wu and C. N. Yang, *Phys. Rev.* **D12**, 3845 (1975).
4. P. Bocchieri and A. Loinger, *Nuovo Cim.* **47A**, 475 (1978).
5. U. Klein, *Lett. Nuovo Cim.* **25**, 33 (1979); A. Zeilinger, *Lett. Nuovo Cim.* **25**, 333 (1979); D. Bohm and B. J. Hiley, *Nuovo Cim.* **52A**, 295 (1979); D. M. Greenberger, *Phys. Rev. D23*, 1460 (1981); H. J. Lipkin, *Phys. Rev. D23*, 1466 (1981).
6. V. L. Lyuboshits and Y. A. A. Smorodinskii, *Zh. ERSP and Teor. Fiz. (U.S.S.R.)* **75**, 40 (1978).
7. C. G. Kuper, *Phys. Lett.* **794**, 413 (1980).
8. D. M. Greenberger *et al.*, *Phys. Rev. Lett.* **47**, 751 (1981).
9. A. Tonomura *et al.*, *Phys. Rev. Lett.* **48**, 1443 (1982).
10. D. Gabor, *Proc. Roy. Soc. (London)* **A197**, 454 (1949).
11. A. Tonomura *et al.*, *J. Electron Microsc.* **28**, 1 (1979).
12. A. Tonomura *et al.*, *Phys. Rev.* **B25**, 6799 (1982).
13. O. Costa de Beauregard and J. M. Vigoureux, *Lett. Nuovo Cim.* **33**, 79 (1982).
14. A. Tonomura *et al.*, *Phys. Rev. Lett.* **51**, 331 (1983).
15. P. Bocchieri *et al.*, *Lett. Nuovo Cim.* **35**, 370 (1982).
16. A. Tonomura *et al.*, in *Proceedings of the International Symposium on Foundations of Quantum Mechanics*, Tokyo, 1983 (Physical Society of Japan, Tokyo, 1984), p. 20.
17. D. Home and S. Sengupta, *Am. J. Phys.* **51**, 942 (1983).

ANGULAR MOMENTUM AND ROTATIONAL PROPERTIES OF A CHARGED PARTICLE ORBITING A MAGNETIC FLUX TUBE

Mark P. Silverman

Department of Physics, Trinity College, Hartford, CT 06106

Investigations of the composite quantum system comprising a charged spinless particle orbiting, but not penetrating, a magnetic flux tube, have led to disagreements among theorists concerning properties of the system. Of particular significance is the problem of the flux dependence of the angular momentum and rotational behavior of the particle. At issue are fundamental questions concerning the single-valuedness of the wave function and the role, spectrum, and observability of certain basic quantum dynamical variables.

I. INTRODUCTION

Quantum mechanics is a subject of such richness and subtlety that despite the innumerable applications to systems as diverse as elementary particles nuclei, atoms, molecules, and condensed matter, its consequences continue to surprise us when weighed against our classical expectations. The Aharonov–Bohm (AB) effect provides one example. More than a quarter century ago the possibility was first raised of a quantum interference effect engendered by the local interaction of charged particles with an electromagnetic potential in a region where the electric and magnetic fields, themselves, are strictly null.[1,2] The topic has since been a subject of much interest and controversy because of the questions of fundamental physical principle to which it has given rise— questions, for example, concerning the primacy or equivalence of fields vis-à-vis potentials; the single- or multiple-valuedness of wave functions in non-simply connected spaces; the actual existence of the proposed— and tested—AB effect, and therefore the correct application and internal consistency of quantum mechanics, itself.

Over the last few years another facet to the problem of a particle in the presence of long-range magnetic flux (i.e. a particle interacting locally with a potential, but subject to no classical force) has come under scrutiny. At issue is not the flux dependence of a diffraction pattern or scattering cross section, but rather the angular momentum spectrum, rotational behavior, and statistics of particles in rotational motion about magnetic flux tubes. It has been claimed[3,4] that a composite charged particle-solenoid system can have a flux-dependent—and therefore arbitrary—angular momentum; the wave function of the particle would then acquire a flux-dependent phase factor upon rotation about the solenoid, and an ensemble of such systems would follow a statistics that interpolate between boson and fermion statistics. These conclusions have been challenged by a number of theorists[5-7] who maintain instead that the orbital angular momentum spectrum remains integer-valued and the wave function of the rotated particle unaffected by the flux. The systems would then obey the traditional quantum statistics determined by the spin of the particle.

As with the earlier discussions of the AB effect, these, too, raise questions of fundamental import that transcend in scope the particularities of the individual systems under study. These questions concern, for example, the role, spectrum, and observability of various gauge invariant and non-invariant dynamical quantities in quantum mechanics. Underlying the above disagreement is the issue of identifying correctly the generator of rotation—a problem not unique to consideration of the AB effect, but one certainly brought into sharper focus by the unusual nature of the systems involved.

In this paper I will try to clarify some of the features of these composite quantum systems by examining the dynamics of a single spinless charged particle in circular orbit about an infinite flux tube. I will discuss the question of identifying the generator of rotation and consider an experimental procedure for testing whether or not the confined magnetic field affects the wave function of the orbiting particle.

II. BOUND STATES OF A SPINLESS CHARGED PARTICLE IN THE PRESENCE OF LONG-RANGE MAGNETIC FLUX

The symmetry of the planar rotator has made it an ideal system for a number of model studies. It is one of few quantum systems for which exact solutions to the Stark effect[8] and Zeeman effect[9] can be obtained. Similarly, a number of studies have been made of the planar rotator in long-range magnetic flux in support or denial of the existence of the AB effect,[10-13] or in consideration of the single-valuedness of wave functions,[14] or in illustration of a general gauge invariant method of analysis.[15] The present study is undertaken with emphasis on the current controversy concerning the rotational properties of a particle-flux tube composite.

Since there is a certain measure of arbitrariness in the specification of a gauge field, it is pertinent to comment briefly at this point on the question of gauge invariance and observability of a dynamical variable. To be an observable, a dynamical variable must be representable by a hermitian operator with complete set of eigenstates.[16] For a mechanical system coupled to a gauge field, e.g. a charged particle to the scalar and vector potentials, a further criterion must be met so that physically meaningful quantities do not depend on the arbitrary choice of gauge. All theoretical expressions, e.g. quantum mechanical expectation values and transition matrix elements, representing experimentally measurable quantities must be invariant under a gauge transformation. In the case of a charged particle minimally coupled to the electromagnetic potentials $\mathbf{A}, \phi$, a gauge transformation entails

$$\mathbf{A} \rightarrow \mathbf{A}' = \mathbf{A} + \nabla\Lambda \tag{1a}$$

$$\phi \rightarrow \phi' = \phi - \partial\phi/\partial t \tag{1b}$$

$$\Psi \rightarrow \Psi' = \exp(iq\Lambda)\Psi \tag{1c}$$

where the gauge function $\Lambda(x, t)$, to be admissible, must not lead to an altered configuration of the electric and magnetic fields, $\mathbf{E}, \mathbf{B}$. For a particle orbiting a flux tube the vector potentials outside the tube must satisfy

$$\oint_c \mathbf{A} \cdot d\mathbf{s} = \oint_c \mathbf{A}' \cdot d\mathbf{s} = \text{Total magnetic flux} \tag{2a}$$

where the contour circumscribes the solenoid. This requires that

$$\oint_c \nabla\Lambda \cdot d\mathbf{s} = \oint d\Lambda = 0, \tag{2b}$$

i.e. that Λ be single-valued.[11]

Under a gauge transformation the hamiltonian

$$H = (\mathbf{p} - q\mathbf{A})^2/2m + q\phi \tag{3a}$$

and Schrödinger equation

$$H\psi = i\,\partial\psi/\partial t \tag{3b}$$

are form-invariant, the transformed hamiltonian being given by

$$H' = \exp(iq\Lambda)H\exp(-iq\Lambda) - q\frac{\partial\Lambda}{\partial t} = \frac{1}{2m}(\mathbf{p} - q\mathbf{A}')^2 + q\phi'. \tag{3c}$$

Thus, the time evolution of the physical system is unaffected by a gauge transformation|ψ' evolves under H' in the same way that ψ evolves under H. The eigenvalue spectrum of H, however, is not invariant under a gauge transformation. If $\psi_E = \psi_E^0\exp(-iEt)$ is a stationary state of H with eigenvalue E, then $\psi'_E = \exp(iq\Lambda)\psi_E$ is an eigenstate, but not a stationary state, of H' with eigenvalue $E' = E - q\,\partial\Lambda/\partial t$. Thus E' is both gauge- and time-dependent unless Λ is independent of time. If the latter condition holds, the hamiltonians H' and H are simply related by a unitary transformation and necessarily have the same eigenvalue spectrum.

Since the eigenvalues of H are not gauge invariant, does this imply that energy is not an observable? The answer is no, for the reason that in quantum mechanics, as in classical mechanics, the hamiltonian need not represent the energy of the system. A suitable hamiltonian that *does* represent the energy must have a time-independent scalar potential $\phi(x)$. Under an arbitrary gauge transformation, the corresponding energy operator, whose eigenvalues represent the energy of the system via

$$W\psi = E\psi \tag{4a}$$

is given by[17]

$$W = H' + q\frac{\partial\Lambda}{\partial t} = \exp(iq\Lambda)H\exp(-iq\Lambda). \tag{4b}$$

The eigenvalue spectrum and expectation values of W are independent of gauge and correspond to observable quantities.

The natural gauge for characterizing the field of an infinite solenoid of radius a is the Coulomb or cylindrical gauge

$$A_\varphi = \begin{cases} \Phi r/2\pi a^2 & (r \leqslant a) \\ \Phi/2\pi r & (r \geqslant a) \end{cases} \tag{5}$$

$$\phi = 0$$

where $\Phi = \pi a^2 B$ is the total flux through the solenoid. With this choice the hamiltonian and energy operator of the planar rotator are identical; the resulting energy eigenvalues and transition probabilities are gauge invariant. The hamiltonian of the rotator can be expressed as

$$H = (\mathbf{p} - q\mathbf{A})^2/2m = K^2\varepsilon = (L - \alpha)^2\varepsilon \tag{6a}$$

where

$$\varepsilon = (2mr^2)^{-1} \qquad (r = \text{orbital radius} > a) \tag{6b}$$

$$\alpha = q\Phi/2\pi \tag{6c}$$

and

$$\mathbf{L} = \mathbf{r} \times \mathbf{p} \Rightarrow L_z = -i\,\partial_\varphi \tag{7a}$$

is the canonical orbital angular momentum,

$$\mathbf{K} = \mathbf{r} \times (\mathbf{p} - q\mathbf{A}) \Rightarrow K_z = -i\,\partial_\varphi - \alpha \tag{7b}$$

is the kinetic angular momentum of the planar rotator.

The AB effect is frequently perceived as a uniquely quantum mechanical phenomenon for which there is no classical analogue. If this is construed to mean that the particle interacts locally only with the gauge field and not with fields for which the resulting effects are interpretable in terms of classical forces, then one must consider a system with constant flux Φ; a time-varying flux would produce at the particle a local electric field by means of Faraday's law of induction.

With α constant the Schrödinger equation

$$\varepsilon(L_z - \alpha)^2\psi = i\frac{\partial\psi}{\partial t} \tag{8}$$

readily admits of two stationary state solutions whose wave functions and energy and angular momentum eigenvalues are given as follows:

(A) Solution I

$$\psi_m^1(\varphi, t) = (2\pi)^{-1/2} \exp[i(m + \alpha)\varphi] \exp(-iE_m^0 t) \tag{9a}$$

Energy
$$E_m^0 = m^2\varepsilon \qquad (m = 0, \pm 1, \pm 2, \ldots) \tag{9b}$$

Canonical angular momentum $l_m = m + \alpha$ (9c)

Kinetic angular momentum $k_m = m$ (9d)

(B) Solution II

$$\psi_m^{II}(\varphi, t) = (2\pi)^{-1/2} \exp(im\varphi) \exp(-iE_m t) \tag{10a}$$

Energy $E_m = (m - \alpha)^2 \varepsilon$ (10b)

Canonical angular momentum $l_m = m$ (10c)

Kinetic angular momentum $k_m = m - \alpha.$ (10d)

(Note: units with $\hbar = c = 1$ are being used.)

A cursory examination of the solutions shows that if the magnetic flux is quantized in units of $\Phi_0 = hc/e$, then α is integer-valued and the two solutions are entirely equivalent; i.e. to each state of Solution I labelled by m corresponds a state of Solution II with identical eigenvalues although different label. If, as experiment seems to indicate,[18] the flux through a current ring is not quantized, then α can take on a continuum of values, and the two solutions are not equivalent. Solution I has the energy spectrum of the flux-free planar rotator, an integer kinetic angular momentum spectrum, and a flux-dependent canonical angular momentum spectrum; the wave function is multiple-valued for non-integer α. Solution II has integer-valued canonical angular momentum and flux-dependent kinetic angular momentum and energy; the wave function is single-valued.

Both solutions satisfy a correspondence principle. In the limit of large quantum number m, the phase of the "classical" rotator is given by

$$\Delta E/\Delta m \rightarrow d\delta/dm = \varphi - w_{\text{rot}} t \tag{11a}$$

where δ is the phase of the wave function

$$\delta^{(I)} = (m + \alpha)\varphi - m^2 \varepsilon t \tag{11b}$$

$$\delta^{(II)} = m\varphi - (m - \alpha)^2 \varepsilon t \tag{11c}$$

and the angular speed of rotation is given by

$$w_{\text{rot}}^{(I)} = 2m\varepsilon = k^{(I)}/mr^2 \tag{11d}$$

$$w_{\text{rot}}^{(II)} = 2(m - \alpha)\varepsilon = k^{(II)}/mr^2. \tag{11e}$$

The mechanical rotation frequency in both cases is related to the appropriate mechanical angular momentum.

An understanding of the physical content of the solutions may be sought by examining them as special cases of the general solution to the rotator in the presence of a time-dependent flux. This solution is given by

$$\psi(t) = \exp\left(-i \int^t H(t')\, dt'\right)\psi(0)$$

$$= \exp\left(-i\varepsilon \int^t [L_z - \alpha(t')]^2\, dt'\right)\psi(0). \tag{12}$$

A basic quantum mechanical theorem[19] states that if $\psi^{(0)}(t)$ is the solution to the Schrödinger equation with $\mathbf{A} = 0$, then

$$\psi(\mathbf{x}, t) = \exp\left(iq \int^{\mathbf{x}} \mathbf{A} \cdot d\mathbf{s}\right)\psi^{(0)}(\mathbf{x}, t) \tag{13}$$

is the solution to the Schrödinger equation for specified $\mathbf{A}$. If at $t = 0$ the potential-free rotator is in an energy-angular momentum eigenstate with quantum number m, then the state of the rotator in the presence of a constant potential $A_{0\varphi} = \Phi_0/2\pi r$ is given by the wave function

$$\psi_m(0) = (2\pi)^{-1/2} \exp[i(m + \alpha_0)\varphi] \tag{14a}$$

where evaluation of the phase integral has led to

$$q \int^{\mathbf{x}} \mathbf{A}_0 \cdot d\mathbf{s} = \int^\varphi (q\Phi_0/2\pi r) r\, d\varphi' = (q\Phi_0/2\pi)\varphi = \alpha_0\varphi. \tag{14b}$$

Thus, from Eq. (12) one has

$$\psi_m(\varphi, t) = (2\pi)^{-1/2} \exp[i(m + \alpha_0)\varphi] \exp\left\{-i\varepsilon \int_0^t [m + \alpha_0 - \alpha(t')]^2\, dt'\right\}. \tag{14c}$$

Consider the following cases for the time dependence of $\alpha(t)$.

Case A

The flux Φ_0 is initially null and rapidly brought to the constant value Φ at $t = 0_+$. Then $\alpha_0 = 0$ $(t \leqslant 0)$, $\alpha(t) = \alpha = \text{const.}$ $(t > 0)$. The resulting wave function is $\psi_m^{(\mathrm{II})}$. The initiation of the flux generates an electric field

$$\mathscr{E}_\varphi = -(2\pi r)^{-1}\, \partial\Phi/\partial t \tag{15a}$$

which produces a torque on the particle

$$N_z = qr\mathscr{E}_\varphi = -(q/2\pi)\,\partial\Phi/\partial t. \tag{15b}$$

The torque changes the initial angular momentum $L_z = m$ by the amount

$$\Delta L_z = \int_0^\infty N_z\,dt = -q\Phi/2\pi = -\alpha \tag{15c}$$

thereby giving the state a final mechanical angular momentum

$$K_z = L_z + \Delta L_z = m - \alpha. \tag{15d}$$

The induced electric field also does work on the particle at the rate

$$dW/dt = w_{\mathrm{rot}}N_z = (L_z/mr^2)N_z = -2\varepsilon(m-\alpha)\,\partial\alpha/\partial t, \tag{16a}$$

leading to a change in energy by the amount

$$\Delta E = \int (dW/dt)\,dt = -2\varepsilon\alpha(m - \tfrac{1}{2}\alpha), \tag{16b}$$

and thereby giving the state a final energy

$$E_m = E_m^0 + \Delta E = (m-\alpha)^2\varepsilon. \tag{16c}$$

The properties of Solution II can therefore be thought to derive from the local interaction of the particle with the induced electric field over the period of initiation of the magnetic flux.

Case B

If the flux remains for all time at its initial value, then $\alpha_0 = \alpha(t) = \alpha =$ constant. The resulting wave function is ψ_m^1. The particle-flux tube composite may be thought to be of "cosmological" origin, for at no time has the system been created (either by initiation of the solenoidal current while the particle is in orbit, or by capture of the particle about a current-carrying solenoid). In this case there is no effect of the flux on initial energy and kinetic angular momentum, although the canonical angular momentum spectrum is shifted by α.

From the form of the wave function and eigenvalue spectra one might conclude that Solution I can be made equivalent to the solution of the flux-free rotator by a gauge transformation with gauge function $\Lambda = -(q\Phi/2\pi)\varphi$. Upon transforming the planar rotator hamiltonian

$H = \varepsilon L_z^2$ (with $\mathbf{A} = \phi = 0$) one obtains

$$H' = (L_z - \alpha)^2 \varepsilon - \varphi\, \partial\alpha/\partial t \tag{17a}$$

$$W = (L_z - \alpha)^2 \varepsilon \tag{17b}$$

$$\psi'_m = \exp(iq\Lambda)\psi_m = \psi^1_m(\alpha(t)) \tag{17c}$$

with potentials

$$A'_\varphi = \Phi/2\pi r; \qquad \phi' = -(\varphi/2\pi)\,\partial\Phi/\partial t. \tag{17d}$$

For constant α one has the hamiltonian of the rotator in constant long-range magnetic flux with stationary state solution ψ^1_m. Such a transformation has been used in the past[20] as proof of the non-existence of the Aharonov–Bohm effect. The conclusion is not valid, however, because Λ is not an admissible gauge function. Since $\Lambda(2\pi) \neq \Lambda(0)$, the function Λ alters the initial field configuration (namely $\mathbf{E} = \mathbf{B} = 0$) by giving rise to a magnetic flux where none had existed before. The vector potential is singular at the origin and corresponds to the introduction of a singular current distribution.[11]

The question of the multiple-valuedness of the wave function ψ^1_m is a somewhat thorny one. Whereas the imposition of single-valuedness as a criterion for the electromagnetic field to be physically meaningful is equivalent to requiring that the classical forces acting on a particle be uniquely specified, the same criterion applied to the quantum mechanical wave function is not so transparent and readily motivated. In a simply connected space, such as that of the flux-free rotator, specification of single-valuedness is necessary to avoid spurious solutions introduced by the arbitrary choice of a particularly convenient coordinate system with the polar axis through the loop. An alternative choice of polar axis outside the loop leads only to single-valued solutions. In the case of a non-simply connected space, however, the situation is less clear.

Some theorists have cited as fallacious the argument that multiple-valued wave functions are to be rejected a priori,[21] others have argued that single-valuedness is a necessary criterion for solutions to be physically meaningful,[11,14] and still others have used the criterion but questioned whether the assumption is necessary in a multiply-connected region.[22] One argument against multiple-valuedness in the case of the particle-solenoid composite has been based on the assertion that the two-dimensional system is an idealization of a system with high, but finite, potential barrier and finite return flux.[14] This sidesteps the problem by converting it to a three-dimensional one. The problem as posed is a conceptually valid one, and diverse theoretical studies have shown that the physics of two-dimensional systems can differ qualitatively from the physics of three. Another argument for the rejection of multiple-valued wave functions is that they fail to regenerate the expected single-valued solutions upon adiabatic extinction of the magnetic flux.[11] This argument would not be relevant, however, to the composite system of cosmological origin for which the particle experiences an invariable flux beyond the control of the experimenter. General current algebraic considerations[23] have also led to the conclusion that the possibility of two-dimensional particle-flux composites with undiscoverable history cannot be ruled out.

For the present the boundary conditions to be imposed on the wave function to eliminate nonphysical solutions is an actively considered matter.[24]

III. OBSERVABLE CONSEQUENCES OF PARTICLE ROTATION

For a system in a simply connected space, it has long been the practise to identify the rotation generator G with the canonical angular momentum L. This association is generally motivated by the intimate connection between rotational symmetry and angular momentum conservation. Some theorists have referred to it as the "usual symmetry theory assumption" in relating L and G for a quantum system in a multiply connected space as well.[25] At root, however, L is called a generator because it induces an infinitesimal rotational transformation on a coordinate basis state or coordinate operator as a result of the commutation relations

$$[L_i, x_j] = ie_{ijk} x_k \tag{18a}$$

$$[L_i, L_j] = ie_{ijk} L_k . \tag{18b}$$

Eq. (18a) leads to the transformation

$$\langle \mathbf{r}|(1 + i\mathbf{n} \cdot \mathbf{L} \, d\theta) = \langle \mathbf{r} + (\mathbf{n} \times \mathbf{r}) \, d\theta| \tag{19}$$

identified as a rotation by $d\theta$ about $\mathbf{n}$.[26] Eq. (18b) expresses the group closure property of a rotational transformation.

The matrix elements of **L**, however, are not invariant under a gauge transformation, but transform as

$$\langle \mathbf{L} \rangle_{\psi'} = \langle \mathbf{L} \rangle_{\psi} + \langle \mathbf{r} \times q\nabla\Lambda \rangle_{\psi} \tag{20}$$

where the bracket represents either a transition moment or expectation value. **L**, therefore, ceases in general to be an observable in a mechanical system coupled to an electromagnetic potential. This result does not depend on the connectivity of the space. It has been demonstrated that in the Coulomb gauge **L** is identical to the total angular momentum **J** (the sum of kinetic and electromagnetic angular momenta) which is an observable.[12] While this may be of some heuristic value, the equivalence of **L** and **J** is not gauge invariant and does not generally hold in other gauges.

The dynamical variable **K** represents the mechanical angular momentum of the particle; it has gauge-invariant matrix elements and the commutation relations

$$[K_i, x_j] = ie_{ijk}x_k \tag{21a}$$

$$[K_i, K_j] = ie_{ijk}[K_k + qx_k(\mathbf{x} \cdot \text{curl } \mathbf{A})]. \tag{21b}$$

For a particle in the field-free region outside the solenoid (**B** = curl **A** = 0) the commutation relations of **K**, Eqs. 21a,b), reduce to those of **L**, Eqs. (18a,b).

What operator generates the rotation of the particle about the flux tube (i.e. the source of the magnetic flux is not co-rotated with the particle)? If this generator must be an observable, then K_z is a possibility—but only if curl **A** strictly vanishes. **J** the angular momentum of the total particle-electromagnetic field system, effects a rotation of the total system; such a global rotation has no observable consequences, a point stressed previously in the case of spinor rotation.[27] If this generator need not be an observable (and it has been shown in Section II that the hamiltonian H, the generator of time translation, is not necessarily an observable), then it can be identified with L_z.

The selection of either **K** or **L** as generator of particle rotation leads to experimental consequences that can be manifested, for example, by a split-beam quantum interference experiment[28,29] such as the one diagrammed in Fig. 1. A collimated beam of spinless charged particles is split into two coherent beams; in the presence of uniform time-independent background magnetic fields of equal strength and opposite

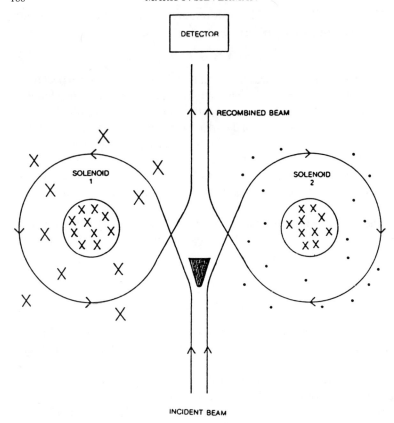

Figure 1   Schematic representation of the experiment.

orientations the resulting two beams circulate in orbits of equal radius
and opposite sense about similar solenoids generating fluxes Φ_1, Φ_2,
respectively. [Flux through the orbital plane of a particle is positive if the
particle circulates about the field in a right-hand sense.] The beams are
recombined and the forward beam intensity is examined as a function of
magnetic flux. [The background fields do not produce a relative phase
shift between the wave functions of the two beams.]

   With idealized experimental conditions—typically characterizing
analyses of the Aharonov–Bohm effect—of a time-independent particle
beam orbiting infinitely long solenoids producing constant magnetic
flux, it follows from the interpretations given in Section II that the
quantum state of the particles are described by the Solution I (multiple-

valued) wave functions (Eqs. (9a–d)). The wave function of a particle in a sharp angular momentum state that has been rotated by an angle θ about a flux tube in a region where curl $\mathbf{A} = 0$ then incurs a flux-dependent phase factor according to

$$\psi^1(r, \varphi + \theta) = \exp(-iL_z\theta)\psi^1(r, \varphi) = \exp[-i(m + \alpha)\theta]\psi^1(r, \varphi) \quad (22a)$$

if L_z is generator; it incurs a flux-independent phase factor according to

$$\psi^1(r, \varphi + \theta) = \exp(-iK_z\theta)\psi^1(r, \varphi) = \exp(-im\theta)\psi^1(r, \varphi) \quad (22b)$$

if K_z is generator. Upon recombination of the two beams after an integral number n of revolutions the forward beam intensity is given by

$$I(2\pi n) \propto I_0 \cos^2[nq(\Phi_1 - \Phi_2)/2\pi\hbar c] \quad (23a)$$

if L_z is generator, and by

$$I(2\pi n) \propto I_0 \quad \text{(independent of } \Phi\text{)} \quad (23b)$$

if K_z is generator. The results are unchanged if the particles are in a linear superposition of angular momentum eigenstates, since the flux-dependent factor is independent of state quantum numbers and the time evolution of each coherent component is the same, since the energy is independent of magnetic flux. Moreover, charged spin-$\frac{1}{2}$ particles (electrons, protons) can be employed since the spin variables do not affect the test of the points at issue concerning orbital angular momentum. For the Gedankenexperiment under consideration, the selection of $\mathbf{L}$ as the rotation generator leads to the same result as would be obtained by determining the time evolution of the system by means of the planar rotator propagator.[30]

IV. CONCLUSION

Nearly six decades after the codification of the basic principles of quantum mechanics there still remain unsettled questions of a fundamental nature for which seemingly straightforward application of quantum mechanical procedures may result in theoretically different and experimentally distinguishable results. A study of the angular momentum and rotational properties of charged particle-magnetic flux tube composites provides one such example. All the more remarkable is that these systems focus attention on issues of conceptual importance such as the role, spectrum, and observable consequences of basic dynamical variables and the boundary conditions to be imposed on physically significant solutions to the quantum equations of motion. Their investigation has provided physicists insight into the content of quantum mechanics, the relationship between quantum and classical physics, and the profound connections between physics and geometry.

Acknowledgments

The author gratefully acknowledges helpful conversations with Akira Inomata, Roman Jackiw, Charles Miller, and Harvey Picker. Support was provided by a Trinity College Faculty Research Grant.

References

1. W. Ehrenberg and R. E. Sidey, Proc. Phys. Soc. **62B**, 8 (1949).
2. Y. Aharonov and D. Bohm, *Phys. Rev.* **115**, 485 (1959).
3. F. Wilczek, *Phys. Rev. Lett.* **48**, 1144 (1982).
4. F. Wilczek, *Phys. Rev. Lett.* **49**, 957 (1982).
5. A. Goldhaber, *Phys. Rev. Lett.* **49**, 905 (1982).
6. H. Lipkin and M. Peshkin, *Phys. Lett.* **118B**, 385 (1982).
7. R. Jackiw and A. N. Redlich, *Phys. Rev. Lett.* **50**, 555 (1983).
8. M. P. Silverman, *Phys. Rev.* **A24**, 339 (1981).
9. M. P. Silverman, *Phys. Rev.* **A24**, 342 (1981).
10. P. Bocchieri and A. Loinger, *Nuovo Cim.* **47A**, 475 (1978).
11. D. Bohm and B. J. Hiley, *Nuoco Cim.* **52A**, 295 (1979).
12. M. Peshkin, *Phys. Repts.* **80**, 375 (1981).
13. D. H. Kobe, *J. Phys. A: Math. Gen.* **15**, L543 (1982).
14. E. Merzbacher, *Am. J. Phys.* **30**, 237 (1962).
15. D. H. Kobe, *J. Phys. A: Math. Gen.* **16**, 737 (1983).
16. C. Cohen Tannoudji, B. Diu and F. Laloe, *Quantum Mechanics*, Vol. I (Wiley, New York, 1977), p. 136.
17. K. Yang, *Ann. Phys. (N.Y.)* **101**, 62 (1976).
18. A. Tonomura, H. Umezaki, T. Matsuda, N. Osakabe, J. Endo and Y. Sugita, *Phys. Rev. Lett.* **51**, 331 (1983).
19. G. Baym, *Lectures on Quantum Mechanics* (Benjamin, New York, 1969), pp. 80–81.
20. P. Bocchieri, A. Loinger and G. Siragusa, *Nuovo Cim.* **51A**, 1 (1979).
21. J. M. Blatt and V. F. Weisskopf, *Theoretical Nuclear Physics* (Wiley, New York, 1952), pp. 783, 787.
22. M. Peshkin, I. Talmi and L. J. Tassie, *Ann. Phys. (N.Y.)* **12**, 426 (1961).
23. G. A. Goldin and D. H. Sharp, *Phys. Rev.* **D28**, 830 (1983).
24. W. C. Henneberger, *Phys. Rev. Lett.* **52**, 573 (1984).
25. L. J. Tassie and M. Peshkin, *Ann. Phys. (N.Y.)* **16**, 177 (1961).
26. G. Baym, Reference 19, pp. 148–153.
27. M. P. Silverman, *Eur. J. Phys.* **1**, 116 (1980).
28. M. P. Silverman, *Phys. Rev. Lett.* **51**, 1927 (1983).
29. M. P. Silverman, *Phys. Rev.* **D29**, 2404 (1984).
30. C. Bernido and A. Inomata, *Phys. Lett.* **77A**, 394 (1980); C. C. Gerry and V. A. Singh, *Nuovo Cim.* **73B**, 161 (1983); M. P. Silverman (unpublished).

A DYNAMICAL FORMULATION OF THE AHARONOV–BOHM EFFECT

Mark D. Semon

Department of Physics, Bates College, Lewiston, ME 04240, and Department of Physics, Amherst College, Amherst, MA 01002

John R. Taylor

Department of Physics, University of Colorado, Boulder, CO 80309

In 1969 Aharonov, Pendleton and Petersen introduced a new dynamical variable, the modular momentum, and discussed how several nonlocal interactions could be characterized as resulting from its exchange. In this paper we review and extend their approach by applying it to the Aharonov–Bohm effect with an electromagnetic scalar potential. First we review the nonclassical nature of the effect, showing that the interaction involved does not change the linear momentum of any particle creating the interference pattern. We then use the modular momentum to describe the interaction in the Heisenberg picture. Studying the equation of motion for the modular momentum, we prove the assertion of Aharonov et al. that modular momentum provides a dynamical description of the Aharonov–Bohm effect.

I. INTRODUCTION

In 1959 Aharonov and Bohm[1] proposed two experiments that called into question our understanding of what constitutes an interaction. In both cases an interference pattern was predicted to shift even though the particles creating it were never acted upon by any force. Since then equivalent effects have been predicted to occur in gauge fields,[2] Yang–Mills fields,[2,3] gravitational fields,[2,4] and inertial fields.[5-7] Nonetheless, our understanding of *how* these effects occur has remained incomplete. Explanations have been offered in terms of local effects of potentials,[8] nonlocal effects of field strengths,[9-11] and field–field interactions.[12,13] The effects even have been attributed to the time-dependent processes involved in setting up the experimental conditions.[14,15] Although each of these explanations illuminates certain aspects of the effect, a complete picture has yet to emerge from any one of them.

In 1969, Aharonov, Pendleton and Petersen[16] introduced a new dynamical variable, the modular variable, and discussed how several nonlocal interactions could be characterized as resulting from its exchange. Their work indicated that modular variables could provide a new formalism for describing the Aharonov–Bohm effect, and more generally, could give a new approach to nonlocality. Because their paper introduced the variables, it developed them in the context of several examples and only sketched proofs of the main results. After studying their work we felt that modular variables were interesting enough to review and expand upon in a simpler context. What we present here is a discussion of our approach and a summary of what we have proved so far.

II. REVIEW OF THE AB EFFECT WITH AN ELECTROMAGNETIC SCALAR POTENTIAL

In order to better understand modular variables, we decided to study them in the context of the simplest Aharonov–Bohm effect, the one involving the electromagnetic scalar potential. Following Aharonov *et al.*[16] we refer to this as the potential effect. In this section we give a brief review of the effect and its nonclassical nature.[16]

Consider a charged particle inside a conducting cylinder. The Schrödinger equation in this case is

$$H_0\psi_0 = i\, \partial\psi_0/\partial t \qquad (2.1)$$

where H_0 is the free particle Hamiltonian, and we take $\hbar = 1$. If we begin putting charge on the outside of the cylinder then a time-dependent potential is created inside, but no **E** or **B** fields. In this case the Schrödinger equation becomes

$$H\psi = i\, \partial\psi/\partial t \tag{2.2}$$

with

$$H = H_0 + q\phi(t) \tag{2.3}$$

and ϕ the time-dependent potential, which is independent of position inside the cylinder. It is easily shown that solutions to this equation have the form

$$\psi = \psi_0 \exp(-i\alpha) \tag{2.4}$$

where ψ_0 is any solution to Eq. (2.1), and α is a phase given by

$$\alpha = q \int^t \phi(t')\, dt'. \tag{2.5}$$

Since α is independent of position, the state of any particle whose wavefunction lies entirely within the cylinder is changed only by an overall phase factor. Thus, the presence of $\phi(t)$ inside the conducting cylinder has no observable consequences, which is what we would expect classically, since the region is force-free.

In 1959, however, Aharonov and Bohm showed that, contrary to our classical expectations, there are situations in which the presence of potentials in force-free regions can result in observable effects. They proposed placing a conducting cylinder behind each slit in a two-slit interference experiment. As the particle passes through the slits its state is given by

$$\psi_0 = \psi_1 + \psi_2 \tag{2.6}$$

where ψ_1 is the probability amplitude for going through slit one, and ψ_2 the probability amplitude for going through slit two. If no potentials are applied during the particle's passage through the cylinders, the interference pattern on the screen is described by

$$\psi_0^* \psi_0 = |\psi_1|^2 + |\psi_2|^2 + 2|\psi_1||\psi_2|\cos\delta \tag{2.7}$$

where δ is the relative phase between ψ_1 and ψ_2. However, if a potential is applied to the second cylinder during the particle's time within the cylinders, and removed before the particle exits, a new situation results. When the particle hits the screen its wavefunction can be written as

$$\psi_\alpha = \psi_1 + \exp(-i\alpha)\psi_2 \tag{2.8}$$

with α given by

$$\alpha = q \int_{-\infty}^{+\infty} \phi(t)\, dt. \qquad (2.9)$$

The interference pattern on the screen is then described by

$$\psi_\alpha^* \psi_\alpha = |\psi_1|^2 + |\psi_2|^2 + 2|\psi_1||\psi_2| \cos(\delta + \alpha). \qquad (2.10)$$

Thus, the interference pattern is predicted to shift even though the particles creating it are never acted upon by any force. The shift is gauge invariant, as we would expect, but shows that some type of interaction has occurred, which, from a classical point of view, is a surprise.

The approach that Aharonov *et al.* take to this is as follows: Normally we describe an interaction by finding some variable that it changes. In classical physics, for example, it is the change in a particle's momentum that characterizes its interaction with a source. In light of this we ask: is it possible to describe the interaction in the potential effect in terms of the change of some variable belonging to a particle contributing to the interference pattern? For example, we can ask if $\langle p \rangle$, where p is the linear momentum of a particle along the screen, depends on α. If not, then the interaction leaves the momentum unchanged, and hence is nonclassical.

We have proved what Aharonov *et al.* suggested to be the case, that the interaction leaves $\langle p^n \rangle$ unchanged, for n a positive integer. That is, we have proved that the interaction does not affect the linear momentum, kinetic energy, or any higher moment of the particle's momentum. Our proof depends on the fact that while the particle is in the cylinders ψ_1 and ψ_2 are nonoverlapping, and requires the cylinders to be short enough and wide enough so that during the time the particle is inside the spreading of ψ_1 and ψ_2 is unimportant.

Having proved that $\langle p^n \rangle$ is unaffected for $n = 1, 2, 3, \ldots$, we can then use this result to prove that $\langle x^n \rangle$ is also unaffected. In this way we have proved that, although the interference pattern shifts when the potential is applied, the interaction that causes the shift does not affect any moment of the position or momentum of any particle contributing to the pattern.

III. MODULAR MOMENTUM

Aharonov *et al.* indicated that there is a variable whose expectation value does change in the potential effect. If we assume that the two slits are centered at $x = 0$ and $x = l$, and take p to be the x component of the momentum operator, then the variable they ask us to consider is

$$A = (p_0/2\pi) \sin 2\pi p/p_0 \qquad (3.1)$$

with $p_0 = h/l$. Knowledge of A determines p within a multiple of p_0, so A is related to the momentum modulo p_0. For convenience, we simply call A the modular momentum.

We can give two heuristic arguments that a variable like A should be important in the potential effect.[16] From a physical standpoint, we know that a particle which lands at a maximum in the interference pattern will be in one of the directions θ for which

$$l \sin \theta = n\lambda \qquad (3.2)$$

where n is an integer and λ the wavelength of the incident particle. This equation is easily rewritten using the de Broglie relation to read

$$p = h \sin \theta/\lambda = np_0. \qquad (3.3)$$

Thus, a particle landing in a maximum has had its x component of momentum shifted to some multiple of p_0 by the two-slit barrier. If a potential difference is applied to the cylinders and the experiment run again, then the maxima are created by particles with momentum

$$p = np_0 + \bar{p} \qquad (3.4)$$

where $0 < \bar{p} < p_0$. Thus, the interaction between a charged particle and the source of potential causes a shift of momentum in *fractions* of p_0. This suggests that the interaction will be described by a variable simply related to the x momentum *modulo* p_0.

From a mathematical standpoint, we note that if we express the modulo variable as a sine or cosine function, then its expectation value in the momentum representation will involve quantities like

$$\int \tilde{\psi}(p, t)^* \, e^{-ilp} \tilde{\psi}(p, t) \, dp \qquad (3.5)$$

since $2\pi p/p_0 = lp$. If we let l be a variable, rather than fix its value as the slit separation, then the integral (3.5) is simply the Fourier transform of the momentum probability density $|\tilde{\psi}|^2$. This means that knowledge of $\sin lp$ or $\cos lp$ for all values of l will be equivalent to knowledge of the momentum distribution itself, and if the interaction changes $|\tilde{\psi}|^2$ at all, it should also change its Fourier transform, and hence $\langle A \rangle$. These considerations lead us to believe that A, as defined in Eq. (3.1), will be important in describing the interaction.

We have been able to prove the following results about A:

(1) In the Heisenberg picture, A evolves according to

$$\frac{d}{dt} A(t) = -\frac{1}{2l} \{ [V(x + l, t) - V(x, t)]e^{ilp}$$

$$+ [V(x, t) - V(x - l, t)]e^{-ilp} \} \quad (3.6)$$

where $V(x, t)$ is the potential energy of the particle and all operators are in the Heisenberg picture. Thus, the time evolution of $A(t)$ depends upon the potential energy at several different points in space at the same time. This means that the change in $A(t)$ is determined by a nonlocal equation, even though the Hamiltonian

$$H = p^2/2m + V(x, t)$$

is local.

(2) If we multiply Eq. (3.6) on the left by $\psi(x, 0)^*$, on the right by $\psi(x, 0)$, and integrate, we find

$$\frac{d}{dt} \langle A \rangle = -\int_{-\infty}^{+\infty} \left[\frac{V(x + l, t) - V(x, t)}{l} \right]$$

$$\times \left[\frac{\psi(x, 0)^* \psi(x + l, 0) + \psi(x + l, 0)^* \psi(x, 0)}{2} \right] dx. \quad (3.7)$$

This equation was derived by Aharonov *et al.* in the Schrödinger picture, is true for any wavefunction, and gives the time evolution of $\langle A \rangle$. As Aharonov *et al.* note, in the limit as l goes to zero, Eq. (3.7) reduces to Ehrenfest's theorem

$$\frac{d}{dt} \langle p \rangle = \left\langle -\frac{\partial V}{\partial x} \right\rangle. \quad (3.8)$$

This means that Eq. (3.7) is a generalization of Ehrenfest's theorem, and that we can identify the potential difference inside the integral as a generalized, nonlocal force. If we confine ourselves to two-slit interference experiments, then $\psi(x, 0)$ is very well peaked about the slits at $x = 0$ and $x = l$, and Eq. (3.7) becomes

$$\frac{d}{dt} \langle A \rangle = -\left[\frac{V(l, t) - V(0, t)}{l} \right] \langle \cos lp \rangle. \quad (3.9)$$

This shows clearly that it is finite differences of the potential energy, and not forces, that lead to a change in $\langle A \rangle$.

(3) If we evaluate $\langle A \rangle$ in the Schrödinger picture,

$$\langle A \rangle = \frac{1}{2i} \int_{-\infty}^{+\infty} \psi(x, t)^*[(e^{ilp} - e^{-ilp})]\psi(x, t)\,dx \qquad (3.10)$$

we find that when the particle hits the screen

$$\langle A \rangle = (\hbar/2l)\sin\alpha. \qquad (3.11)$$

Thus, $\langle A \rangle$ *is* changed by the interaction.

(4) The results of (2) and (3) together lead to an interesting mathematical question. Suppose that, rather than expressing the sine function in Eq. (3.10) as the sum of two exponentials, we had used instead its power series representation. In this case we might have argued that, since none of the moments of the momentum is affected by the interaction, the modular momentum should also remain unaffected.

The problem with this argument is that the power series representation of the sine function cannot be applied to all wavefunctions. At the very least it can only be used on wavefunctions that are infinitely differentiable, and for which the resulting power series is uniformly convergent. On the other hand, a wavefunction need only be $\mathscr{L}_2(\mathbf{R})$ to be a member of the Hilbert Space. In our case, even though our two-slit wavefunctions may be infinitely differentiable, they are not analytic. This is because the physical conditions require them to be identically zero in an interval along the x-axis, and the only analytic function for which this is true is the trivial function. Since our two-slit wavefunctions are not analytic, $\sin lp$ acting on them does not produce the same result as its power series, and the power series argument is not valid.

The same objection does not apply when we represent the sine function as the difference of two exponentials. The exponentials can be defined by their power series on a set of vectors dense in $\mathscr{L}_2(\mathbf{R})$, called analytic vectors, and their properties established in the usual way.[17,18] Then, because the exponentials are continuous unitary operators their extension to the rest of the Hilbert Space is unique, and their properties apply to any vector in the space, including two-slit wavefunctions. The interesting point here is that because of the nonanalyticity of the two-slit wavefunctions, which is required by the physical situation, the interaction in the potential effect can change the modular momentum while leaving all the moments of the linear momentum unchanged.

IV. SUMMARY

Equations (3.9) and (3.11) show that modular momentum gives a new formalism and language for describing the potential effect. According to Eq. (3.9) the modular momentum of the particle is conserved unless a generalized nonlocal force acts. This generalized nonlocal force can cause a change in the modular momentum even when none of the moments of the linear momentum are affected. This is in fact what happens since the generalized nonlocal force is nonzero even though the local force vanishes wherever the particle can be found. We note that this description of the potential effect, in terms of the change in modular momentum, is quite similar to the classical description of interactions. The work of Aharonov *et al.* indicates that such an approach can be generalized to other variables, such as energy, angular momentum, and position, and that not only will it lead to a new understanding of known quantum effects, but it also could provide a way to make new predictions of effects peculiar to quantum mechanics which have no classical analog.

Acknowledgements

We would like to thank Professor Aharonov for making available to us some lecture notes on this subject, and for several helpful conversations. We also would like to thank Professor K. Jagannathan and Professor R. H. Romer for many helpful discussions.

Note Added in Proof

For proofs of the results described here, and their extension to the magnetic A–B effect, see our paper, 'Expectation Values in the Aharonov-Bohm Effect', submitted to *Phys. Rev.* D**15**.

References

1. Y. Aharonov and D. Bohm, *Phys. Rev.* **115**, 485 (1959).
2. D. Wisnivesky and Y. Aharonov, *Ann. Phys.* (*NY*) **45**, 479 (1967).
3. T. T. Wu and C. N. Yang, *Phys. Rev.* D**12**, 3845 (1975).
4. J. Stachel, *Phys. Rev.* D**26**, 1281 (1982).
5. Y. Aharonov and G. Carmi, *Found. Phys.* **3**, 493 (1973).
6. J. H. Harris and M. D. Semon, *Found. Phys.* **10**, 151 (1980).
7. M. D. Semon, *Found. Phys.* **12**, 49 (1982).
8. Y. Aharonov and D. Bohm, *Phys. Rev.* **123**, 1511 (1962); **125**, 2192 (1963); **130**, 1625 (1963).
9. B. S. DeWitt, *Phys. Rev.* **125**, 2189 (1962).
10. P. D. Noerdlinger, *Nuovo Cim.* **23**, 158 (1962).
11. F. J. Belinfante, *Phys. Rev.* **128**, 2832 (1962).
12. E. J. Konopinski, *Am. J. Phys.* **46**, 499 (1978).
13. H. Erlichson, *Am. J. Phys.* **38**, 162 (1970).
14. D. Home and S. Sengupta, *Am. J. Phys.* **51**, 942 (1983).
15. P. Frolov and V. D. Skarzhinsky, *Nuovo Cim.* **76B**, 35 (1983).
16. Y. Aharonov, H. Pendleton and A. Petersen, *Int. J. Theor. Phys.* **2**, 213 (1969).
17. T. F. Jordan, *Linear Operators for Quantum Mechanics* (Wiley, New York, 1969), p. 112.
18. M. Reed and B. Simon, *Methods of Modern Mathematical Physics*, Vol. II (Academic Press, New York, 1975), pp. 200ff.

TOPOLOGICAL EFFECTS IN QUANTUM MECHANICS

Christopher C. Gerry

Department of Physics, St. Bonaventure University, St. Bonaventure, NY 14778

Akira Inomata

Department of Physics, State University of New York at Albany, Albany, NY 12222

.

The winding numbers associated with the Aharonov–Bohm experiment, flux quantization and the Josephson oscillation are considered with the aid of Feynman's path integral. Possible tests of the topological effects due to the higher winding numbers are discussed. Some remarks are also made on the recent double ring experiment by Deaver and Donaldson.

I. INTRODUCTION

The purpose of this paper is to discuss the possibility of observing nontrivial effects in quantum mechanics when the configuration space of the system is multiply connected. These effects which we shall refer to as topological effects may be characterized by the so-called winding number. Specifically we shall be concerned with the winding numbers related to the Aharonov–Bohm effect, flux quantization and the Josephson oscillation which are all taking place in a multiply connected space.

The Aharonov–Bohm (AB) effect[1] indicates that the magnetic vector potential has observable consequences unique in quantum mechanics.

The topological nature of the AB effect, though not immediately obvious, has been well recognized.[2,3] In fact, it becomes more transparent if Feynman's path integral is used to perform summation over all possible paths in the configuration space where a region containing magnetic flux is inaccessible to electrons.[4-6] The path integral analysis allows us to classify the paths according to the number of entanglements about the inpenetrable region. This number is the winding number.

The proper mathematical tool for classifying paths in a multiply connected space is known as homotopy theory. Since the homotopical understanding of the winding number is useful for our discussion, we start with a brief résumé of the homotopy theory in a fashion relevant to the problems to be discussed here. This is followed in Section III by an analysis of the AB effect with the winding number explicitly introduced into the path integral.

While the lowest two winding numbers are responsible for the type of interference described by Aharonov and Bohm[1] and verified by various experiments,[7] the higher winding numbers are also expected to lead to observable consequences. Section IV deals with the related phenomena of magnetic flux quantization in a superconducting ring. In Section V, we propose that the higher winding numbers will give rise to higher frequencies in the a.c. Josephson oscillation. Remarks are also made on a recent experiment by Deaver and Donaldson[8] which claims to have exhibited the winding number dependence of flux quantization.

II. HOMOTOPICAL CLASSIFICATION OF PATHS

To begin with, we briefly outline the concept of homotopy[9] and explain its relation to the winding number.

A particle in motion leaves a trace in its configuration space M (arcwise connected). Suppose it occupies a point $x \in M$ at time t. As t varies from t' to t'', $x = f(t)$ defines a path $f: T \to M$ which starts at $x' = f(t')$ and ends at $x'' = f(t'')$. For simplicity, we can parameterize the path $f(t)$, without loss of generality, by $t \in I = [0, 1]$, so that $x' = f(0)$ and $x'' = f(1)$.

Any two paths, say $f_0(t)$ and $f_1(t)$, with the same starting and ending points are said to be *homotopic* and expressed as $f_0 \simeq f_1$ is they can be continuously deformed into each other. This means that for $f_0 \simeq f_1$ there is a family of paths $\phi(t, s)$ continuous with respect to a parameter $s \in I = [0, 1]$ such that

$$\phi(t, 0) = f_0(t), \qquad \phi(t, 1) = f_1(t)$$

$$\phi(0, s) = x', \qquad \phi(1, s) = x''. \tag{2.1}$$

For the homotopic relation, the following rules apply:

(i) $f \simeq f$

(ii) If $f_0 = f_1$, then $f_1 \simeq f_0$

(iii) If $f_0 \simeq f_1$ and $f_1 \simeq f_2$, then $f_0 \simeq f_2$. $\qquad$ (2.2)

Let $f(t)$ be a path from x' to x'' and $g(t)$ be a path from x'' to x'''. Here, $f(1) = g(0) = x''$. A function $h(t)$ such that

$$h(t) = \begin{cases} f(2t) & 0 \leqslant t \leqslant \frac{1}{2} \\ g(2t - 1) & \frac{1}{2} \leqslant t \leqslant 1 \end{cases} \tag{2.3}$$

defines a path from x' to x''' via x''. We write this composition of two paths as $h = f \cdot g$. For a path $f(t)$ from x' to x'', $f^{-1}(t) = f(1 - t)$ gives a path which traces the path $f(t)$ reversely from x'' to x'. Apparently, $(f^{-1})^{-1} = f$. Furthermore, $f \cdot f^{-1} = x'$ and $f^{-1} \cdot f = x''$.

The path $f(t)$ is *closed* with reference to x' if $f(0) = f(1) = x'$. Let $P(M, x')$ be a set of all closed paths on the arcwise connected space M which start at $x' \in M$. Let f be a closed path belonging to $P(M, x')$. We write a set of paths homotopic to f as $\{f\}$. If we choose two of such sets, $\{f\}$ and $\{g\}$, then we have either $\{f\} = \{g\}$ or $\{f\} \cap \{g\} = 0$. Thus, we can classify every path belonging to $P(M, x')$ into one of mutually exclusive homotopy classes, $\{f\}, \{g\}, \{h\}, \ldots$. All loops homotopic to the constant $f(0)$, that is, those which can be shrunk to the point x', belong to the identity class $\{e\}$. For each $\{f\}$, there is $\{f^{-1}\}$ such that $\{f \cdot f^{-1}\} = \{f^{-1} \cdot f\} = \{e\}$. Since $f_1 \simeq g_1$ and $f_2 \simeq g_2$ imply $f_1 \cdot g_1 \simeq f_2 \cdot g_2$, we have the composition rule,

$$\{f\} \cdot \{g\} = \{f \cdot g\}. \tag{2.4}$$

Therefore, the entire set $\pi(M, x')$ of the homotopy classes forms a group with respect to the group composition (2.4). This is called the *fundamental group* of M in reference to $x' \in M$. The structure of this group is independent of the choice of the reference point x'.

In a simply connected space, the identity class $\{e\}$ is the only element of π_1. For example, a spherical surface S^2 is simply connected. All closed paths starting from a point $x \in S^2$ are contractable to the point x, so that $\pi_1(S^2)$ has $\{e\}$ alone. The simplest example of multiply connected spaces is a circle $S^1 = \{P(\theta); 0 \leqslant \theta < 2\pi\}$. Closed loops starting at $P(0) = 0$ are given by $g_n(t) = P(2\pi n t)$ where $t \in I = [0, 1]$ and $n = 0, \pm 1, \pm 2, \ldots$. Each of $g_n(t)$ indicates a path which encircles counterclockwise n times if $n \geqslant 0$ and clockwise $|n|$ times if $n < 0$. Obviously, $g_n \cdot g_m = g_{n+m}$. Thus, the fundamental group of S^1 is $\pi_1(S^1) = \{g_n\} \simeq Z$ where Z is the set of integers.

In the case of our interest, the configuration space M is a two-dimensional plane R^2 with origin $(0, 0)$ removed, i.e. $M = R^2 - (0, 0)$. Any path $f(t)$ can be decomposed as $f(t) = f_0(t) \cdot g_n(t)$ where $f_0(t)$ is a path from x' to x'' without encircling the origin and $g_n(t)$ a closed path starting at x' and encircling the origin n times. All paths from x' to x'', moving around the origin less than 2π, are homotopic to $f_0(t)$. Any loop $g_0(t)$ not encircling the origin $(0, 0)$ can be contracted into x', and hence belongs to the identity class $\{e\} = 0$. A loop of n entanglements $g_n(t)$, counterclockwise if $n > 0$ and clockwise if $n < 0$, belongs to class $\{n\}$. Thus, paths from x' to x'' on M are characterized by the homotopy classes, $\ldots \{-2\}, \{-1\}, \{0\}, \{1\}, \{2\}, \ldots$. It is easy to see that for these classes the composition rule $\{n\} \cdot \{m\} = \{n + m\}$ holds. The fundamental group of our configuration space M is apparently the set of integers, that is, $\pi_1(M) \simeq Z$. The integer $n \,(= 0, \pm 1, \pm 2, \ldots)$ associated with paths from x' to x'' on $M = R^2 - (0, 0)$ is of course the winding number.

Feynman[10] asserted that the propagator of the particle from x' to x'' is given by the sum of the contributions from all possible paths from x' to x''. If the paths are classified into different homotopy classes, then the propagator must be a linear combination of the partial propagators belonging to those classes. It has been shown that the coefficients of the partial propagators form a one-dimensional unitary representation of the fundamental group.[11] For $\pi_1(M) \simeq Z$, the coefficient of the nth partial propagator is given by $e^{i\xi n}$ where ξ is a real constant. With $\xi = 0$, we have a trivial representation of $\pi_1(M)$.

III. AHARONOV–BOHM EFFECT

The usual experimental setup for the AB effect consists of the source S, the detector D, and a long solenoid containing magnetic flux Φ placed right behind a double slit. Charged particles emitted from S are to arrive

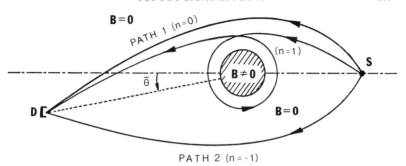

PATH 2 (n = -1)

Figure 1   Idealized Aharonov–Bohm experiment.

at D through field-free regions outside the solenoid. For the present discussion, however, as shown in Fig. 1 we shall ignore the double slit which mechanically selects two possible paths of the particles. We also take the solenoid to be ideally thin so that it can be treated as a line singularity. The vector potential outside the solenoid is given in cylindrical coordinates (r, θ, z) by

$$\mathbf{A} = (\Phi/2\pi)\,\nabla\theta \qquad (3.1)$$

so that $\mathbf{B} = \nabla \times \mathbf{A} = 0$. The various possible paths of a particle around the solenoid can be classified into equivalent homotopy classes and delineated by the winding number n. Since the particle starts from the source position $\mathbf{r}'$ and ends at the detector position $\mathbf{r}''$, their paths are not really closed. Nevertheless, the winding number n can be defined through the relation,[4]

$$\varphi(t) = \int^{t} \dot{\theta}\, dt = \theta + 2\pi n, \qquad n = 0, \pm 1, \pm 2, \dots \qquad (3.2)$$

where θ is the usual polar variable with range $[0, 2\pi)$ and φ is the covering variable with range $(-\infty, \infty)$.

The propagator of the particle with mass μ and charge q going from the source at $\mathbf{r}'$ to the detector at $\mathbf{r}''$ in the time τ can be calculated in the covering space M^* of the configuration space M by using the periodic constraint (3.2) in the path integral, the result being a sum of the partial propagators, each of which belongs to a class of homotopically equivalent paths,[4]

$$K(\mathbf{r}'', \mathbf{r}': \tau) = \sum_{n=-\infty}^{\infty} K_n(\mathbf{r}'', \mathbf{r}'; \tau). \qquad (3.3)$$

The partial propagator K_n can further be decomposed into the flux-dependent factor and the flux-independent propagator,[5,6]

$$K_n(\mathbf{r}'', \mathbf{r}' : \tau) = \exp[i(q\Phi/hc)(\theta'' - \theta' + 2\pi n)]\tilde{K}_n(\mathbf{r}'', \mathbf{r}' : \tau) \quad (3.4)$$

where

$$\tilde{K}_n = (\mu/2\pi i h\tau) \exp[i(r'' - r')^2 \mu/2h\tau + ih\tau/(8r'r''\mu)]$$
$$\times \exp[\tfrac{1}{2}i(r'r''\mu/h\tau)(\theta'' - \theta' + 2\pi n)^2]. \quad (3.5)$$

The flux-dependent factor in (3.4) is nothing but the contribution from the so-called nonintegrable phase factor of Wu and Yang,[12]

$$\exp\left[i(q/hc)\left(\int_{\mathbf{r}'}^{\mathbf{r}''} \mathbf{A} \cdot d\mathbf{r} + n \oint \mathbf{A} \cdot d\mathbf{r} \right) \right]. \quad (3.6)$$

The experimental basis for detecting the effect of the isolated flux is of course the interference of two propagators belonging to two different winding numbers. The propagators corresponding to the winding numbers n and m ($n \neq m$) will give rise to an interference at $\theta'' = \theta' + \pi$ as[6]

$$\text{Re}(K_n^* K_m) = \left(\frac{\mu}{2\pi h\tau} \right)^2 \cos\left[2\pi(m - n)\left\{ \frac{q\Phi}{2\pi hc} + \frac{r'r''}{h\tau}(m + n + 1) \right\} \right]. \quad (3.7)$$

For the double slit experiment, in particular, the paths with $m = 0$ and $n = -1$ contribute to the interference, resulting in the traditional flux dependence,[13]

$$\text{Re}(K_0^* K_{-1}) = (\mu/2\pi\tau)^2 \cos[q\Phi/hc]. \quad (3.8)$$

Now notice that (3.7) contains a term depending on the winding number as well as the term involving the flux Φ. Even in the absence of flux in the solenoid, we can expect an interference shift dependent on the winding number. This indicates that in general there is an interference shift of the topological origin in addition to the proper flux-dependent AB effect. Such topological shifts disappear if $m + n + 1 = 0$. Apparently, the topological effect slips off from detection by the double slit experiment. How one can detect these additional shifts is as yet an open question.

IV. FLUX QUANTIZATION

The quantization of magnetic flux trapped in a superconducting ring is often described as a consequence of the singlevaluedness requirement imposed on the wave function of a charged carrier (a Cooper pair) in the ring.[13] It has also been argued that flux quantization is a manifestation of the AB effect in a bound state.[14] Here we wish to point out that these two views are not compatible with each other.

For simplicity, we consider an ideally thin superconducting ring of radius R centered about the flux as in Fig. 2. Let $\psi(\theta)$ be the wave function satisfying the Schrödinger equation for a carrier in the ring. Then, it is apparent that

$$\psi(\theta + 2\pi) = \exp\left[(iq/\hbar c) \oint \mathbf{A} \cdot d\mathbf{r} \right] \psi(\theta) \qquad (4.1)$$

where $\mathbf{A}$ is the vector potential in the ring. Certainly, the singlevaluedness requirement $\psi(\theta) = \psi(\theta + 2\pi)$ leads to the quantization on the flux trapped,

$$\Phi = \oint \mathbf{A} \cdot d\mathbf{r} = n\Phi_0, \qquad (4.2)$$

where $\Phi_0 = hc/q$ is the fundamental fluxoid and $n = 0, 1, 2, \ldots$.

However, such a naive application of the singlevaluedness is questionable. The flux quantization (4.2) resulted from the singlevaluedness requirement is independent of whether the ring around

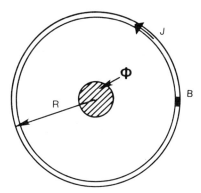

Figure 2   Bloch's ring.

the flux is of normal conductivity or of superconductivity. If the singlevaluedness is a universal requirement for any axially symmetric system, then all magnetic flux, whenever and wherever found, would have to be in quantized states contrary to the experimental fact.[15] If the flux is always quantized, there should be no observable AB effect.

For our idealized superconducting ring, the propagator follows from (3.3) as a special case,

$$K(\theta'', \theta'; \tau) = \sum_{n=-\infty}^{\infty} \exp[i(\theta'' - \theta' + 2\pi n)\Phi/\Phi_0]\tilde{K}_n(\theta'', \theta'; \tau) \quad (4.3)$$

where

$$\tilde{K}_n(\theta'', \theta'; \tau) = (2\pi i\hbar\tau/I)^{-\frac{1}{2}} \exp[i(I\tau/\hbar)(\theta'' - \theta' + 2\pi n)^2].$$

Apparently, the nth partial propagator $\tilde{K}_n$ in (4.3) equals that of an axial rotator [16] of $I = \mu R^2$. Using the theta function,

$$\vartheta_3(z, \zeta) = \sum_{m=-\infty}^{\infty} \exp[2miz + m^2\pi\zeta]$$

and its transformation,

$$\vartheta_3(z, \zeta) = (-i\zeta)^{-\frac{1}{2}} \exp(-iz^2/\pi\zeta)\vartheta_3(z/\zeta, -1/\zeta), \quad (4.4)$$

the full propagator (4.3) may be converted into

$$K(\theta'', \theta'; \tau) = (2\pi)^{-1} \sum_{m=-\infty}^{\infty} \exp[im(\theta'' - \theta')]$$

$$\times \exp[-i(\hbar\tau/2I)(m - q\Phi/\hbar c)^2]. \quad (4.5)$$

This result can of course be obtained directly from the evolution operator $U = \exp[-iH\tau/\hbar]$ with the Hamiltonian,[14] $H = (2I)^{-1}(L_z - qRA_\theta/c)^2$. The integer m in (4.5) is the eigenvalue of the angular momentum L_z. The singlevaluedness, $K(\theta'', \theta') = K(\theta'' + 2\pi, \theta')$, of (4.5) is self-evident. It does not lead to the flux quantization (4.2). However, once the flux is quantized, the propagator (4.5) takes the form,

$$K(\theta'', \theta'; \tau) = (2\pi)^{-1} \sum_{m=-\infty}^{\infty} \exp[im(\theta'' - \theta')] \exp[-i(\hbar\tau/2I)m^2],$$

so that the charge carrier becomes insensitive to the flux enclosed by the ring. Similarly, the flux-dependent shift in (3.7) and (3.8) also disappears. Therefore, we cannot expect any observable AB effect. The

singlevaluedness of (4.5) is independent of whether the flux is quantized
or not, and hence plays no role in flux quantization.

At this point, it is interesting to compare the two expressions (4.3) and
(4.5) for the same propagator, connected by the theta transformation
(4.4). We note that the winding number n and the angular quantum
number m are complementary to each other in the sense that the two
numbers cannot be specified at the same time. A partial propagator with
a fixed angular momentum is a superposition of propagators belonging
to all possible winding numbers, while a propagator with a specific
winding number is a superposition of partial propagators belonging to
all possible angular quantum numbers.

In flux quantization, actually the flux enclosed by the superconducting
ring consists of flux Φ_{ex} due to external fields and flux Φ_i due to induced
surface currents. The total flux Φ will adjust itself so as to minimize the
free energy of charge carriers in the ring.[17] The adjusted flux is quantized
by (4.2). To see this, we construct the partition function for the charge
carriers. The single particle partition function can be obtained from the
propagator (4.3) by the relation

$$Z(\Phi, \beta) = \text{Tr } K(\theta'', \theta'; -i\hbar\beta).$$

Namely, we find[18,19]

$$Z(\Phi, \beta) = \gamma \left[1 + 2 \sum_{n=1}^{\infty} \cos(2\pi n\Phi/\Phi_0) \exp(-\pi n^2 \gamma^2) \right] \quad (4.6)$$

where $\gamma = (2\pi\mu R^2/\hbar^2\beta)^{\frac{1}{2}}$. For the system of $\bar{N}$ carriers, the partition
function is

$$Q(\Phi, \beta) = Z^{\bar{N}}/\bar{N}! \quad (4.7)$$

and the corresponding free energy is

$$F = -\beta^{-1}(\bar{N} \ln Z - \ln \bar{N}!). \quad (4.8)$$

Now it is readily seen that the flux quantization $\Phi = k\Phi_0$ ($k = 0, 1, 2,$
...) occurs when the free energy (4.8) is a minimum or when the partition
function (4.6) is a maximum. Indeed, the flux quantization is a
manifestation that the charge carriers in the ring is in an equilibrium
state where the body current $J = -c \, \partial F/\partial \Phi$ is nil.

What would then the role of the winding number be in flux
quantization? Recently, Deaver and Donaldson[8] have performed an
experiment on a superconducting ring which twice encircles a long
solenoid containing flux. They have shown that the trapped flux is

quantized as $\Phi = \frac{1}{2}k\Phi_0$ ($k = 0, \pm 1, \pm 2, \ldots$) so that the change in flux is a half of the fundamental fluxoid, i.e. $\Delta\Phi = \frac{1}{2}\Phi_0$. An experiment with an N-turn ring would then presumably show the fractional quantization $\Phi = (k/N)\Phi_0$. They have interpreted the result as a demonstration of the winding number dependence of the AB effect suggested in Section III. However, the number of turns N is not quite the same as the winding number n. A suggestion made by Yang[20] is that doubling the number of turns effectively doubles the trapped flux.

For an N-turn ring the winding number about the structure is limited to $n = n'N$ where n' is an integer. Thus the partition function (4.6) has to be modified as

$$Z_N(\Phi, \beta) = \gamma\left[1 + 2\sum_{n'=1}^{\infty}\cos(2\pi n'N\Phi/\Phi_0)\exp(-\pi n'^2N^2\gamma^2)\right]. \quad (4.9)$$

In comparison with (4.6), we can write (4.9) in the form,

$$Z_N(\Phi, \beta) = N^{-1}Z(N\Phi, \beta/N^2). \quad (4.10)$$

This implies that at a fixed temperature the N-turn ring of radius r encircling flux Φ is equivalent to a single turn ring of radius Nr encircling flux $N\Phi$.

A maximum of (4.9), which minimizes the corresponding free energy F_N, takes place when $\Phi = (k/N)\Phi_0$ ($k = 0, 1, 2, \ldots$). This is the fractional quantization Deaver and Donaldson[8] have expected.

If the terms higher than $n' = 1$ are suppressed, then the winding number n coincides with the configurational turning number N. The contribution from $|n| = N$ for a single-turn ring is therefore equivalent to the dominant contribution from an N-turn ring with the same radius.[19] In this sense, the result of Deaver and Donaldson[8] may be considered as a demonstration of the winding number dependence of the AB effect.

V. JOSEPHSON EFFECT

Here we shall discuss an alternative method for detecting the effect of the winding number n, particularly for $|n| > 1$. Again, we consider a superconducting ring centered on a solenoid. This time, however, we place a barrier or a weak link in the ring so as to block the current flow as shown in Fig. 2. If the flux in the solenoid varies linearly with time as

$\Phi = cVt$ where V is an effective d.c. bias, then alternating currents are induced in the ring. The frequency of such a current oscillation is given by $v_1 = qVh$. This setup was originally suggested by Bloch[21] as a simple explanation of the a.c. Josephson effect.

For an N-turn ring, from (4.7) and (4.9), we obtain the oscillatory body current,[18,19]

$$J_N = \sum J_{Nn'} \sin(2\pi v_{Nn'} t) \tag{5.1}$$

where $J_{Nn'}$ are amplitudes containing damping factors due to the barrier[19] and $v_{Nn'}$ are the Josephson frequencies

$$v_{Nn'} = Nn'v_1. \tag{5.2}$$

Detection of higher harmonics $v_{Nn'}$ such that $v_{Nn'}/N = n'v_1$ ($n' = 1, 2, 3, \ldots$) must be a confirmation of the winding number $n = Nn'$.

Naturally, a question arises as to whether detecting the higher harmonics in the Josephson junction is experimentally feasible, or not. A proposal is to seek a departure from the current-phase relation $I = I_0 \sin \delta$ where δ is the phase difference across the junction. In the a.c. Josephson effect, $\delta(t) = \delta(0) - 2\pi v_1 t$. Deaver and Donaldson[22] have suggested a static measurement of the current-phase relation by measuring the flux induced in the ring with a given external field applied. Such an experiment would indicate a nonsinusoidal behavior of the current because of the possible depairing of carriers across the junction. In the case of the external flux varying with time as cVt, the higher harmonic terms are expected to attenuate by P^n where P is the tunneling probability for Cooper pairs at the junction. Now $P \sim 10^{-8}$. Consequently, the higher harmonics may not be observable for a weak link. On the other hand, for a stronger coupling, the junction itself will contribute to the current oscillations and hence the effect we are seeking may not be discernable. The possibility of higher harmonics in the Josephson effect has also been discussed by Widom *et al.*[23] in a somewhat different context. However, the experimental feasibility of finding the higher harmonics is as yet unclear.

If the higher terms ($|n'| > 1$) in (5.1) are strongly damped, the dominant contribution will be

$$I_N \sim \sin(2\pi v_N t)$$

where $v_N = Nv_1$. Apparently this is identical with the Nth harmonic of a single-turn ring. Therefore, the nth harmonic of a single turn ring can be determined in principle by the dominant oscillation of an n-turn ring.

References

1. Y. Aharonov and D. Bohm, *Phys. Rev.* **115**, 485 (1959).
2. D. Bohm and B. J. Hiley, *Nuovo Cim.* **52A**, 295 (1978).
3. D. M. Greenberger, *Phys. Rev.* *D23*, 1460 (1981).
4. A. Inomata and V. A. Singh, *J. Math. Phys.* **19**, 2318 (1978); A. Inomata, Lecture Notes at SUNY-Albany, unpublished (1978).
5. C. C. Gerry and V. A. Singh, *Phys. Rev.* *D20*, 2550 (1979).
6. C. Bernido and A. Inomata, *Phys. Lett.* **77A**, 384 (1980); *J. Math. Phys.* **22**, 715 (1981).
7. R. G. Chambers, *Phys. Rev. Lett.* **5**, 3 (1960); H. A. Fowler *et al.*, *J. Appl. Phys.* **32**, 1153 (1961); G. Moellenstedt and W. Bayh, *Phys. Bl.* 299 (1962); H. Boersch *et al.*, *Z. Phys.* **165**, 79 (1961); R. C. Jaklevic *et al.*, *Phys. Rev.* **140**, A1628 (1965); A. Tonomura *et al.*, *Phys. Rev. Lett.* **48**, 1443 (1982).
8. B. S. Deaver and G. B. Donaldson, *Phys. Lett.* **89A**, 178 (1982).
9. W. Mariano and H. Pagels, *Phys. Rep.* **36**, 137 (1978), where a readable account of homotopy theory can be found.
10. R. P. Feynman and A. R. Hibbs, *Quantum Mechanics and Path Integrals* (McGraw-Hill, 1965).
11. C. DeWitt-Morette *et al.*, *Phys. Rep.* **50**, 255 (1979).
12. T. T. Wu and C. N. Yang, *Phys. Rev.* *D12*, 3845 (1975).
13. See, e.g., J. J. Sakurai, *Advanced Quantum Mechanics* (Addison-Wesley, 1967).
14. M. Peshkin, *Phys. Rev.* *A23*, 360 (1981).
15. A. Tonomura *et al.*, *Phys. Rev. Lett.* **51**, 331 (1983).
16. W. Langguth and A. Inomata, *J. Math. Phys.* **20**, 499 (1979).
17. N. Byers and C. N. Yang, *Phys. Rev. Lett.* **7**, 46 (1961).
18. C. C. Gerry and V. A. Singh, *Nuovo Cim.* **73B**, 161 (1983).
19. A. Inomata, *Phys. Lett.* **95A**, 176 (1983).
20. C. N. Yang, as quoted in ref. 8.
21. F. Bloch, *Phys. Rev. Lett.* **21**, 1241 (1968); *Phys. Rev.* *B2*, 109 (1970).
22. B. S. Deaver and G. B. Donaldson, Private communication.
23. A. Widom, T. D. Clark and G. Meagaloudis, *Phys. Lett.* **76A**, 163 (1980).

QUANTUM INTERPRETATION USING CONSISTENT HISTORIES

Robert B. Griffiths

Department of Physics, Carnegie–Mellon University, Pittsburgh, PA 15213

A probabilistic interpretation for a closed quantum-mechanical system (time development determined solely by Schrödinger dynamics) is developed in a form which does not rely on the notion of a measurement, but instead employs a mathematical condition of consistency to select sequences of events to which probabilities can be assigned. The result is a generalization of the "orthodox" interpretation of quantum mechanics. It removes certain conceptual difficulties while retaining, and sometimes sharpening other difficulties.

I. INTRODUCTION

There are two basic approaches to the well-known conceptual difficulties of non-relativistic quantum mechanics. One is to replace "standard" quantum theory with something else (hidden variables, for example), the other is to try and re-interpret the standard theory in a way which gets rid of some of the problems. The attempt represented in this paper belongs to the second category: for the most part it is the standard machinery of standard quantum mechanics as set forth in the standard textbooks (or by von Neumann[1]) with, however, a *generalization* of the Born probability interpretation which seems to make physical sense in some simple cases. This generalization, which goes under the name of "consistent histories", resolves certain conceptual dilemmas (see Table I), but leaves others in place and perhaps makes them even more

Table I   *Comparison of classical mechanics and the "orthodox" quantum interpretation with consistent histories*

	Classical	Orthodox quantum mechanics	Consistent histories
1. Describes closed system	Yes	No	Yes
2. Needs measurements	No	Yes	No
3. Time reversible	Yes	No	Yes
4. All physics is in *the* wave function		?	No
5. Deterministic	Yes	No	No
6. Agrees with experiment	No	Yes	Yes
7. Answers all "reasonable" questions	Yes	No	No
8. Standard logic	Yes	No	No

annoying. The reader must judge which he prefers. My own preference is of course obvious, and the supporting polemic, along with considerable mathematical detail, is in a lengthy and difficult article published elsewhere.[2] In what follows I have at some points introduced a few simplifications in the interests of good pedagogy. More precise statements and a number of technical details will be found in the article just mentioned.

The main respects in which consistent histories differ from the "orthodox" quantum interpretation, to use Wigner's term,[3] are indicated in the first three items in Table I. The consistent history interpretation refers to *closed* systems whose time development is determined solely by Schrödinger's equation *without* any outside interaction or intervention. (Strictly speaking, the time development is determined by the unitary transformations which correspond to solutions of Schrödinger's equation.) If there is a measuring apparatus (or observer, or whatever), this (he) must be part of the closed system described quantum mechanically. The basic interpretive unit is an "event", which means a particular state of affairs, macroscopic or microscopic, which exists at a certain time: a pointer reading, mark on a photographic plate, etc., *or* the position of a particle, the z-component of spin, or some other microscopic event. A crucial difference from the orthodox interpretation is that the consistent history approach is *not* restricted to talking about the "results of measurements", and it is *most* essential that the reader, especially the reader familiar with orthodox terminology, try *not* to think of these events as the "results of measurement". He will be much closer to the consistent history approach if he tries to imagine these events in the way one thinks of such things in *classical* mechanics: something which is taking place because it is taking place, and which could be observed to take place (in principle, at least) by bringing up a suitably designed piece of apparatus. All of which is, of course, too good to be true, and will have to be qualified later on. Nonetheless, line 2 of Table I identifies a crucial difference between quantum orthodoxy and the consistent history approach. The reader who simply asserts, as an article of faith, that "you can't talk about it unless it is measured" is missing the point that it is precisely this statement which is being challenged by suggesting an alternative approach.

Compared with what has just been mentioned, lines 3 and 4 of Table I are minor, but still worth noting. The consistent history *interpretation* of quantum mechanics is explicitly invariant under a reversal of the sense of time, in contrast to von Neumann's approach,[1] and this is not unrelated to the fact, as noted by Aharonov *et al.*,[4] that one cannot think of the quantum system as represented at each point in time by "the" wave function. From the consistent history perspective, wave functions are basically mathematical devices for calculating physical quantities. Thus, for example, the "collapse" of a wave function is a mathematical operation carried out to calculate something of physical interest; there is no corresponding physical "collapse". (There seem to be differences of opinion on this topic among the orthodox, which is why I put a question mark on line 4.)

The remaining items (5–8) in Table I are those in which consistent histories and the orthodox approach are similar, though not completely identical. In particular, both yield probabilities rather than certainties, and both seem to be in agreement with experiment. Indeed, the consistent history approach gives the same probabilities for experimental results as does standard quantum mechanics—or at least I believe this is true, though the point is not altogether trivial.[5] Both approaches share a certain agnosticism in refusing to answer, and thus dismissing as "meaningless" certain questions which make perfectly good sense in classical mechanics. In the orthodox interpretation this agnosticism is expressed through a refusal to answer questions unrelated to measurements. But in consistent histories, measurements do not play a fundamental role, and whether a question is answerable is decided by a *mathematical* condition of *consistency*. When this condition is not satisfied, the consistent history approach refuses to answer the question even in the sense of assigning a probability. Since unanswerable questions are easily constructed by a conjunction of answerable questions ("what is the probability of *A and B* given . . . ?"), the resulting conceptual apparatus has a non-standard logic. Whether it is identical with any of the non-standard logics hitherto employed as an aid in interpreting quantum mechanics is something I do not know. On this point (as on many others), my own understanding is quite limited, and I welcome advice from the real experts.

II. WEIGHTS FOR TRAJECTORIES; CONSISTENCY

A classical analogy is useful for understanding the consistent history approach. Figure 1 is a schematic drawing of an orbit (curved line) of a closed system through its phase space (vertical axis) as a function of time t. The vertical lines are "gates", or barriers-with-holes, corresponding to different events taking place at different times: those orbits and only

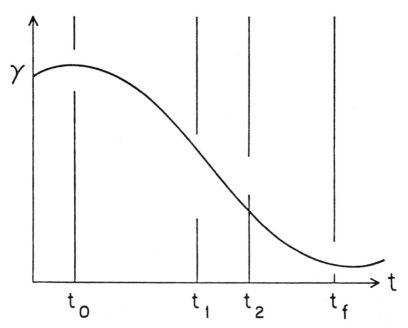

Figure 1   Orbit in a classical phase space (represented schematically by the vertical axis) as a function of time.

those orbits which correspond to the event in question (e.g., "the particle is in the left third of the box") pass through the hole, while the others are blocked. These gates are *not* part of the physical system being described. Instead, they correspond to questions which interest the theoretician who is describing the system.

Consider the set of all orbits which pass through the initial gate D at time t_0 and the final gate F at time t_f (whether or not they pass through any of the intermediate gates), and assign them equal weights using a uniform measure at some fixed time. (The reader is invited to supply the technical conditions.) Now the *fraction* of this weight associated with those orbits which *in addition* pass through the intermediate gates $E_1, E_2,$ ... at times $t_1, t_2,$ ... can be interpreted as a conditional probability

$$P(E_1 \wedge E_2 \wedge \ldots E_n \mid D \wedge F) \qquad (2.1)$$

of these intermediate events given the initial and final events. Read " $\wedge$ " as "and". In particular, if E_j' denotes the complement of E_j—the event corresponding to E_j not occurring, obtained by interchanging the barrier and the hole—then it is true that

$$P(E_1 \wedge \ldots E_j \ldots E_n \mid D \wedge F) + P(E_1 \wedge \ldots E_j' \ldots E_n \mid D \wedge F)$$
$$= P(E_1 \wedge \ldots E_n \mid D \wedge F), \qquad (2.2)$$

where ... on the right side denotes the same sequence of events as on the left, except that the jth event has been omitted.

In the quantum mechanical case the events are represented by orthogonal projection operators, for which I shall also use the symbol E_j, acting on an appropriate Hilbert space, and

$$E_j' = 1 - E_j \qquad (2.3)$$

corresponds to the complementary event. (The classical counterpart of E_j is a function equal to 1 everywhere on the hole and 0 everywhere on the barrier for the gate in question.) The quantum analog of the set of classical orbits passing through all the holes is an "amplitude" operator

$$a = FU(t_f, t_n)E_n U(t_n, t_{n-1}) \cdots U(t_2, t_1)E_1 U(t_1, t_0)D, \qquad (2.4)$$

where D and F are orthogonal projections corresponding to the initial and final states, and $U(t', t)$ is the unitary time transformation from t to t' corresponding to the Schrödinger equation.

Since we know that quantum probabilities come about from squaring amplitudes, it (perhaps) makes sense to think about the weight

$$w(D \wedge E_1 \wedge \dots F) = \text{Tr}[a^\dagger a], \qquad (2.5)$$

where Tr is a trace and $a^\dagger$ denotes the Hermitian conjugate. Unsophisticated people like me may find the following remark helpful. If the quantum system could be described by a finite n-dimensional Hilbert space, you and I would introduce an orthonormal basis in which the operator a would correspond to an n by n complex matrix, which we could then (if we chose) think of as a "vector" with n^2 complex components, or counting real and imaginary parts of each matrix element separately, $2n^2$ real components. The right side of (2.5) is the square of the length of this vector. (Sophisticated people who feel comfortable in infinite-dimensional Hilbert spaces should, I am told, insert "trace class" at appropriate points in the exposition.)

The final step is to define the quantum analog of (2.1):

$$P(E_1 \wedge \dots E_n \mid D \wedge F) = w(D \wedge E_1 \wedge \dots E_n \wedge F)/w(D \wedge F), \qquad (2.6)$$

where the denominator is, of course, obtained from (2.5) by replacing all of the E_j's by 1, that is, a by

$$F U(t_f, t_0) D. \qquad (2.7)$$

But now we are in serious trouble because the right side of (2.6) does *not* behave like a probability: it can be larger than one, (2.2) is in general not satisfied, etc. What can we do? The consistent history approach (this is the *central* point) is to use (2.6) *only for certain families of* sequences-of-events, or *histories*, for which the right side of the equation *does* have the properties which good (classical) probabilities should have. Such families, or the individual histories of which they are composed, are called *consistent histories*, and any history which does not belong to a consistent family is called inconsistent. Probabilities are only assigned to consistent histories. As to what goes on in other (inconsistent) cases, the consistent history approach maintains a discreet silence (the agnosticism referred to earlier).

The technical definition of consistency is given in detail elsewhere,[2] but the following may be a useful perspective. A family of histories with a given D and F corresponds to various ways of writing (2.7) as a sum of operators of the type (2.4), with various different Es. Consistency is then the demand that in each of these sums the summands, regarded as $2n^2$-dimensional real "vectors" in the sense indicated below (2.5), are mutually orthogonal. Such a condition seems sensible if squares of lengths are to be interpreted as probabilities.

Note that consistency is a *mathematical* condition. Given a set of orthogonal projections corresponding to the different events, and given the unitary time transformations appropriate to the (closed!) system, the question of consistency can be settled, in principle, by honest (though perhaps arduous) computation. The mathematical condition makes no reference to "measurements", and hence whether some or all or none of the events in question are related to some measurement is irrelevant. Of course we as theoretical physicists may be interested in events corresponding to the position of the pointer on some measuring apparatus and, if so, we can put this event along with others which interest us into a history and ask whether the history is consistent. If it is, then the consistent history approach will assign a (conditional) probability to it. But measurements *as such* play no fundamental role in the consistent history approach, and for this very reason consistent histories can help interpret measurement processes without leading to circular reasoning.

The reader has no doubt been wondering why all the probabilities discussed above have been conditional on both an initial event D and a final event F. This is because including both of them shows explicitly that the interpretive scheme of consistent histories is unaltered if the sense of time is reversed.One can certainly replace F (or D) with the identity operator—in the classical analog, the hole becomes the whole phase space—so as to obtain probabilities conditional on a single initial (or final) event.

In particular, the usual Born formula for transition probabilities is obtained by letting $F = 1$, choosing D corresponding to a pure initial state, and considering a family in which each history consists of D and some event at a later time t_1. Any single history of this type is automatically consistent, and the family is consistent provided the corresponding projections at time t_1 all commute with one another.

III. EXAMPLES

A crucial test of any interpretive scheme is to see what it predicts for particular examples. Here are two which, to my thinking, illustrate both the pleasant and unpleasant aspects of the consistent history approach. Technical details will be found in my longer paper.[2]

The first is a scattering problem, Fig. 2, in which a particle represented

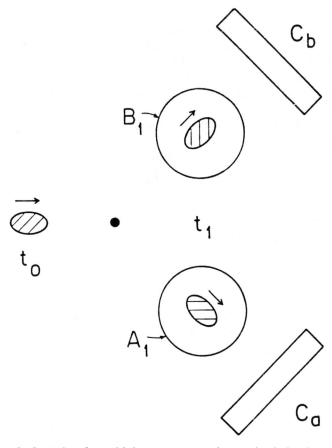

Figure 2   Scattering of a particle into two counters; the cross-hatched regions represent the wave function at two successive times t_0 and t_1.

by a wave packet at time t_0 is scattered in such a way that the wave packet at time t_1 consists of two pieces traveling towards two different counters C_a and C_b. Note that the counters are quantum objects which, along with the particle itself, are part of a closed system. Thus the initial state D corresponds to the initial particle wave packet and untriggered states for both counters. Let t_f be some time after the particle has reached and triggered one of the counters, and suppose that the final state F corresponds to counter C_a in a triggered state indicating arrival of the particle. Question: given D and F, where was the particle at time t_1, i.e., was it in region A_1 or in region B_1?

The physically reasonable answer (at least for someone who has not become confused through studying too much quantum mechanics!) is that the particle was in region A_1: if a particle has just triggered a counter, then surely it is the case that an instant earlier the particle was traveling towards the counter and not away from it. And this is the answer provided by the consistent history approach, in the following way. First one checks that the history represented schematically by

$$D \rightarrow A_1 \rightarrow F, \tag{3.1}$$

where A_1 is the event that the particle is in this region, Fig. 2, at time t_1, satisfies the consistency condition, and next, using (2.6), that

$$P(A_1 \mid D \wedge F) = 1. \tag{3.2}$$

Similarly one can show that

$$P(B_1 \mid D \wedge F) = 0, \tag{3.3}$$

as expected.

It is "retrodictions" of this sort which are particularly troublesome for standard quantum mechanics as represented in von Neumann's well-known scheme,[1] which makes a sharp difference between past and future. By contrast, the consistent history interpretation is formally invariant if the sense of time is reversed, so it is not surprising if the latter puts prediction and retrodiction on an equal footing. There is, however, a price to be paid, and the next example illustrates it very well.

Let two analyzers measuring the z and x components of the spin polarization of a spin $1/2$ particle be placed in series, Fig. 3. Thus if a particle with $S_z = 1/2$ enters the analyzer Z on the left it will emerge on the right with $S_z = 1/2$ and the analyzer will be in a state Z^+, whereas a particle with $S_z = -1/2$ entering on the left emerges on the right with

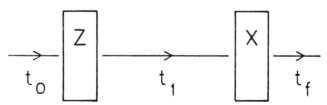

Figure 3   A particle passing successively through analyzers of the z and x components of its spin.

$S_z = -1/2$ and leaves the analyzer in a state Z^-. Here Z, Z^+ and Z^- are macroscopically distinct states (a pointer in different positions). A similar discussion applies to the analyzer X, with z and Z in the preceding discussion replaced by x and X, respectively.

At $t = t_0$ let the initial state D correspond to a particle approaching from the left with $S_x = 1/2$ and both analyzers in their initial untriggered states Z and X. At time t_f the particle has passed through both analyzers, and let us suppose the final state F is $Z^+ \wedge X^+$. Let t_1 be a time at which the particle has passed through Z but not yet through X, and let A_1 and Γ_1 be the events $S_z = 1/2$ and $S_x = 1/2$, respectively, at time t_1. One can show that

$$D \to A_1 \to F \tag{3.4}$$

is consistent, and

$$P(A_1 \,|\, D \wedge F) = 1. \tag{3.5}$$

What this means is that one can be sure that at time t_1 the particle had a polarization $S_z = 1/2$, which is not surprising given the final state F. One can also show that

$$D \to \Gamma_1 \to F \tag{3.6}$$

is consistent, and that

$$P(\Gamma_1 \,|\, D \wedge F) = 1, \tag{3.7}$$

which is to say that on the basis of the same information one can be sure (probability 1) that at time t_1 the particle had a polarization $S_x = 1/2$. Note that in physical terms (3.5) is a "prediction" of the polarization of the particle based on an earlier measurement, and (3.7) a retrodiction based on a later measurement.

The problem, of course, is that (3.5) and (3.7) together then yield, by the usual laws of (classical) probability

$$P(A_1 \wedge \Gamma_1 \,|\, D \wedge F) = 1, \tag{3.8}$$

i.e., we are sure that $S_x = 1/2$ and $S_z = 1/2$ at the same time. This is contrary to standard quantum mechanics—and to the consistent history approach as well. Because if an event is to be included in a (possibly consistent) history, it must be represented by an appropriate orthogonal projection operator. But there is no operator for the "event" $A_1 \wedge \Gamma_1$.

One strategy to avoid this difficulty is to make the events non-simultaneous. Let $t_{1.1}$ be a time slightly later than t_1, but before the particle has reached the second analyzer, and let $A_{1.1}$, $\Gamma_{1.1}$ be the projections corresponding to $S_z = 1/2$ and $S_x = 1/2$, respectively, at this time. Of course (3.4) and (3.6) remain consistent and (3.5) and (3.7) are valid with 1 replaced by 1.1. In addition,

$$D \rightarrow A_1 \rightarrow \Gamma_{1.1} \rightarrow F \qquad (3.9)$$

is consistent, and

$$P(A_1 \wedge \Gamma_{1.1} \mid D \wedge F) = 1, \qquad (3.10)$$

which seems almost as good as the nonsensical (3.8). But any illusion that the real quantum conceptual difficulties have somehow marvelously disappeared is dispelled when one notes that the history with intermediate events interchanged,

$$D \rightarrow \Gamma_1 \rightarrow A_{1.1} \rightarrow F \qquad (3.11)$$

is *inconsistent*, and thus cannot even be assigned a probability!

This example illustrates line 8 of Table I: the logic (or at least the probability theory) implicit in the consistent history approach is not standard. Sequences of events are only "meaningful", i.e., can be discussed within this interpretation, when they form part of a consistent history. Events can always be deleted from such a history, with the (possible) exception of D and F, and the result will be consistent (by the definition of consistency). However, an attempt to interleave events from two separate histories, each of them consistent, in order to form a single history will generally lead to inconsistency, as illustrated by (3.11).

IV. SCHRÖDINGER'S CAT

The literature of quantum interpretation contains a number of standard paradoxes. Let us see what the consistent history approach says about one of these: *Schrödinger's cat*.[6] Consider a closed system whose initial wave function ψ_0 at time t_0 develops, by integrating Schrödinger's equation, into

$$\psi_g = U(t_1, t_0)\psi_0 = (\psi_l + \psi_d)/\sqrt{2} \qquad (4.1)$$

at time t_1, where ψ_l and ψ_d correspond to macroscopically very different states of affairs: a live cat and a dead cat, to use Schrödinger's example. What does the consistent history interpretation say about this "grotesque" wave function ψ_g, and the question as to whether the cat is "really" dead or alive at time t_1?

First let us define the initial event

$$D = |\psi_0\rangle\langle\psi_0| \tag{4.2}$$

in the usual dyad notation. Next, let E^l and E^d be projection operators corresponding to a live cat and a dead cat, respectively, at time t_1. Since these are macroscopic states of affairs, there is some ambiguity in the definition, but it is at least plausible that we can choose them so that

$$E^l\psi_l = \psi_l, \qquad E^d\psi_d = \psi_d, \tag{4.3}$$

$$E^l E^d = 0 = E^d E^l. \tag{4.4}$$

In addition, at least as an aid to discussion, define

$$E^g = |\psi_g\rangle\langle\psi_g| \tag{4.5}$$

corresponding to a "grotesque event" at time t_1.

A straightforward calculation shows that

$$P(E^l \,|\, D) = 1/2 = P(E^d \,|\, D), \tag{4.6}$$

where we assume $F = 1$, and the corresponding histories are (trivially) consistent. In addition, see (4.4),

$$P(E^l \wedge E^d \,|\, D) = 0. \tag{4.7}$$

In words, the probability is $1/2$ that the cat is alive and $1/2$ that it is dead at time t_1, and we can be certain that it is *not* simultaneously dead and alive. (In interpreting what these probabilities mean, it may help to think of the analogy of flipping a coin.)

Note that the conclusions (4.6) and (4.7) are based on weights of the form (2.5) which make no reference, or at least no explicit reference, to the grotesque event E^g or the corresponding wave function ψ_g. To be sure, the reader who has actually derived the first equality in (4.6) using the method he was taught as a university student will probably have computed

$$\langle\psi_g|E^l|\psi_g\rangle = \mathrm{Tr}[E^l E^g]. \tag{4.8}$$

However, an alternative route (in principle!) is to first calculate

$$\tilde{E}^l = U(t_0, t_1)E^l U(t_1, t_0) \tag{4.9}$$

and then

$$\langle\psi_0|\tilde{E}^l|\psi_0\rangle. \tag{4.10}$$

The event $\tilde{E}^l$ at time t_0 is fully as grotesque as E^g at time t_1, and probably worse. The point is that when there are two different mathematical routes for getting to what is manifestly the same answer, it seems prejudicial to ascribe "physical reality" to one rather than the other. The consistent history approach avoids ascribing physical significance to those items which are only a matter of calculational convenience, through the use of a formal expression (2.5) in which they do not appear.

The events which do appear in (2.4), and thus (2.5), are those chosen by the theoretical physicist because he is interested in considering them. He can, if he wishes, think about histories with one or more grotesque events in them, test these for consistency, etc. In particular it is easy to show that

$$P(E^g \mid D) = 1. \qquad (4.11)$$

Note that E^g does not commute with E^l or E^d, so that one cannot combine (4.11) with (4.6) in the way permitted by classical probability theory. The other danger in interpreting (4.11) is to suppose that E^g can somehow be intuitively thought of as corresponding to a cat which is simultaneously dead and alive. That seems to me analogous to thinking of a spin $1/2$ atom with $S_x = 1/2$ as having $S_z = 1/2$ and $S_z = -1/2$ simultaneously. I personally do not find that very helpful. But in these matters, as in all else having to do with interpreting quantum mechanics, the reader must exercise his own judgment and make up his own mind.

Acknowledgments

This research has been supported by the National Science Foundation through grant DMR 8108310. I am also indebted to numerous individuals and institutions as indicated at the end of my longer paper.[2]

References

1. J. von Neumann, *Mathematical Foundations of Quantum Mechanics* (Princeton University Press, Princeton, 1955), translated from *Mathematische Grundlagen der Quantenmechanik* (Springer, Berlin, 1932).
2. R. B. Griffiths, "Consistent Histories and the Interpretation of Quantum Mechanics", *J. Stat. Phys.* **36**, 219 (1984).
3. E. P. Wigner, *Amer. J. Phys.* **31**, 6 (1963).
4. Y. Aharonov, P. G. Bergmann and J. L. Lebowitz, *Phys. Rev.* **134**, B1410 (1964); see also Y. Aharonov and D. Z. Albert, *Phys. Rev.* **D29**, 223 (1984).
5. See Sec. VII.A of ref. 2.
6. E. Schrödinger, *Naturwiss.* **23**, 807 (1935); English translation in *Proc. Amer. Phil. Soc.* **124**, 323 (1980) and in *Quantum Theory and Measurement*, eds. J. A. Wheeler and W. H. Zurek (Princeton University Press, Princeton, NJ, 1983), p. 152.

MULTIPLE-TIME MEASUREMENTS ON QUANTUM MECHANICAL SYSTEMS

Susan S. D'Amato

Department of Physics, Furman University, Greenville, SC 29613

Many surprising regularities emerge when multiple-time measurements (measurements that involve the sum of the value of one observable at one time and the value of another observable at another time) are carried out on a quantum mechanical system. Under certain circumstances, the behavior of a quantum system within a given time interval is completely determined by the outcome of two-time measurements conducted at the boundaries of the interval. This result and many others can be explained in the context of a formalism which assigns eigenstates and eigenvalues to so-called multiple-time operators.

I. INTRODUCTION

Recent investigations in nonrelativistic quantum theory[1-3] indicate that, in some cases, the traditional characterization of quantum systems in terms of states whose forward evolution in time is governed by the Schrödinger equation is not rich enough to convey all that is known about the system in question. For example, consider a spin-one-half particle in circumstances in which its Hamiltonian is independent of spin. Suppose that a measurement of the x-component of the particle's spin is made at a time t_i and it is found that $\sigma_x(t_i) = +1$, where $\sigma_x \equiv (2/\hbar)S_x$ is one of the Pauli spin matrices. Suppose, moreover, that the z-

225

component of the particle's spin is measured at a later time t_f, with the result that $\sigma_z(t_f) = +1$. What can be said about the result of a spin measurement carried out at an intermediate time t_1 $(t_i < t_1 < t_f)$?

$$t_f \text{\textemdash} \sigma_z = +1$$

$$t_1 \text{\textemdash} \sigma_x = ? \quad \sigma_z = ?$$

$$t_i \text{\textemdash} \sigma_x = +1$$

If a measurement of σ_x is made at t_1, the result must, of course, be that $\sigma_x(t_1) = +1$. But what if σ_z is measured instead? According to the traditional approach, which takes into account only the particle's forward evolution in time from t_i, the results $+1$ and -1 are equally likely if σ_z is measured at t_1, since the particle's state at t_i (and therefore at t_1) is a linear combination of the eigenstates of σ_z: $|x\uparrow\rangle = 2^{-1/2}(|z\uparrow\rangle + |z\downarrow\rangle)$.

But this analysis is incorrect. If σ_z is measured at t_1, the probability that the measurement will yield the result $\sigma_z(t_1) = +1$ is *unity*. We can easily see this if we calculate the probability that a measurement of σ_z at t_1 will yield -1. The probability that the particle will be in the state $|x\uparrow\rangle$ at t_i, then in the state $|z\downarrow\rangle$ at t_1, then in the state $|z\uparrow\rangle$ at t_f, is given by:[4]

$$P\{\sigma_z = -1 \text{ at } t_1\} = N|\langle z\uparrow|z\downarrow\rangle\langle z\downarrow|x\uparrow\rangle|^2, \tag{1}$$

where N is a constant such that

$$N\{|\langle z\uparrow|z\downarrow\rangle\langle z\downarrow|x\uparrow\rangle|^2 + |\langle z\uparrow|z\uparrow\rangle\langle z\uparrow|x\uparrow\rangle|^2\} = 1.$$

Since the hypothetical intermediate state is orthogonal to the known final state of the particle, the probability that the value of σ_z at t_1 is -1 is zero. Thus it must be the case that a measurement of σ_z at t_1 will yield $+1$.

We see from this example that a description of a particle in the interval between two complete measurements on the particle depends on the outcome of the final measurement, as well as on the outcome of the initial measurement. Indeed, the prescription for calculating the probability that an intermediate measurement will yield a given result is symmetric with respect to the initial and final states of the particle.[4] That a time-symmetric description of a quantum mechanical system is appropriate becomes even more apparent when we consider the results of multiple-time measurements carried out on the system.

II. MULTIPLE-TIME MEASUREMENTS: THEORY AND EXAMPLES

Aharonov and Albert have shown in reference 1 that it is possible to determine the value of a multiple-time quantity, such as the sum of the z-component of a particle's spin at one time and the x-component of its spin at another time, without measuring separately the values of single-time quantities such as σ_z and σ_x. Their method will briefly be reviewed here.

Suppose we have two measuring devices, one which we plan to switch on at t_1 to measure the x-component of a particle's spin, and another which we plan to switch on at t_2 to measure the z-component of the particle's spin. An appropriate interaction Hamiltonian[5] for such a pair of measurements is that given by Eq. (2).

$$H_{\text{int}} = g_1(t)q_1\sigma_x + g_2(t)q_2\sigma_z. \tag{2}$$

Here, q_1 and q_2 are internal coordinates associated with the first and second measuring devices; $g_1(t)$ is a coupling function which is large and non-zero only during the brief interval about t_1 in which the first measuring device is switched on ($g(t)$ could be a delta function about $t = t_1$, for example); and $g_2(t)$ is a similar coupling function which is large and non-zero only during a brief interval about t_2.

Using the Heisenberg formalism, we can obtain equations of motion for q_1, q_2, σ_x and σ_z, and also for Π_1 and Π_2, the momenta that are canonically conjugate to the measuring device coordinates q_1 and q_2. The result is

$$\left. \begin{array}{l} dq_1/dt = d\sigma_x/dt = 0 \\ d\Pi_1/dt = -g_1(t)\sigma_x \end{array} \right\} \quad \text{for } t_1 - \varepsilon < t < t_1 + \varepsilon,$$

and

$$\left. \begin{array}{l} dq_2/dt = d\sigma_z/dt = 0 \\ d\Pi_2/dt = -g_2(t)\sigma_z \end{array} \right\} \quad \text{for } t_2 - \varepsilon < t < t_2 + \varepsilon.$$

Assuming that the coupling functions are delta functions of unit magnitude about t_1 and t_2, these equations can be integrated immediately to give

$$q_1(t) = q_{10}; \qquad q_2(t) = q_{20}$$

$$\Pi_1(t_1 + \varepsilon) - \Pi_1(t_1 - \varepsilon) = \Delta\Pi_1 = -\sigma_x(t_1)$$

$$\Pi_2(t_2 + \varepsilon) - \Pi_2(t_2 - \varepsilon) = \Delta\Pi_2 = -\sigma_z(t_2).$$

Now suppose that we arrange for the initial conditions on the two measuring devices to be such that $q_{10} - q_{20} = \Pi_{10} + \Pi_{20} = 0$. Then after the measurements are completed, neither Π_1 nor Π_2 will have a well-defined value, since neither quantity will commute with the final condition on the coordinates q_1 and q_2. The *sum* of Π_1 and Π_2 will have a well-defined value, however, and this value will be proportional to the quantity $\sigma_x(t_1) + \sigma_z(t_2)$. Thus this method will enable us to carry out a two-time measurement—a measurement of the *two-time observable* $\sigma_x(t_1) + \sigma_z(t_2)$.

Many surprising regularities emerge when two-time measurements are carried out in the interval between two complete single-time measurements on a quantum system. For example, consider a spin-half particle which is known to be in the state $|x\uparrow\rangle$ at t_i and in the state $|z\downarrow\rangle$ at t_f. Suppose that we wish to "predict" (that is, to make definite statements about) the result of a two-time measurement of $\sigma_z(t_1) + \sigma_x(t_2)$ carried out in the interval between t_i and t_f. If two *single-time* measurements of σ_z and σ_x were to be made at t_1 and t_2, the four pairs of results given below would be equally likely:

$$\begin{cases} \sigma_z(t_1) = +1 \\ \sigma_x(t_2) = +1 \end{cases} \quad \begin{cases} \sigma_z(t_1) = +1 \\ \sigma_x(t_2) = -1 \end{cases} \quad \begin{cases} \sigma_z(t_1) = -1 \\ \sigma_x(t_2) = +1 \end{cases} \quad \begin{cases} \sigma_z(t_1) = -1 \\ \sigma_x(t_2) = -1. \end{cases}$$

One might naively guess from this fact that the probability of finding that $\sigma_z(t_1) + \sigma_x(t_2) = 0$ is 50%.

Direct calculation shows otherwise, however. The probability of finding that $\sigma_z(t_1) + \sigma_x(t_2) = 0$ is given by[4]

$$P\{\sigma_z(t_1) + \sigma_x(t_2) = 0\} = N|\langle z\downarrow|x\uparrow\rangle\langle x\uparrow|z\downarrow\rangle\langle z\downarrow|x\uparrow\rangle$$
$$+ \langle z\downarrow|x\downarrow\rangle\langle x\downarrow|z\uparrow\rangle\langle z\uparrow|x\uparrow\rangle|^2 = 0. \quad (3)$$

Notice that the probability amplitudes associated with the two histories for which $\sigma_z(t_1) + \sigma_x(t_2) = 0$ must be aded before the absolute square is taken. This is because the two histories produce indistinguishable results in the measuring devices[6] (i.e., both histories lead to $\Delta\Pi_1 + \Delta\Pi_2 = 0$).

How are we to interpret the fact that there is zero probability of finding that $\sigma_z(t_1) + \sigma_x(t_2) = 0$ if $\sigma_x(t_i) = +1$ and $\sigma_z(t_f) = -1$? It is well known that when a particle may proceed from one spatial point to another by more than one physical path (when an electron, for example, may proceed from a hot filament to a detecting screen via more than one slit in an intermediate screen), there may be constructive or destructive interference among the alternative trajectories. We see here that when a particle may proceed from one state to another through *time* by more than one *history*, there may be constructive or destructive (in this case, destructive) *interference among the possible histories*.

Let us now turn to cases in which more than one multiple-time measurement is carried out on a particle whose initial and final states are known. As might be expected, the probability that one such measurement will yield a given result depends on the results of other multiple-time measurements performed in the interval between t_i and t_f. For example, suppose that we are told that a particle in the state $|x\uparrow\rangle$ at t_i and in the state $|z\uparrow\rangle$ at t_f is subjected to a two-time measurement of $\sigma_z(t_1) + \sigma_x(t_4)$, with the result that $\sigma_z(t_1) + \sigma_x(t_4) = 0$.

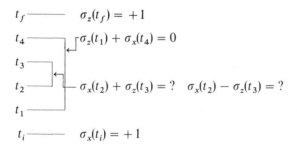

Then we can show by means of calculations similar to that performed in Eq. (3) that, given the information cited above, there is zero probability that a measurement of $\sigma_x(t_2) + \sigma_z(t_3)$ conducted in the interval between t_1 and t_4 will yield zero as its result. On the other hand, the probability that a measurement of $\sigma_x(t_2) - \sigma_z(t_3)$ will yield zero is unity.

As a final example, let us examine a slightly more complicated series of two-time measurements, which will serve as a basis for our introduction of the notion of "two-time eigenstates." Suppose that we have a spin-half particle in some arbitrary spin state $|\phi_i\rangle$ at t_i and in some other (or possibly the same) arbitrary state $|\phi_f\rangle$ at t_f. Suppose that we know that two-time measurements carried out on the particle at t_1/t_6 and at t_2/t_5 yield the results $\sigma_z(t_1) + \sigma_x(t_6) = 0$ and $\sigma_x(t_2) - \sigma_z(t_5) = 0$. What can we say about the result of a measurement of $\sigma_z(t_3) + \sigma_x(t_4)$ carried out in the interval between t_2 and t_5? In particular, what is the probability that the result of this measurement will be zero?

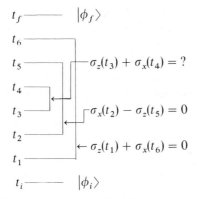

$t_f \text{———} |\phi_f\rangle$

t_6

$t_5 \text{———} \sigma_z(t_3) + \sigma_x(t_4) = ?$

t_4

$t_3 \text{———} \sigma_x(t_2) - \sigma_z(t_5) = 0$

t_2

$t_1 \text{———} \sigma_z(t_1) + \sigma_x(t_6) = 0$

$t_i \text{———} |\phi_i\rangle$

To calculate the probability that the measurement at t_3/t_4 will have one particular result, we must consider all possible outcomes of the measurement. The two-time measurement of $\sigma_z(t_3) + \sigma_x(t_4)$ might yield $+2$ as its result, or it might yield -2, or it might yield 0. Let us calculate the probability that the result of the measurement will be $+2$. First, we note that the arbitrary initial and final states of the system can be expanded in terms of the eigenstates of σ_z and σ_x, respectively:

$$|\phi_i\rangle = a|z\uparrow\rangle + b|z\downarrow\rangle \qquad |\phi_f\rangle = c|x\uparrow\rangle + d|x\downarrow\rangle.$$

Next, we observe that there are four contributions to the probability amplitude that $\sigma_z(t_3) + \sigma_x(t_4)$ will yield $+2$, corresponding to the four ways in which the given conditions $\sigma_z(t_1) + \sigma_x(t_6) = \sigma_x(t_2) - \sigma_z(t_5) = 0$ may be satisfied. Thus we can write

$$P\{\sigma_z(t_3) + \sigma_x(t_4) = +2\} = N|A_1 + A_2 + A_3 + A_4|^2,$$

where

$$A_1 = (\langle x\uparrow|c^* + \langle x\downarrow|d^*)|x\uparrow\rangle\langle x\uparrow \,|\, z\uparrow\rangle\langle z\uparrow \,|\, x\uparrow\rangle\langle x\uparrow \,|\, z\uparrow\rangle$$
$$\times \langle z\uparrow \,|\, x\uparrow\rangle\langle x\uparrow \,|\, z\downarrow\rangle\langle z\downarrow \,|\, (a|z\uparrow\rangle + b|z\downarrow\rangle),$$

$$A_2 = (\langle x\uparrow|c^* + \langle x\downarrow|d^*)|x\uparrow\rangle\langle x\uparrow \,|\, z\downarrow\rangle\langle z\downarrow \,|\, x\uparrow\rangle\langle x\uparrow \,|\, z\uparrow\rangle$$
$$\times \langle z\uparrow \,|\, x\downarrow\rangle\langle x\downarrow \,|\, z\downarrow\rangle\langle z\downarrow \,|\, (a|z\uparrow\rangle + b|z\downarrow\rangle),$$

$$A_3 = (\langle x\uparrow|c^* + \langle x\downarrow|d^*)|x\downarrow\rangle\langle x\downarrow \,|\, z\uparrow\rangle\langle z\uparrow \,|\, x\uparrow\rangle\langle x\uparrow \,|\, z\uparrow\rangle$$
$$\times \langle z\uparrow \,|\, x\uparrow\rangle\langle x\uparrow \,|\, z\uparrow\rangle\langle z\uparrow \,|\, (a|z\uparrow\rangle + b|z\downarrow\rangle),$$

$$A_4 = (\langle x\uparrow|c^* + \langle x\downarrow|d^*)|x\downarrow\rangle\langle x\downarrow \,|\, z\downarrow\rangle\langle z\downarrow \,|\, x\uparrow\rangle\langle x\uparrow \,|\, z\uparrow\rangle$$
$$\times \langle z\uparrow \,|\, x\downarrow\rangle\langle x\downarrow \,|\, z\uparrow\rangle\langle z\uparrow \,|\, (a|z\uparrow\rangle + b|z\downarrow\rangle).$$

Performing the scalar product calculations, we find that

$$A_1 + A_2 + A_3 + A_4 = (32)^{-1/2}(bc^* - bc^* - ad^* + ad^*) = 0,$$

so that the probability that a two-time measurement of $\sigma_z(t_3) + \sigma_x(t_4)$ will yield $+2$ is zero. A similar calculation shows that the probability that the result of the measurement will be -2 is also zero. Therefore the information given about the particle at t_1/t_6 and at t_2/t_5 implies that a measurement of $\sigma_z(t_3) + \sigma_x(t_4)$ must find that $\sigma_z(t_3) + \sigma_x(t_4) = 0$. (In the special case for which $bc^* = -ad^*$, the probability of finding each of the results $+2$, -2 and 0 is zero. This is because the "given" result $\sigma_x(t_2) - \sigma_z(t_5) = 0$ cannot be obtained if $bc^* = -ad^*$.)

What is remarkable about this result is the fact that the (zero) results of the "outer" two two-time measurements guarantee the (zero) result of the "innermost" measurement, *independently of our choice of the initial and final states of the particle*. (This can be shown to be true even if other measurements are carried out in the interval between t_3 and t_4.[1]) The behavior of the particle in the interval between t_i and t_f is completely determined by the outcome of the two-time measurements conducted at t_1/t_6 and at t_2/t_5: it is independent of the outcome of any single-time measurements (such as those performed to determine the state of the particle at t_i and t_f) carried out outside this interval.

How are we to interpret this result? We can proceed by analogy with a familiar example. Consider two spin-half particles in circumstances in which their Hamiltonian is independent of spin. Let the initial state of the two-particle system be some arbitrary state $|\psi_i\rangle_{1,2}$ (which may or may not be expressible as a product state). Suppose that a measurement of the x-component of the total spin of the particles[7] is performed, with the result that $\sigma_x \equiv \sigma_x^{(1)} + \sigma_x^{(2)} = 0$. Suppose that a measurement of the y-component of the total spin then determines that $\sigma_y = 0$. Then we can say with certainty that any further measurements of σ_x or σ_y (or, incidentally, σ_z) must yield zero, regardless of the identity of the initial state of the particles. This is because the experimental results $\sigma_x = \sigma_y = 0$ correspond to a collapse of $|\psi_i\rangle_{1,2}$ to the singlet state, which is an eigenvalue-zero eigenstate of σ_x, σ_y and σ_z.

The similarities between this case and the case of a one-particle system on which two-time measurements are performed suggest that it might be possible to extend the eigenstate formalism in such a way as to account for the observed regularities in the one-particle case. Such an extension is, in fact, possible, as will be shown more fully in the next section.

III. MULTIPLE-TIME EIGENSTATES

Consider once again a spin-half particle in some arbitrary spin state $|\phi_i\rangle$ at t_i and in some other arbitrary spin state $|\phi_f\rangle$ at t_f. Let two two-time measurements be carried out on the particle at t_1/t_6 and at t_2/t_5, with the result that $\sigma_z(t_1) + \sigma_x(t_6) = \sigma_x(t_2) - \sigma_z(t_5) = 0$. Our goal is to predict the outcome of a two-time measurement of $\sigma_z(t_3) + \sigma_x(t_4)$ using a language of "two-time states" that is analogous to the conventional language of states-at-a-given-time.

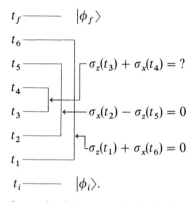

$$t_f \quad\text{——}\quad |\phi_f\rangle$$

$$\sigma_z(t_3) + \sigma_x(t_4) = ?$$

$$\sigma_x(t_2) - \sigma_z(t_5) = 0$$

$$\sigma_z(t_1) + \sigma_x(t_6) = 0$$

$$t_i \quad\text{——}\quad |\phi_i\rangle.$$

Let us begin by tentatively defining the *two-time state* of the particle at t_i/t_f to be

$$|\phi_f\rangle\cdots\langle\phi_i|.$$

(The usefulness of this notation will become apparent later.) Since the initial and final states of the particle can be expanded in terms of the eigenstates of σ_z and σ_x, respectively, the two-time state of the particle at t_i/t_f can be written as

$$(c|x{\uparrow}(t_f)\rangle + d|x{\downarrow}(t_f)\rangle)\cdots(a^*\langle z{\uparrow}(t_i)| + b^*\langle z{\downarrow}(t_i)|)$$

$$= a^*c|x{\uparrow}(t_f)\rangle\cdots\langle z{\uparrow}(t_i)| + a^*d|x{\downarrow}(t_f)\rangle\cdots\langle z{\uparrow}(t_i)|$$

$$+ b^*c|x{\uparrow}(t_f)\rangle\cdots\langle z{\downarrow}(t_i)| + b^*d|x{\downarrow}(t_f)\rangle\cdots\langle z{\downarrow}(t_i)|.$$

Now, since the values of $\sigma_x(t_f)$ and $\sigma_z(t_i)$ (whatever they are) are the same as the values of $\sigma_x(t_6)$ and $\sigma_z(t_1)$, we can easily write this two-time state as

$$a^*c|x{\uparrow}(t_6)\rangle\cdots\langle z{\uparrow}(t_1)| + a^*d|x{\downarrow}(t_6)\rangle\cdots\langle z{\uparrow}(t_1)|$$

$$+ b^*c|x{\uparrow}(t_6)\rangle\cdots\langle z{\downarrow}(t_1)| + b^*d|x{\downarrow}(t_6)\rangle\cdots\langle z{\downarrow}(t_1)|.$$

It is very tempting to look upon these four terms as a linear combination of four two-time eigenstates of $\sigma_z(t_1) + \sigma_x(t_6)$, corresponding to the eigenvalues $+2, 0, 0$ and -2, respectively. If we take this view, then the fact that the two-time measurement at t_1/t_2 yields zero implies that $|\phi_f\rangle \cdots \langle\phi_i|$ "collapses" at t_1/t_6 to

$$a^*d|x\!\downarrow(t_6)\rangle \cdots \langle z\!\uparrow(t_1)| + b^*c|x\!\uparrow(t_6)\rangle \cdots \langle z\!\downarrow(t_1)|.$$

In order to analyze the result of the next inner two-time measurement on the particle (that of $\sigma_x(t_2) - \sigma_z(t_5)$), let us re-express the two-time state of the particle at t_1/t_6 using the relationships:

$$|x\!\updownarrow(t_6)\rangle = |x\!\updownarrow(t_5)\rangle = 2^{-1/2}(|z\!\uparrow(t_5)\rangle \pm |z\!\downarrow(t_5)\rangle)$$

$$|z\!\updownarrow(t_1)\rangle = |z\!\updownarrow(t_2)\rangle = 2^{-1/2}(|x\!\uparrow(t_2)\rangle \pm |x\!\downarrow(t_2)\rangle).$$

We then have for the two-time state of the particle at t_2/t_5 (omitting the factors of $2^{-1/2}$):

$$a^*d(|z\!\uparrow(t_5)\rangle - |z\!\downarrow(t_5)\rangle) \cdots (\langle x\!\uparrow(t_2)| + \langle x\!\downarrow(t_2)|)$$

$$+ b^*c(|z\!\uparrow(t_5)\rangle + |z\!\downarrow(t_5)\rangle) \cdots (\langle x\!\uparrow(t_2)| - \langle x\!\downarrow(t_2)|)$$

$$= (a^*d + b^*c)(|z\!\uparrow(t_5)\rangle \cdots \langle x\!\uparrow(t_2)| - |z\!\downarrow(t_5)\rangle \cdots \langle x\!\downarrow(t_2)|)$$

$$+ (a^*d - b^*c)|z\!\uparrow(t_5)\rangle \cdots \langle x\!\downarrow(t_2)| - (a^*d - b^*c)|z\!\downarrow(t_5)\rangle \cdots \langle x\!\uparrow(t_2)|.$$

Again, it is tempting to interpret this expression as a statement that a measurement of $\sigma_x(t_2) - \sigma_z(t_5)$ may yield $0, -2$ or $+2$, respectively. If this is so, and if the result of the measurement is that $\sigma_x(t_2) - \sigma_z(t_5) = 0$, then we may say that, as a consequence of the t_2/t_5 measurement, the two-time state of the system "collapses" to

$$(a^*d + b^*c)(|z\!\uparrow(t_5)\rangle \cdots \langle x\!\uparrow(t_2)| - |z\!\downarrow(t_5)\rangle \cdots \langle x\!\downarrow(t_2)|).$$

Re-expressing this two-time state in terms of $|x\!\updownarrow(t_4)\rangle$ and $|z\!\updownarrow(t_3)\rangle$ and omitting, once again, factors of $2^{-1/2}$, we have for the two-time state of the particle at t_3/t_4:

$$(a^*d + b^*c)(|x\!\uparrow(t_4)\rangle \cdots \langle z\!\downarrow(t_3)| + |x\!\downarrow(t_4)\rangle \cdots \langle z\!\uparrow(t_3)|),$$

which, if we are to be consistent, we must interpret as the two-time eigenstate of $\sigma_z(t_3) + \sigma_x(t_4)$ corresponding to the eigenvalue zero. The formalism correctly predicts, then, what we know to be true: that the (zero) results of the two outer two-time measurements determines the (zero) result of the innermost measurement, independently of our choice of the initial and final states of the particle. (The formalism also shows us that the history pictured is impossible for the special case in which $a^*d = -b^*c$, because in that case the two-time measurement at t_2/t_5 cannot yield zero.)

As the reader may easily verify, the two-time eigenstate formalism introduced here may be used to analyze the results of other series of two-time measurements, such as those given on pages 228 and 229. Refer, for example, to the situation described on page 228, for which it was shown that there is zero probability of finding that $\sigma_z(t_1) + \sigma_x(t_2) = 0$, given that the initial state of the particle is $|x\uparrow\rangle$ and that the final state of the particle is $|z\downarrow\rangle$. We can derive this result using the two-time formalism by writing:

$$|\phi_f\rangle \cdots \langle \phi_i| = |z\downarrow(t_f)\rangle \cdots \langle x\uparrow(t_i)|$$

$$= |x\uparrow(t_2)\rangle \cdots \langle z\uparrow(t_1)| - |x\downarrow(t_2)\rangle \cdots \langle z\downarrow(t_1)|$$

$$+ |x\uparrow(t_2)\rangle \cdots \langle z\downarrow(t_1)| - |x\downarrow(t_2)\rangle \cdots \langle z\uparrow(t_1)|.$$

Recall that we have postulated that the two-time eigenstate of $\sigma_z(t_1) + \sigma_x(t_2)$ corresponding to the eigenvalue zero is $|x\uparrow(t_2)\rangle \cdots \langle z\downarrow(t_1)| + |x\downarrow(t_2)\rangle \cdots \langle z\uparrow(t_1)|$, not the combination of these two terms that appears in the expression above for $|\phi_f\rangle \cdots \langle \phi_i|$. Thus the formalism tells us that a two-time measurement of $\sigma_z(t_1) + \sigma_x(t_2)$ is certain *not* to find that $\sigma_z(t_1) + \sigma_x(t_2) = 0$.

The question naturally arises as to why the eigenstates and eigenvalues of the two-time operators $\sigma_z(t) + \sigma_x(t')$ and $\sigma_x(t) - \sigma_z(t')$ should be the ones which we have stated, namely:

$$
\left.
\begin{aligned}
&|x\uparrow(t')\rangle \cdots \langle z\uparrow(t)| && +2 \\
&|x\uparrow(t')\rangle \cdots \langle z\downarrow(t)| && \\
&\quad + |x\downarrow(t')\rangle \cdots \langle z\uparrow(t)| && 0 \\
&|x\downarrow(t')\rangle \cdots \langle z\downarrow(t)| && -2
\end{aligned}
\right\} \quad \text{for } \sigma_z(t) + \sigma_x(t'),
$$

and

$$
\left.
\begin{aligned}
&|z\downarrow(t')\rangle \cdots \langle x\uparrow(t)| && +2 \\
&|z\uparrow(t')\rangle \cdots \langle x\uparrow(t)| && \\
&\quad - |z\downarrow(t')\rangle \cdots \langle x\downarrow(t)| && 0 \\
&|z\uparrow(t')\rangle \cdots \langle x\downarrow(t)| && -2
\end{aligned}
\right\} \quad \text{for } \sigma_x(t) - \sigma_z(t').
$$

This question is best answered by means of an illustrative example.

Suppose we project the two-time state of a particle at t_i/t_f onto one of the two-time eigenstates listed above. In particular, suppose we project $|z\downarrow(t_f)\rangle \cdots \langle x\uparrow(t_i)|$ onto the eigenvalue-zero eigenstate of $\sigma_z(t_1) + \sigma_x(t_2)$, as shown in expression (4).

$$\langle z\downarrow(t_f)|\{|x\uparrow(t_2)\rangle \cdots \langle z\downarrow(t_1)| + |x\downarrow(t_2)\rangle \cdots \langle z\uparrow(t_1)|\}|x\uparrow(t_i)\rangle. \tag{4}$$

Now let us replace the ellipsis dots in each term by the scalar product of the bra and ket in that term. Then expression (4) becomes

$$\langle z\downarrow(t_f)|x\uparrow(t_2)\rangle\langle x\uparrow(t_2)|z\downarrow(t_1)\rangle\langle z\downarrow(t_1)|x\uparrow(t_i)\rangle$$
$$+ \langle z\downarrow(t_f)|x\downarrow(t_2)\rangle\langle x\downarrow(t_2)|z\uparrow(t_1)\rangle\langle z\uparrow(t_1)|x\uparrow(t_i)\rangle. \tag{5}$$

Notice that the absolute square of (5) is identical (to within a constant) to the right-hand-side of Eq. (3). Herein lies the rationale for choosing the eigenvalue-zero eigenstate of $\sigma_z(t) + \sigma_x(t')$ to be $|x\uparrow(t')\rangle \cdots \langle z\downarrow(t)| + |x\downarrow(t')\rangle \cdots \langle z\uparrow(t)|$: the form of the eigenstate must be such that its projection onto the state of the system at a time prior to t and a time subsequent to t' gives the correct probability amplitude (as prescribed in references 4 and 6) for finding that $\sigma_z(t) + \sigma_x(t') = 0$.

This requirement also explains why the eigenvalue-zero eigenstate of $\sigma_x(t) - \sigma_z(t')$ has the form $|z\uparrow(t')\rangle \cdots \langle x\uparrow(t)| - |z\downarrow(t')\rangle \cdots \langle x\downarrow(t)|$. Since the probability amplitude associated with a history of the system in which the particle's state is $|\phi_i\rangle$ at t_i and $|\phi_f\rangle$ at t_f, with $\sigma_x(t) - \sigma_z(t') = 0$ in the interval, is given by

$$\langle \phi_f(t_f)|z\uparrow(t')\rangle\langle z\uparrow(t')|x\uparrow(t)\rangle\langle x\uparrow(t)|\phi_i(t_i)\rangle$$
$$+ \langle \phi_f(t_f)|z\downarrow(t')\rangle\langle z\downarrow(t')|x\downarrow(t)\rangle\langle x\downarrow(t)|\phi_i(t_i)\rangle,$$

it follows that the "expanded" form of the eigenvalue-zero eigenstate of $\sigma_x(t) - \sigma_z(t')$ is

$$|z\uparrow(t')\rangle\langle z\uparrow(t')|x\uparrow(t)\rangle\langle x\uparrow(t)| + |z\downarrow(t')\rangle\langle z\downarrow(t')|x\downarrow(t)\rangle\langle x\downarrow(t)|.$$

When the scalar product terms are evaluated, then replaced by ellipsis dots, the fact that $\langle z\uparrow(t')|x\uparrow(t)\rangle = -\langle z\downarrow(t')|x\downarrow(t)\rangle$ implies that the eigenstate can be written as $|z\uparrow(t')\rangle \cdots \langle x\uparrow(t)| - |z\downarrow(t')\rangle \cdots \langle x\downarrow(t)|$.

We note in closing that two-time measurements can also be carried out on a system of two particles. The measurements that are of interest in this case are those that involve the sum of the value of an observable associated with one particle at one time and the value of an observable associated with the second particle at a different time. Consider, for example, two spin-half particles in some arbitrary spin state $|\psi_i\rangle$ at t_i and in some other (or possibly the same) arbitrary spin state $|\psi_f\rangle$ at t_f.

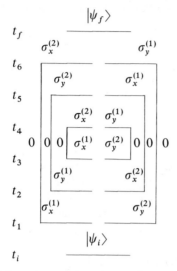

Then it can be shown by means of calculations similar to those performed in the one-particle case that if two-time, two-particle measurements are carried out on the system at times t_1/t_6 and t_2/t_5, with the result that

$$\sigma_x^{(1)}(t_1) + \sigma_x^{(2)}(t_6) = \sigma_y^{(1)}(t_6) + \sigma_y^{(2)}(t_1) = 0$$

$$\sigma_y^{(1)}(t_2) + \sigma_y^{(2)}(t_5) = \sigma_x^{(1)}(t_5) + \sigma_x^{(2)}(t_2) = 0,$$

then the result of a third pair of measurements carried out at t_3/t_4 must be that $\sigma_x^{(1)}(t_3) + \sigma_x^{(2)}(t_4) = \sigma_y^{(1)}(t_4) + \sigma_y^{(2)}(t_3) = 0$, independently of the identity of the initial and final states of the system. We can explain this result using the language of two-time eigenstates simply by stating that the outer measurements (given that their result is zero) place the system in a two-time state that is an eigenstate of certain two-particle, two-time operators. Further applications of two-time measurements on two particles are described in reference 3.

IV. SUMMARY AND CONCLUSION

When a quantum mechanical particle is subjected to one complete measurement that determines its state at t_i and to another complete measurement that determines its state at t_f, a description of the particle in the interval between t_i and t_f depends on the outcome of the final measurement, as well as on the outcome of the initial measurement. In this sense, quantum particles, in contrast to classical particles, seem to experience a "multiple-time existence."

The multiple-time nature of quantum mechanical particles becomes most apparent when a particle is subjected to multiple-time measurements—measurements that involve the sum (or difference or product, etc.) of the value of one observable associated with the particle at one time and the value of another observable associated with the particle at a different time. Many surprising regularities emerge when such measurements are performed. For example, under certain circumstances, the behavior of a particle within some interval of time is completely determined by the outcome of multiple-time measurements (specifically, two-time measurements) conducted at the outer boundaries of the interval: it is completely independent of the outcome of any single-time measurements (such as measurements to determine the state of the particle) conducted before or after the interval. This result and many others can be obtained by utilizing a formalism of multiple-time eigenstates, in which multiple-time operators such as $\sigma_z(t) + \sigma_x(t')$ have eigenvalues and eigenstates that are analogous to the eigenvalues and eigenstates of "ordinary," single-time operators.

It should be mentioned at this point that it is not *necessary* to use the multiple-time eigenstate formalism presented here, either as a calculational device or as an interpretive framework, when analyzing the results of various series of multiple-time measurements carried out on a quantum mechanical system. The probability amplitude associated with a history of the system in which multiple-time measurements yield a particular pattern of results may be calculated in terms of the traditional forward-time-evolution picture associated with the Schrödinger representation. Furthermore, it is possible to describe the regularities that emerge when nested series of two-time measurements involving $\sigma_z(t)$ and $\sigma_x(t')$ (for example) are carried out on a particle in terms of *correlations* between the values of non-commuting components of the particle's spin at different times, without referring at all to "two-time states" of the system. (These correlations can ultimately be traced to the correlation between the particle and the measuring devices to which it is coupled.)

On the other hand, the multiple-time eigenstate formalism does provide a simple and efficient method of calculating the probability that a nested series of two-time measurements on a particle will yield given results. Moreover, it does so in a language which refers only to the particle of interest, in contrast to the Schrödinger approach, which requires us to trace the history of all measuring devices coupled to the particle.[1] Furthermore, the strong analogy that exists between the results of two-time measurements carried out on a single spin-half particle and the results of single-time measurements carried out on two spin-half particles gives a strong esthetic appeal to the notion of single-particle, multiple-time eigenstates. The formalism has already led to new insights into the behavior of a quantum particle subjected to multiple-time measurements—to wit, the fact that the two-time behavior of a particle is completely determined by the outcome of two nested two-time measurements—and it will surely continue to do so.

While the need for a multiple-time characterization of quantum systems subjected to multiple-time measurements may be obvious, it is also true that systems subjected only to single-time measurements may require a multiple-time characterization. As mentioned previously, systems subjected to two single-time measurements of non-commuting observables require a multiple-time description in the interval between the measurements. Such a description can be obtained simply by treating the results of the initial and final measurements on an equal footing. A multiple-time description of a particle for which single-time measurements at t_i and t_f show that $A(t_i) = a$ and $B(t_f) = b$ (where $[A, B] \neq 0$), for example, consists of the following statements:

(1) If A is measured at a time t such that $t_i < t < t_f$, the outcome of the measurement is certain to be that $A(t) = a$.

(2) If B, rather than A, is measured at a time t between t_i and t_f, the outcome of the measurement is certain to be that $B(t) = b$.

The validity of this description can be verified by carrying out a pair of *compensating measurements*[3] on the system—measurements designed to reveal both what the outcome of a measurement of A would have been, had not a measurement of B occurred, *and* what the outcome of a measurement of B would have been, had not a measurement of A occurred.

A multiple-time description can thus be profitably applied in a wide variety of situations, offering new insights into the behavior of even the simplest of systems over time. In the present work, we have set forth the basic elements of a multiple-time characterization of quantum mechanical systems. It is to be hoped that future investigations into the multiple-time nature of such systems will contribute significantly to our understanding of the role of time in the quantum domain.

Acknowledgement

This work was carried out in collaboration with Yakir Aharonov and David Z. Albert of the University of South Carolina.

References

1. Y. Aharonov and D. Z. Albert, *Phys. Rev.* **D29**, 223 (1984).
2. R. B. Griffiths, *J. Stat. Phys.* **36**, 219 (1984).
3. Y. Aharonov, D. Z. Albert and S. S. D'Amato, *Phys. Rev.* **D32**, 1975 (1985).
4. Y. Aharonov, P. G. Bergmann and J. L. Lebowitz, *Phys. Rev.* **134**, B1410 (1964).
5. J. Von Neumann, *Mathematical Foundations of Quantum Mechanics* (Princeton University Press, Princeton, NJ, 1955), pp. 437–445; D. Bohm, *Quantum Theory* (Prentice Hall, New York, 1951), pp. 583–623.
6. R. P. Feynman and A. R. Hibbs, *Quantum Mechanics and Path Integrals* (McGraw-Hill, New York, 1965), pp. 13–21.
7. Y. Aharonov and D. Z. Albert, *Phys. Rev.* **D21**, 3316 (1980); *Phys. Rev.* **D24**, 359 (1981).

A PHASE-SPACE MODEL FOR QUANTUM MECHANICS IN WHICH ALL OPERATORS COMMUTE

Itamar Pitowsky

Department of Philosophy, Brown University, Providence, RI 02912

With every physical system we associate a "state function" defined on phase space at every given moment. Measurement of position and momentum are represented by the operators "multiplying the state function by position" and "multiplying the state function by momentum" respectively, so that all operators commute. The strange mathematical character of the state function allows for the recovery of the usual expectation values and probability densities of quantum mechanics. Also the uncertainty relations are shown to be valid. An analogue of the Schrödinger and Hamilton equations is formulated and a "recipe" for translating the model back to the language of quantum mechanics is provided. Some philosophical points are discussed in the conclusion.

I. INTRODUCTION

In previous articles[1,2] I have noted the similarity between some peculiarities of spin-statistics and certain "pathological" mathematical constructions that one can obtain in set theory. The purpose of this article is to extend these observations to the case of position-momentum and the operators in their algebra. A few results in that direction have already been obtained by S. Gudder.[3]

The idea is as follows: We represent a physical system, say a one-dimensional system, by a complex function $\Phi_t(q, p)$ of the position q and momentum p, which is defined on phase space at every moment t. The operators corresponding to position measurement and momentum measurement are just "multiplying Φ_t by q" and "multiplying Φ_t by p" respectively, so that all the operators in their algebra commute. Still it is shown that the expectation value of every traditional quantum mechanical observable coincides with the expectations given in the model. In particular, the probability densities in configuration space and momentum space are obtained as marginals from $|\Phi_t|^2$. Also the uncertainty relation $\Delta p \, \Delta q \leqslant 1/2$ is valid.

The reason why all this is possible has to do with the mathematical character of the function Φ_t. Its real part and modulus are smooth but its imaginary part is so extremely discontinuous that the operations "integrating in configuration space" and "integrating in momentum space" do not commute for certain operators. Thus, for example, the equality $\iint qp\Phi_t(q, p) \, dp \, dq - \iint qp\Phi_t(q, p) \, dq \, dp = i$ corresponds to the quantum mechanical relation $[Q, P] = i$. In spite of this one can obtain an analogue of the Schrödinger equation and the Hamilton equations in the framework of the model. Moreover, it is shown that the model can be "translated" back into the conventional language of quantum mechanics.

The existence of functions like Φ_t is a result of one single set theoretical construction which is formulated at the outset and proved in the Appendix. The esoteric nature of that result may disturb some readers, so I should note that my purpose is not so much to provide a full-fledged alternative to quantum mechanics as it is to stir some scepticism with respect to certain interpretations of the theory, Bohr's views in particular. It seems to me that those interpretations draw far-reaching and unwarranted conclusions largely on the basis of one particular mathematical formulation.

The model presented below accounts for one-dimensional systems but, as I note in the end, it can easily be extended to any number of space dimensions.

Throughout this paper, I shall assume $\hbar = 1$.

II. MATHEMATICAL PRELIMINARIES

Let $\mathbf{R}^{(2)}$ denote the Euclidean plane and fix a system of Cartesian coordinates (q, p) in this plane. We shall say that a line in $\mathscr{R}^{(2)}$ is "horizontal" if it is parallel to the q-axis and "vertical" if it is parallel to the p-axis. Let $F = F(q, p)$ be a complex function defined on $\mathbf{R}^{(2)}$. F is *locally integrable* if the integrals: $E(F \mid q) = \int_{-\infty}^{\infty} F(q, p)\, dp$ and $E(F \mid p) = \int_{-\infty}^{\infty} F(q, p)\, dq$ are well defined for all values of q and p respectively. F is *globally integrable* if the integrals $E_v(F) = \int_{-\infty}^{\infty} E(F \mid q)\, dq$ and $E_h(F) = \int_{-\infty}^{\infty} E(F \mid p)\, dp$ are well defined. $E_v(F)$ and $E_h(F)$ are the vertical and horizontal integrals of F, respectively, and they need not be equal. If $E_v(F) = E_h(F)$, we say that F is *totally integrable* and put $E(F) = E_v(F) = E_h(F)$. The model to be presented here depends on the following set theoretical construction:

Main Theorem: There exists a subset M of $\mathbf{R}^{(2)}$ with the following properties:

(i) If l is a vertical line $M \cap l$ is a countable set.

(ii) If l is a horizontal line $M \cap l$ is the whole line l save perhaps countably many points.

The proof is given in the Appendix where I shall also note the set theoretical assumptions involved in the proof of such "strange" results. Let $\phi = \phi(q)$ be a complex function and assume that ϕ together with all its derivatives up to a sufficiently high order are integrable, square integrable and vanish at $\pm \infty$. Also assume that ϕ is normalized so that $\|\phi\|^2 = \int |\phi(q)|^2\, dq = 1$. Let $\hat{\phi} = \hat{\phi}(p)$ be the normalized Fourier transform of ϕ, that is

$$\hat{\phi}(p) = \frac{1}{\sqrt{2\pi}} \int \phi(q) e^{-iqp}\, dq.$$

We have

$$\phi(q) = \frac{1}{\sqrt{2\pi}} \int \hat{\phi}(p) e^{ipq}\, dp$$

and $\|\phi\|^2 = \|\hat{r}\|^2$ (Plancherel's theorem). Now put

$$\Phi(q, p) = \frac{1}{\sqrt{2\pi}} [\chi_M(q, p)\phi(q)\hat{\phi}^*(p) e^{-ipq} + \chi_{\bar{M}}(q, p)\phi^*(q)\hat{\phi}(p) e^{ipq}].$$

$$(2.1)$$

Where M is the set whose existence is guaranteed in the Main Theorem, $\bar{M} = \mathbf{R}^{(2)} \backslash M$, $\chi_M(q, p) = 1$ for $(q, p) \in M$ and zero otherwise so that $\chi_{\bar{M}} = 1 - \chi_M$, and $\phi^*, \hat{\phi}^*$ are the complex conjugates of $\phi, \hat{\phi}$ respectively. Φ is an extremely discontinuous function (in fact it is non-measurable in terms of the Lebesque measure on $\mathbf{R}^{(2)}$) but nevertheless its real part is smooth $\mathrm{Re}\,\Phi(q, p) = \mathrm{Re}(\phi^*(q)\hat{\phi}(p)e^{ipq})/\sqrt{2\pi}$ and so is its absolute value $|\Phi|^2 = \Phi\Phi^* = |\phi(q)|^2|\hat{\phi}(p)|^2/2\pi$.

Consider the integral $E(\Phi\,|\,q) = \int \Phi(q, p)\,dp$. When q is fixed the line $l = \{(q, p)\,|\,-\infty < p < \infty\}$ is vertical so that $M \cap l$ is only a countable set (Main Theorem, part i) and thus has measure zero on l. Hence $\bar{M} \cap l$ is the whole line l save perhaps countably many points. We therefore conclude

$$E(\Phi\,|\,q) = \phi^*(q)\frac{1}{\sqrt{2\pi}}\int \hat{\phi}(p)e^{ipq}\,dp = |\phi(q)|^2 \qquad (2.2)$$

Similar considerations are involved in the calculation of all the integrals that follow. The rule is simple: If we integrate dq first, then we consider only the component which contains $\chi_M(q, p)$, and if we integrate dp first then we consider the component which contains $\chi_{\bar{M}}(q, p)$. Hence we have

$$E(\Phi\,|\,p) = \hat{\phi}^*(p)\frac{1}{\sqrt{2\pi}}\int \phi(q)e^{-ipq}\,dq = |\hat{\phi}(p)|^2. \qquad (2.3)$$

Therefore Φ is totally integrable and we have

$$E(\Phi) = \|\phi\|^2 = \|\hat{\phi}\|^2 = 1. \qquad (2.4)$$

More generally if

$$\Psi(q, p) = \frac{1}{\sqrt{2\pi}}[\chi_M(q, p)\psi(q)\hat{\psi}^*(p)e^{-ipq} + \chi_{\bar{M}}(q, p)\psi^*(q)\hat{\psi}(p)e^{ipq}],$$

we see that $\Phi\Psi^*$ is totally integrable and

$$E(\Phi\Psi^*) = \frac{1}{2\pi}|(\phi, \psi)|^2 \qquad (2.5)$$

where (ϕ, ψ) is the usual inner product in $L^2(\mathbf{R})$, this result follows directly from Plancherel's theorem: $(\phi, \psi) = (\hat{\phi}, \hat{\psi})$.

Consider now the function $q^m p^n \Phi(q, p)$ for $m, n = 0, 1, 2, \ldots$. If ϕ satisfies the conditions cited above we have

$$\frac{\partial^n \phi}{\partial q^n} = \frac{1}{\sqrt{2\pi}}\int (ip)^n \hat{\phi}(p)e^{ipq}\,dp. \qquad (2.6)$$

Let us introduce the operator notation $(Q\phi)(q) = q\phi(q)$ and $(P\phi)(q) = -i\,\partial\phi/\partial q$. Then, from (2.6), we obtain

$$E(q^m p^n \Phi \,|\, q) = \phi^*(q)q^m \frac{1}{\sqrt{2\pi}} \int p^n \hat{\phi}(p)e^{ipq}\,dp = \phi^*(q)(Q^m P^n \phi)(q).$$

Therefore the vertical integral of $q^m p^n \Phi$ is

$$E_v(q^m p^n \Phi) = (Q^m P^n \phi, \phi). \tag{2.7}$$

Similarly,

$$E(q^m p^n \Phi \,|\, p) = \hat{\phi}^*(p)p^n \frac{1}{\sqrt{2\pi}} \int q^m \phi(q)e^{-ipq} = \hat{\phi}^*(p)p^n Q^m \hat{\phi}(p).$$

Now, since $(\psi_1, \psi_2) = (\hat{\psi}_1, \hat{\psi}_2)$ and since the inverse Fourier transform of $p^n Q^m \hat{\phi}(p)$ is

$$\frac{1}{i^n} \frac{\partial^n}{\partial q^n}(Q^m \phi) = P^n Q^m \phi$$

we conclude that the horizontal integral of $q^m p^n \Phi$ is

$$E_h(q^m p^n \Phi) = (P^n Q^m \phi, \phi). \tag{2.8}$$

Thus for $m, n \geq 1$ the function $q^m p^n \Phi$ is not necessarily totally integrable and in fact we have in this case

$$E_v(q^m p^n \Phi) - E_h(q^m p^n \Phi) = ([Q^m, P^n]\phi, \phi) \tag{2.9}$$

where $[\,,\,]$ denotes the commutator. In particular, for $m = n = 1$, we get

$$E_v(qp\Phi) - E_h(qp\Phi) = ([Q, P]\phi, \phi) = i. \tag{2.10}$$

The expression (2.1) defines the function Φ in terms of ϕ (and $\hat{\phi}$). The question is whether we can "recover" the values of ϕ from those of Φ. We can indeed do that (up to a phase factor) in the following way. Note that

$$E(e^{-ipq}\Phi \,|\, q) = \phi^*(q) \frac{1}{\sqrt{2\pi}} \int \hat{\phi}(p)\,dp = \phi^*(q)\phi(0).$$

Thus for $\phi(0) \neq 0$ we have

$$\phi(q) = c \frac{E^*(e^{-ipq}\Phi \,|\, q)}{\|E(e^{-ipq}\Phi \,|\, q)\|} \qquad \phi(0) \neq 0 \quad \text{and} \quad |c| = 1. \tag{2.11}$$

In the case $\phi(0) = 0$ we can always take a point q_0 for which $\phi(q_0) \neq 0$ and recover ϕ in terms of $E(e^{-ip(q-q_0)}\Phi \,|\, q)$. I shall call (2.11) the "inversion formula".

All of the above considerations seem at first glance to depend heavily upon the particular coordinate system in $\mathbf{R}^{(2)}$ which was fixed at the outset. This, however, is not the case. It is easy to see that all of the above integrals remain well defined under any "translation" of the form

$$(q, p) \rightarrow (q + q_0, p + p_0).$$

The reason is that such a transformation maps vertical and horizontal lines into vertical and horizontal lines, respectively. There are other transformations, such as a $90°$ "rotation" of the space, q-reflection, p-reflection, etc., which map well-defined quantities into well-defined quantities and it is not difficult to calculate the transformation rules for them. It is evident, however, that a general canonical transformation may map the function Φ into a function which is not even locally integrable. As long as we do not impose further constraints on the set M, this is inevitable.

III. A PHASE-SPACE MODEL FOR QUANTUM MECHANICS

The state of a one-dimensional quantum mechanical system is usually described by a complex function $\phi = \phi(q, t)$ of position and time, which is a solution of the time-dependent Schrödinger equation with the appropriate boundary and initial conditions. As before, denote by $\hat{\phi}(p, t)$ the normalized Fourier transform of ϕ (with respect to q, the time t remains intact) and consider the representation of one-dimensional systems which is based on the following principles:

(a) *State Functions:* With every particle at every given moment t we associate a phase space state function $\Phi_t(q, p)$ of the form,

$$\Phi_t(q, p) = \frac{1}{\sqrt{2\pi}} [\chi_M(q, p)\phi(q, t)\hat{\phi}^*(p, t)e^{-ipq}$$
$$+ \chi_{\bar{M}}(q, p)\phi^*(q, t)\hat{\phi}(p, t)e^{ipq}]. \quad (3.1)$$

The magnitudes $E(\Phi_t \,|\, q)$ and $E(\Phi_t \,|\, p)$ are the probability densities in configuration and momentum space, respectively. Therefore, for example, $\int_a^b E(\Phi_t \,|\, q)\, dq$ is the probability that the particle is in the interval $[a, b]$ at time t. The function $2\pi|\Phi_t|^2$ is the probability density in phase space. From (2.2) and (2.3), it is clear that $E(\Phi_t \,|\, q) = 2\pi E(|\Phi_t|^2 \,|\, q)$

and $E(\Phi_t \mid p) = 2\pi E(|\Phi_t|^2 \mid p)$ and more generally that $2\pi E(\Phi_t \Psi_t^*)$ in (2.5) is the transition probability from the state Ψ_t to the state Φ_t at time t.

(b) *Position Momentum and Uncertainty:* By analogy with conventional quantum mechanics we introduce operators which correspond to the measurement of physical magnitudes. Defining the position operator $\bar{q}$ to be $(\bar{q}\Phi_t)(q, p) = q\Phi_t(q, p)$ and the momentum operator $\bar{p}$ by $(\bar{p}\Phi_t)(q, p) = p\Phi_t(q, p)$, we see that the position and momentum operators commute. More generally let

$$F(q, p, t) = \sum a_{mn}(t) q^m p^n \qquad (3.2)$$

be a complex analytic function of the variables q, p (and possibly the time t). The operator $\bar{F}$ associated with it is simply $(\bar{F}\Phi_t)(q, p) = F(q, p, t)\Phi_t(q, p)$. Call the magnitudes $\operatorname{Re} E_v(F\Phi_t)$ and $\operatorname{Re} E_h(F\Phi_t)$ the *vertical and horizontal expectations* of $\bar{F}$ (in the state Φ_t). The operator $\bar{F}$ is said to be *definite* if its vertical and horizontal expectations are identical. Thus, for example, whenever F is a real function, $\bar{F}$ is definite since in this case the coefficients $a_{mn}(t)$ in (3.2) are real and thus from (2.7) and (2.8) we obtain

$$E_v(F\Phi_t) = \sum a_{mn}(Q^m P^n \phi, \phi) = \sum a_{mn}(\phi, P^n Q^m \phi) = \sum a_{mn}(P^n Q^m \phi, \phi)^*$$
$$= (\sum a_{mn} P^n Q^m \phi, \phi)^* = E_h^*(F\Phi_t).$$

Thus, $\operatorname{Re} E_v(F\Phi_t) = \operatorname{Re} E_h^*(F\Phi_t) = \operatorname{Re} E_h(F\Phi_t)$. In particular, if F is real and $F\Phi_t$ is totally integrable we have $\operatorname{Re} E_v(F\Phi_t) = \operatorname{Re} E_h(F\Phi_t) = E(F\Phi_t)$. There are, however, definite operators which correspond to complex functions F. Thus, for example, take $F = -iq + q^2 p$ we have $E_v(F\Phi_t) = ((-iQ + Q^2 P)\phi, \phi) = (QPQ\phi, \phi)$ and $E_h(F\Phi_t) = ((-iQ + PQ^2)\phi, \phi) = -2i(Q\phi, \phi) + (QPQ\phi, \phi)$ and thus, $\operatorname{Re} E_v(F\Phi_t) = \operatorname{Re} E_h(F\Phi_t) = (QPQ\phi, \phi)$. In fact, every self-adjoint operator in the Q, P algebra corresponds to some definite operator $\bar{F}$ in our model and the usual quantum mechanical expectation coincides with our definition. Note, however, that functional relations are not necessarily preserved, that is, the operator $(QPQ)^2$ does not correspond with $(-iq + q^2 p)^2$. In the case of functions of the form q^m or p^n no such problem arises and in particular, since $q\phi_t$, $q^2\Phi_t$, $p\Phi_t$ and $p^2\Phi_t$ are all totally integrable, the standard deviations Δq, Δp are given immediately by:

$$\Delta q = [E(q^2\Phi_t) - E^2(q\Phi_t)]^{1/2}, \qquad \Delta p = [E(p^2\Phi_t) - E^2(p\Phi_t)]^{1/2},$$

and thus, $\Delta q \, \Delta p \geqslant 1/2$ as can easily be proved applying standard reasoning to Eq. (2.7).

The uncertainty relations are therefore valid in the present model but the reasons for that are quite different from those cited in the context of traditional quantum mechanics. In the above formulation one can consistently hold the view that both position and momentum have simultaneous well-defined values (though perhaps unknown values) and that $2\pi|\Phi_t|^2$ is their simultaneous probability density. Moreover, in this model every real operator $\bar{F}$ has a well-defined expectation, even in cases when it does not correspond with a self-adjoint operator. The point is that the state function Φ_t is extremely discontinuous and thus for certain magnitudes the mathematical operations: "integrating in configuration space" and "integrating in momentum space" do not commute (e.g., $E_v(qp\Phi_t) \neq E_h(qp\Phi_t)$). This essentially is the reason for "uncertainty". Thus, to a certain extent, we have removed the blame for incommensurability from "Nature" and have reduced the uncertainty relations to a mathematical feature of the state function Φ_t.

(c) *Energy:* Let $H = H(q, p, t)$ be the classical Hamiltonian of the system. We shall assume that H does not contain "mixed" components of the form $q^m p^n$ for $m, n \geqslant 1$. In particular, we shall take the paradigm case of a Hamiltonian of the form $H(q, p) = p^2/2m + V(q, t)$ where V is a reasonably smooth potential. If these conditions obtain, $H\Phi_t$ is a totally integrable function and the expectation

$$\langle \bar{H} \rangle_\Phi = E(H\Phi_t)$$

is the average energy of the system (at time t).

Let $\mathscr{H}$ denote the quantum mechanical Hamiltonian, then using the time-dependent Schrödinger equation, we get (from (2.2))

$$i\frac{\partial E(\Phi_t \mid q)}{\partial t} = i\left[\phi^* \frac{\partial \phi}{\partial t} + \phi \frac{\partial \phi^*}{\partial t} \right] = \phi^* \mathscr{H} \phi - \phi(\mathscr{H}\phi)^*.$$

On the other hand, $E(H\Phi_t \mid q) = \phi^*(\mathscr{H}\phi)$, Hence we obtain

$$\frac{\partial E(\Phi_t \mid q)}{\partial t} = 2\,\mathrm{Im}\,E(H\Phi_t \mid q) \tag{3.3}$$

and similarly

$$\frac{\partial E(\Phi_t \mid p)}{\partial t} = 2\,\mathrm{Im}\,E(H\Phi_t \mid p). \tag{3.4}$$

For $H = p^2/2m + V(q, t)$, these equations are

$$\frac{\partial E(\Phi_t \mid q)}{\partial t} = 2\,\mathrm{Im}\,E\left(\frac{p^2}{2m}\Phi_t \mid q \right), \qquad \frac{\partial E(\Phi_t \mid p)}{\partial t} = 2\,\mathrm{Im}\,E(V\Phi_t \mid p). \tag{3.5}$$

Therefore the rate of change of the position probability density depends on the momentum and the rate of change of momentum density depends on the potential.

Since Φ_t is a smooth function of t (when q, p are fixed), $\partial \Phi_t / \partial t$ is well defined and

$$\frac{\partial E(\Phi_t | q)}{\partial t} = E\left(\frac{\partial \Phi_t}{\partial t} \bigg| q\right),$$

as can easily be verified. Thus Eq. (3.3) can also be written in the form

$$\mathrm{Re}\, E\left(\frac{1}{2}\frac{\partial \Phi_t}{\partial t} + iH\Phi_t \bigg| q\right) = \mathrm{Re}\, E\left(\frac{1}{2}\frac{\partial \Phi_t}{\partial t} + iH\Phi_t \bigg| p\right) = 0. \qquad (3.6)$$

If the densities $E(\Phi_t | q)$, $E(\Phi_t | p)$ and the average energy $E(H\Phi_t)$ are constant in time then Φ_t corresponds with the quantum mechanical stationary state $\phi(q, t) = \phi(q)e^{-i\varepsilon t}$, where ε is the energy. In this case, $\Phi_t = \Phi$ is *independent* of the time variable t. Using standard methods, we obtain for stationary states Φ

$$E((H - \varepsilon)\Phi | q) = E((H - \varepsilon)\Phi | p) = 0, \quad \varepsilon = E(H\Phi), \qquad (3.7)$$

which corresponds to the time-independent Schrödinger equation.

(d) *The Hamilton Equations:* From the classical equations $\partial H/\partial p = \dot{q}$ and $\partial H/\partial q = -\dot{p}$, we can derive a definition for the operators $\dot{q}$ and $\dot{p}$; namely,

$$\dot{q}\Phi = \frac{\partial H}{\partial p}\Phi_t, \qquad \dot{p}\Phi = -\frac{\partial H}{\partial q}\Phi_t.$$

On the other hand, we know that $\dot{Q} = -i[Q, \mathscr{H}]$, $\dot{P} = -i[P, \mathscr{H}]$, hence, from Eq. (2.9) we have

$$E_v(qH\Phi_t) - E_h(qH\Phi_t) = -iE\left(\frac{\partial H}{\partial p}\Phi_t\right) \qquad (3.8)$$

$$E_v(pH\Phi_t) - E_h(pH\Phi_t) = iE\left(\frac{\partial H}{\partial q}\Phi_t\right) \qquad (3.9)$$

which are the analogues of the Hamilton equations.

(e) *Returning to the Conventional Picture:* Using the inversion formula (2.11) we can "retranslate" the above model back to the language of quantum mechanics. Suppose that Φ_0 is the state of the system at the time $t = 0$, then

$$\phi(q, 0) = c \, \frac{E^*(e^{-ipq}\Phi_0 \,|\, q)}{\|E(e^{-ipq}\Phi_0 \,|\, q)\|}$$

where c is a time independent phase factor. From the values of $\phi(q, 0)$ one can easily obtain those of $\phi(q, t)$ using the time-dependent Schrödinger equation.

(f) *Higher Dimensions:* The model is easily generalized to any number of dimensions. Consider the three-dimensional case: Let $\mathbf{q} = (q_1, q_2, q_3)$, $\mathbf{p} = (p_1, p_2, p_3)$. Then a horizontal "line" in $\mathbf{R}^{(6)}$ is a set of the form

$$\{(\mathbf{q}, \mathbf{p}_0) \,|\, -\infty < q_i < \infty, \, i = 1, 2, 3\}$$

and a vertical "line" is the set

$$\{(\mathbf{q}_0, \mathbf{p}) \,|\, -\infty < p_i < \infty, \, i = 1, 2, 3\}$$

where $\mathbf{q}_0, \mathbf{p}_0$ are fixed vectors. Since the intersection of a horizontal and a vertical "line" contains one single point $(\mathbf{q}_0, \mathbf{p}_0)$, the Main Theorem is valid when we replace lines by "lines". The rest of the model is the same as the one-dimensional model with the sole difference that the normalization factor is now $(2\pi)^{-3/2}$.

In the three-dimensional case, we have some additional interesting operators, namely the angular momenta $\bar{J}_j(\bar{\mathbf{q}}, \bar{\mathbf{p}}) = \bar{q}_k \bar{p}_l - \bar{q}_l \bar{p}_k$ $((j, k, l)$ cyclic permutation of $(1, 2, 3))$. Again all these operators commute and we can maintain that they all have definite values at all times. This contention is also valid in my more-detailed spin model.[2]

IV. CONCLUSION

For logicians and philosophers of science the existence of models such as the one proposed above will come as little surprise. Their's has long been the contention that "theories are underdetermined by observation",[4] that is, empirical data, be it as rich and diversified as you wish, does not in itself determine a single theoretical framework. Many, even countless mutually incompatible theories may be provided to account for the same data.

If mathematical elegance and clarity is a criterion for choosing an appropriate theory, then there is little doubt that quantum mechanics supersedes all its present alternatives. But quantum theory had come to the world with a great metaphysical muddle, namely the Copenhagen Interpretation. Position-momentum and other incommensurable pairs, we were told, do not have simultaneous definite values. Apart from its being metaphysically obscure this approach puts into question the very reality of the subject matter of the theory itself. It seems that a growing number of physicists take Bohr's position as unsatisfactory and numerous different and mutually incompatible interpretations have already been proposed. Feynman has aptly summed up the confusing situation when he exclaimed: "I think I can safely say that nobody understands quantum mechanics".[5]

What I have been trying to demonstrate is that one can "play around" with the mathematical apparatus of the theory to obtain a formulation which is, no doubt, mathematically less elegant but still is logically consistent and, from a certain metaphysical angle, perhaps more satisfactory. But surely, neither mathematical elegance nor metaphysical clarity provide the ultimate criterion for choosing among rival theories. It is rather the overall success in predicting and interpolating observable results which counts. In this respect, too, quantum mechanics has an impeccable record. Still the theory has its limitations. Consider the question: "What precisely happens to the position of a particle during a momentum measurement?" Quantum mechanics renders such questions unintelligible. In the above model it is not unintelligible, and future and better theories may provide an intelligible answer which may yield (indirectly) observable predictions.

There are some proofs that "no hidden variable theories are possible" but a closer examination of such theorems reveals that more than establishing the futility of attempts to extend quantum mechanics these results actually impose constraints on any future extension. The above model testifies to just that effect.

In the face of these facts, I believe that Feynman's attitude is indeed the most rational: admit ignorance and remain open-minded.

Acknowledgement

I wish to thank J. Bub for his constant encouragement.

APPENDIX—PROOF OF THE MAIN THEOREM

Let $\mathscr{L}$ be the set of all horizontal and vertical lines. $\mathscr{L}$ has the cardinality of the continuum and thus, using the axiom of choice, we can well order $\mathscr{L}$

$$\mathscr{L} = \{l_\alpha \,|\, \alpha < \Omega\}$$

where Ω is the first ordinal whose cardinality is that of the continuum. Now define sets λ_α by induction on the order:

$$\lambda_1 = l_1.$$

If $l_\alpha = l_\beta$ for some $\beta < \alpha$ put $\lambda_\alpha = \lambda_\beta$ otherwise put $\lambda_\alpha = l_\alpha \backslash \cup_{\beta < \alpha}(l_\alpha \cap l_\beta)$.
Consider the set,

$$\bigcup_{\beta < \alpha} (l_\alpha \cap l_\beta).$$

The cardinality of α is less than that of the continuum and the intersection of any two lines contains at most one point, hence, if we accept the continuum hypothesis it follows that $\cup_{\beta < \alpha}(l_\alpha \cap l_\beta)$ is countable. Now put $M = \cup_{\alpha < \Omega} \{\lambda_\alpha \,|\, l_\alpha$ is horizontal$\}$ and M is the required set. Q.E.D.

One can replace the continuum hypothesis by the weaker axiom: "Every set of reals whose cardinality is less than the continuum has measure zero", and prove a result similar to the Main Theorem only that the word "countable" is replaced by "has measure zero on l". This will not change the model at all. The reader should be warned that the continuum hypothesis and even its weaker alternative are scarcely used in mathematical analysis. Their (relative) consistency have been demonstrated a few decades ago by K. Gödel.[6]

References

1. I. Pitowsky, *Phys. Rev. Lett.* **48**, 1299 (1982); see also discussion in I. Pitowsky, *Phys. Rev. Lett.* **49**, 1214 (1982).
2. I. Pitowsky, *Phys. Rev.* **D27**, 2316 (1983).
3. S. Gudder, "Local Probability Spaces", to be published.
4. See, e.g., W. V. O. Quine, *Erkenntnis*, **9** (1975).
5. R. Feynman, *The Character of Physical Laws* (M.I.T. Press, Cambridge, MA, 1965), p. 129.
5. K. Gödel, *The Consistency of the Axiom of Choice and of the Generalized Continuum Hypothesis*, Annals of Mathematical Studies, Vol. 3 (Princeton University Press, Princeton, NJ, 1938).

LOCAL STRUCTURE AND SYMPLECTIC GAUGE SYMMETRY

Russell Dubisch

Department of Physics, Siena College, Loudonville, NY 12211

In Hawking-type processes, the curvature of spacetime produces an interaction-like effect owing to a transformation of the complex structure attached to the space H *of c-number solutions to the first-quantized field equations. A similar transformation is produced in accelerated coordinate reference frames. It is here suggested that such complex-structure transformations be regarded as gauge transformations, and that a necessary qualification for a quantum theory, general enough to include quantum gravitational effects, should be invariance under some as-yet-undetermined subgroup of the infinite-dimensional symplectic group* Sp(∞, C) *on* H. *The usefulness of this point of view is limited by the commonly-held opinion that the complex structure of a field must be determined by its global properties. An approach to a local description of complex structure is therefore outlined.*

I. INTRODUCTION: QUANTUM FIELDS, COMPLEX STRUCTURE, VACUUM EFFECTS

A quantum Boson field is an object obtained as a summation over basis vectors u_k of the solution space H to a c-number field equation, i.e.,

$$\phi = \sum_k (a_k u_k + a_k^\dagger u_k^*), \tag{1.1}$$

in which the coefficients of the basis vectors are operators a_k, $a_k^\dagger$ satisfying the canonical commutation relations

$$[a_k, a_{k'}] = [a_k^\dagger, a_{k'}^\dagger] = 0$$

$$[a_k, a_{k'}^\dagger] = \delta_{kk'}. \tag{1.2}$$

The quantum field ϕ is thus an element of a vector module Φ over the algebra of operators $\{a_k, a_k^\dagger : k \in K\}$ having the property (1.2) for some appropriate index set K.

Given another basis $\{\bar{u}\}$ for H, we have the alternative expansion

$$\phi = \sum_k (b_k \bar{u}_k + b_k^\dagger \bar{u}_k^*) \tag{1.3}$$

of the same field ϕ. In general, alternative bases for H will be related by the Bogoliubov transformation

$$\bar{u}_j = \sum_i (\alpha_{ji} u_i + \beta_{ji} u_i^*)$$

and   (1.4)

$$u_i = \sum_j (\alpha_{ji}^* \bar{u}_j - \beta_{ji}^* \bar{u}_j^*)$$

provided the bases $\{u_j\}$, $\{\bar{u}_j\}$ are orthonormal. Consequently, the operators a and b are related by

$$a_i = \sum_j (\alpha_{ji} b_j + \beta_{ji}^* b_j^\dagger)$$

(1.5)

$$b_j = \sum_j (\alpha_{ji}^* a_i - \beta_{ji} a_i^\dagger).$$

If there are nonzero β_{ji} for any j, i, then the Fock spaces of the two operator sets $\{a\}$ and $\{b\}$ will differ; in particular, the vacuum state $|\Omega\rangle_a$, defined by $a_k|\Omega\rangle_a = 0$ for all $k \in K$, will not be the b-vacuum, that is, $b_k|\Omega\rangle_a$ will not in general vanish.

The above considerations form the foundation for a set of results that may be described as "particle creation in vacuum spacetimes", or "Hawking-type processes". In the Hawking calculation, production of particles by a black hole is demonstrated by allowing the u_k's and $\bar{u}_k$'s to be the solutions to the Klein–Gordon equation that are plane-wave-like asymptotically as $t \to -\infty$ and $t \to +\infty$, respectively. In flat spacetime, Fulling and others[2] have investigated the consequences of basing the quantum field on solutions u_k and $\bar{u}_k$ obtained in Minkowski and Rindler coordinates, respectively, to illustrate the effects of acceleration to the vacuum. Again a (thermal) flux is conjured up in the accelerated frame by the unaccelerated vacuum. These effects appear to be "real", and not just some computational artifact or failure of renormalization. The Hawking effect, for example, is considered necessary on thermodynamic grounds. The effect due to an accelerated reference frame may already have been observed in particle-accelerator collisions involving the high accelerations produced by the strong interaction.[3]

II. SYMPLECTIC GAUGE SYMMETRY AND COVARIANCE CONDITION

Notwithstanding the evident reality of the effects mentioned above, calculations whose results depend crucially on choice of coordinate system have a repugnant anti-relativistic flavor. One suspects, moreover, that the stress-energy obtained from measurements of the Einstein tensor in an accelerated frame in flat space, ought to yield the same zero result as in an unaccelerated frame in flat space; any other result would appear to violate general covariance and would lead to the paradoxical conclusion that the geometry of the universe is observer-dependent! One would prefer to have a framework for quantum field theory in which general covariance is more manifestly upheld. It is conceivable that such a framework could lead to a covariance principle that would be useful in constructing new physical theories.

I would like to suggest that our present situation resembles that facing a pre-special relativistic user of Maxwell's theory of the electromagnetic field. To such a theorist, electric and magnetic effects would appear to vary according to choice of inertial frame, in clear violation of Galilean relativity. This "paradox", of course, was resolved by learning to regard the description of the electromagnetic field (in terms of B and E), as involving a gauge freedom corresponding to the Lorentz group, rather than the Galilean group. In order to appreciate particle creation by acceleration or spacetime curvature we must, analogously, learn to regard the presence or absence of particles as an aspect of some gauge freedom in the description of a quantum system. Accordingly, we seek the corresponding gauge group, and resolve to consider as truly fundamental only those equations, definitions and relations that are invariant under the action of that group.

It is well known[4] that in the particle production process, a crucial role is played by the so-called almost-complex-structure attached to the space H of solutions to the c-number field equations. This structure is given by a linear mapping $J:H \to H$ having the properties

$$J^2 = -1, \qquad \tilde{G}JG = J, \tag{2.1}$$

where 1 is the identity map and G the Hermitian inner product on H. By virtue of (2.1), J will have two eigenvalues, $+i$ and $-i$. Thus J is specified by the corresponding two subspaces, the positive- and negative-frequency subspaces, respectively. If the pos–freq subspace is spanned by the basis $\{u_k\}$, then

$$Ju_k = +iu_k$$
$$Ju_k^* = -iu_k^*. \tag{2.2}$$

The Bogoliubov transformation (1.4) induces a transformation of the almost-complex structure:

$$J \to JB, \tag{2.3}$$

where

$$B = \begin{pmatrix} (\alpha_{ij}) & (\beta_{ij}) \\ (\beta_{ji}^*) & (\alpha_{ji}^*) \end{pmatrix}.$$

Owing to the properties of the Bogoliubov coefficients,

$$\sum_k (\alpha_{ik}\alpha_{jk} - \beta_{ik}\beta_{jk}) = \delta_{ij}$$

$$\sum_k (\alpha_{ik}\beta_{jk}^* - \beta_{ik}\alpha_{jk}^*) = 0, \tag{2.4}$$

the invariance condition $BJ\tilde{B} = J$ holds, and thus B is an element of the (infinite-dimensional) symplectic group $Sp(\infty, C)$.

Unruh,[5] in discussing the effect of acceleration on a particle-detection apparatus, points out that the effect of the change in the almost-complex structure induced by acceleration is manifested in a change in the absorption–emission processes undergone by the detector. That is, the emission events, seen as occurring when viewed by an unaccelerated observer, are the very absorption events taken by the accelerated observer as a manifestation of the acceleration-induced thermal bath. It therefore seems reasonable, when talking about accelerated systems (or, similarly, about systems in curved spacetime), to regard those particulars of the quantum description of the system that are frame-dependent, as being determined by a choice of gauge. The most general group of such a gauge is the symplectic group $Sp(\infty, C)$.

It may be appropriate to remark at this point that any theory contrived to be gauge-invariant under $Sp(\infty, C)$ (acting in the manner described above) would have properties at least roughly equivalent to the property of general covariance met by the general theory of relativity. (General covariance requires that a classical field theory be expressed in a form invariant under the spacetime diffeomorphism group.) In general relativity, however, the principle of general covariance merely requires that all theories be "tensorial" in character, that is, that all quantities in the equations of the theory be sections of the tangent tensor algebra bundle over the spacetime manifold. General covariance is therefore (at least formally) an easy requirement.

On the other hand, the requirement of "symplectic covariance" is apparently a much harder restriction to apply, in that it concerns global aspects of a theory. This is because the complex structure is obtained by dividing the space H of wave solutions into its positive- and negative-frequency parts (i.e., eigenfunctions of the time-translation vector with positive and negative eigenvalues), and the frequency is not a tensor (that is, local) property, but rather an extended attribute according to the usual treatment of the theory. If, for example, we confine a particle to a box†, so as to deprive its wavefunction of some of its extension, then the uncertainty principle $\Delta t \, \Delta \omega \sim h$ frustrates precise description of its frequency. Indeed, calculations[6] using a variety of accelerated coordinate systems show that the radiation field generated in an accelerated coordinate system from the Minkowski vacuum depends in detail on the properties of the chosen coordinate system either at infinity or at the coordinate "event horizon". These results indicate that we cannot obtain unique results for an accelerated lab by specifying a single acceleration of the lab; we must instead specify the trajectory of each part of an infinitely-extended "lab" in order to obtain an unambiguous result for the acceleration-induced effects. Unfortunately, the situation is confused by the lack of any covariant meaning to the concept of "uniform acceleration" as applied to an extended object†.

The fundamental question that emerges from all this: "Can the symplectic group be used as a gauge symmetry to obtain a covariance

---

† A four-dimensional box, that is, with duration Δt.

† This is a well-known fact in special relativity.

condition?" The answer is not yet known, but, to paraphrase Fred Hoyle, the correct answer is usually the one yielding the most interesting results.[7] A new covariance condition is an exciting prospect because it may be expected to function, as in the case of general relativity, as a guide to a theory or theories of improved generality. It may not even be too much to expect that the sought-after grand unification might hinge on finding the proper covariance condition.

Although there is no assurance that spacetime will be the fundamental arena for such a powerful covariance principle (see, e.g., W. Wootters' contribution to this symposium), the example set by Einstein in obtaining general covariance urges us to look for a local version of a principle of symplectic covariance. That is, even granting the caveats† mentioned two paragraphs ago, we are motivated to see how much of the algebra of complex structure of a field can be embodied in the local structure of that field. Accordingly, in the following example we exploit the idea of a kind of "Taylor-expansion-like" structure of a c-number field at each chosen point in a spacetime manifold.

III. JETS AND LOCAL COMPLEX STRUCTURE OF A FIELD

Consider a scalar Boson field φ on a spacetime M. We may write the field equation covariantly as (minimally coupled case)

$$(g^{ab}\nabla_a\nabla_b + m^2)\varphi = 0, \tag{3.1}$$

where ∇_a is the covariant derivative. Obtaining a complex structure on the manifold of solutions to (3.1) is conveniently done by finding the solutions having positive and negative frequencies, respectively (u_+ and u_-), and specifying condition (2.2), that is,

$$Ju_\pm = \pm iu_\pm. \tag{3.2}$$

Unfortunately, in a general curved spacetime, there is no canonical way

---

† Which should perhaps be tempered with the observation that, while the differentiability class necessary for most purposes in physics is no greater than C^4, that assumed in the usual formulation of general covariance is C^∞. In other words, the gauge group of diffeomorphisms leading to general covariance is smaller than that gauge diffeomorphism group justifiable on purely physical grounds. General covariance has proved nevertheless to be sufficient for its purpose.

to specify the sign of the frequency of a solution to (3.1); the solution set does not generally include plane waves. Nevertheless, if the conditions

$$\left|\frac{k_0}{c}\right| \gg \frac{1}{\text{dimensions of lab}} \gg |\text{local curvature of spacetime}| \quad (3.3)$$

hold, then the solutions $\exp(+ik_a x^a)$ are at least approximate solutions to (3.1) (with $g^{ab}k_a k_b = m^2$), inside the lab. So to at least this extent, the concept of "local frequency" makes sense: for high frequency (large values of the components of the wave-vector k_a), the difference between the second covariant derivative $\nabla_a \nabla_b \varphi$ of the wave-function, and the second partial derivative $\partial_a \partial_b$ is insignificant, since that difference is

$$\partial_a \partial_b \varphi - \nabla_a \nabla_b \varphi = \Gamma^c_{ab} \partial_c \varphi - \Gamma^c_{ab} k_c \varphi,$$

and each term on the lhs is of order

$$\partial_a \partial_b \varphi \sim k_a k_b \varphi.$$

It is indeed the low-frequency end of the spectrum where effects due to acceleration or curvature are the most significant. Therefore we need a formalism that will allow a definition of the wavenumber within an interval small compared with the oscillation period (in time) and to the wavelength (in space). Accordingly, we consider the following ansatz, which resembles a jet-bundle formalism.[8]

Let p be a point in a spacetime M, and let X^a be a timelike vector field in a neighborhood of p. Let φ be a scalar field obeying (3.1) in a neighborhood of p. We may describe the time-dependence of at p by means of its jet-structure: we specify that the j-jet of at p is the equivalence class of functions φ' having the same values for $X^n \varphi(p)$, $n = 0, 1, \ldots, j$, where $X = X^a \nabla_a$. We may give j-jet coordinates as

$$J^j_{(\varphi, p, X)} = [\varphi(p), (X\varphi)(p), (X^2\varphi)(p), \ldots, (X^j\varphi)(p)]. \quad (3.4)$$

In terms of the jet-structure, we may now introduce the concept of local jth-order frequency. Let

$$_p\varphi^j_{X:\omega} = \{\varphi : J^j_{(X\varphi, p, X)} = i\omega J^j_{(\varphi, p, X)}\}. \quad (3.5)$$

This equivalence class can be considered as an "eigen-jet" with "derivational frequency" ω. This is not a true frequency, since it is not a vector component; rather it depends on the extended behavior of the vector field X^a. We may overcome this shortcoming by introducing local geodesic coordinates at p. Let x^i, $i = 0, \ldots, 3$ be geodesic coordinates at

p. Let $X_n = \partial/\partial x^n$. Then, setting $X = X_n$,

$$X^b X^a_{;b} = 0, \tag{3.6}$$

so that

$$\begin{aligned}
X^2 \varphi &= X^b (X^a \varphi_{;a})_{;b} \\
&= X^b X^a_{;b} \varphi_{;a} + X^b X^a \varphi_{;ab} \\
&= X^b X^a \varphi_{;ab}, \tag{3.7}
\end{aligned}$$

and similarly for higher-order derivatives, so that choice of a geodesic X^a yields a derivational frequency that is, in fact, dependent only on the value of X^a at the point p, and therefore a "directional" frequency, that is, a vector component. Furthermore, if $Y = X_m$, then

$$\begin{aligned}
X Y \varphi &= X^a (Y^b \varphi_{;b})_{;a} \\
&= X^a Y^b_{;a} \varphi_{;b} + X^a Y^b \varphi_{;ba} \\
&= Y^a X^b_{;a} \varphi_{;b} + Y^b X^a \varphi_{;ab} \\
&= Y X \varphi \tag{3.8}
\end{aligned}$$

at the point p. The four fields X_n are consequently a set of commuting operators, and therefore admit common eigenfunctions, which are furthermore the same equivalence classes as those of any other set of independent geodesic fields at p. We will label these classes ${}_p\varphi^i_k$. The associated complex structure (the *normal* complex structure) is then defined by

$$\left. {}_p J^j_N \right| {}_p \varphi^i_k \rangle = \pm i \left| {}_p \varphi^j_k \right\rangle \tag{3.9}$$

if $k_0 = \pm |k_0|$.

According to the spirit of the present paper, the normal complex structure is not to be regarded as the only possible choice, but rather as a choice of gauge, which may be in some respects a convenient one, and in others not so convenient. Inasmuch as it is "canonical", it avoids ambiguities in the description of a system, such as those associated with the asymptotic behavior of the coordinate system; furthermore, it is an appealing choice of complex structure for a freely-falling observer in curved spacetime. (In flat spacetime, the $j = \infty$ case is identical to the usual choice of complex structure.) We might even reasonably expect the normal complex structure to be the appropriate one for the calculation of the stress-energy in quantum black-hole theory (see, e.g., Wald[9]).

Of course, various difficulties exist. Each j-jet corresponds to an entire equivalence class of functions φ on the manifold. For finite j, the normal complex structure at p can be regarded only as an approximation to an idealized complex structure (presumably $_pJ_N^\infty$). But allowing j to approach ∞ produces an infinite set of algebraic conditions on the φ_k's; techniques for handling this system have not been worked out. Furthermore, any approach that is local in the meaning assumed herein cannot explicitly handle the effects of "news" in the solutions to the field equation (3.1). (For an approach that can, see Deutsch.[10])

Acknowledgements

I would like to express my gratitude to Edwin Rogers for many stimulating and helpful discussions. Part of this work was supported by a Faculty Research Assistance Grant from Siena College, spring, 1981.

References

1. N. D. Birrell and P. C. W. Davies, *Quantum Fields in Curved Space* (Cambridge University Press, Cambridge, 1982), p. 46.
2. S. Fulling, *Phys. Rev.* **D7**, 2850 (1973); W. G. Unruh, *Phys. Rev.* **D14**, 870 (1976).
3. S. Barshay and W. Troost, *Phys. Lett.* **73B**, 437 (1978).
4. A. Ashtekar and R. Geroch, *Rep. Prog. Phys.* **37**, 1211 (1974); D. Deutsch, *Phys. Rev.* **D28**, 1907 (1983).
5. W. Unruh, op. cit.
6. N. Sanchez, *Phys. Lett.* **87B**, 212 (1979).
7. F. Hoyle, *Third John Danz Lecture Series* (University of Washington Press, Seattle, 1964), p. 41.
8. R. Herman, *Vector Bundles in Mathematical Physics* (Benjamin, New York, 1970); B. A. Kuperschmidt, in *Geometric Methods in Mathematical Physics* (Springer-Verlag, Berlin, 1979), p. 165.
9. R. Wald, *Gen. Rel. and Grav.* **9**, 95 (1978).
10. D. Deutsch, in *Quantum Gravity 2*, ed. Isham *et al.* (Clarendon Press, Oxford, 1981), p. 131.

TIME IN QUANTUM MECHANICS

David Park

Williams College, Williamstown, MA 01267

"Time is defined so that motion looks simple."—Misner, Thorne and
Wheeler

The word "time" is used by physicists in at least three ways: as the reading of a clock, as one of the dimensions in which we perceive the world, and as one of the four coordinates that are used in the mathematical structure of physical theory. The chief contention of this paper is that whereas we can do fairly well in classical physics without troubling to make careful distinctions among these three meanings, quantum physics repays a little more thought. I shall talk first about Newtonian physics, then Einsteinian, and finally about quantum physics.

I. CLASSICAL TIME

"Absolute, True, and Mathematical Time, of itself, and from its own nature flows equably without regard to anything external . . . ."—these words, familiar to every physicist, contain a rather large amount of physical hypothesis. It is not obvious why Newton found it necessary to be so explicit, since this definition and the accompanying definition of space are never mentioned again in the *Principia*; not only that, but they provided a large, vulnerable, and tempting target for Newton's critics on the Continent. To Leibniz, for example, the concepts of space and time refer to relations between solid bodies and between events. Without these relations there would be no space or time. For Newton, time and space emanate from God and are prior to the created universe. Both Newton and Leibniz have clocks; the difference is that Newton's clock exists independently of the world, while Leibniz's *is* the world. I think that the Scholium on absolute time is intended to clarify for the reader Newton's thought and his language rather than his mathematics. What is this time of the Scholium? Clearly it is not a coordinate, since a coordinate does not flow. It is, so to speak, the Platonic Idea of the time that is measured on a clock: it is an Idea because by saying that it flows without relation to anything external he cuts us off from any criterion by which his statement could be tested. But, all the same, I think I know what he means. It is that the revolution of day and night and the march of mechanical clocks seem to us to proceed evenly, after we allow for the roughness of our perceptions. Today we might say that there is only one basic dynamical principle, that the pulse of the universe runs in us all.

In the whole *Principia* Newton represents time only by words, and never by a letter. In 1736 the young Leonhard Euler in his *Mechanica* effortlessly makes the transition to modern notation. There, for the first time that I know of, the Second Law is written in differential form; time no longer flows but has tacitly appeared as a coordinate, and dynamics can be developed as a branch of modern mathematics instead of ancient geometry. Newton is left behind with hardly a backward glance, but I think we cannot fully understand the role of time in physics unless we occasionally recall how Newton thought of it.

The spatial dimensions of a thing can be measured with a rigid rod and, as Einstein emphasized, the process requires that both ends of the thing be observed at the same time. Obviously, to compare the beginning and end of a time interval we have to have some counterpart of a rigid rod to lay beside it, and this is what a clock is. Thus coordinates in space may be operationally assigned in terms of an array of static objects, but time requires a functioning mechanism. The contrast runs very deep, as we shall see.

Because the goal of this discussion is quantum mechanics I shall discuss classical mechanics from the point of view that most resembles it. Let the motion of a material particle be represented by a superposition of waves of the form $e^{ikW(q,t,E)}$, where W is Hamilton's principal function, given by

$$W = S(q, E) - Et, \qquad S(q, E) = \int p_i(q, E)\, dq^i \tag{1}$$

and k is some constant. E represents the energy, and the momentum p is supposed to be given in terms of q and E. Consider the motion of a superposition of waves with energies in a narrow distribution around the value $\bar{E}$. Then, as is well known, the condition for constructive interference is $\partial W / \partial E = 0$, or

$$t(q, \bar{E}) = \frac{\partial S(q, \bar{E})}{\partial \bar{E}}. \tag{2}$$

This derivation, which represents a classical state by a superposition of states of different though closely adjacent energies, is obviously a hybrid of classical and quantum ideas. If we start by assuming Newton's laws we can derive the same result without it, but I want to recall here the remarkable fact that it is possible to get time out of expressions that do not contain time by a superposition of states with slightly different energies. We shall need it later.

Whatever route is followed, time in Eq. (2) emerges naturally from classical dynamics as a parameter, defined in terms of other dynamical quantities, that tells where a particle (or more complicated system) is situated on its path. For reasons that will become clear I think it is extremely important to show how a quantum time can emerge from a spatial description in quantum physics in the same way that we have just produced Newtonian time from classical physics. I have not been able to do this in terms as general as those used here, but at the end of this talk I will show how it can be done in a clear and simple example.

Let us try to introduce in quantum mechanics a time operator analogous to the time quantity that has just emerged in Eq. (2). In classical mechanics we encounter the Poisson-bracket relation

$$\frac{dA}{dt} = [\![A, H]\!]$$

and writing $t(q, p)$ we expect to find, and do find in practice, that

$$[\![t, H]\!] = 1.$$

This scene does not play well in quantum theory. There, dynamical variables are represented by q-numbers and we would expect to represent $t(q, E)$ by an operator $\hat{T}$ satisfying

$$[\hat{T}, \hat{H}] = i.$$

Let us see what it does. Let $|E'\rangle$ be an eigenfunction of $\hat{H}$ belonging to the eigenvalue E', and let $|E'\rangle_\varepsilon = e^{i\varepsilon \hat{T}} |E'\rangle$. Then

$$\hat{H}|E'\rangle_\varepsilon = e^{i\varepsilon \hat{T}} e^{-i\varepsilon \hat{T}} \hat{H} e^{i\varepsilon \hat{T}} |E'\rangle = (E' + \varepsilon)|E'\rangle_\varepsilon$$

and for arbitrary values of ε not only do the eigenvalues of $\hat{H}$ form a continuum but they extend to negative infinity. Pauli[1] wrote in 1933 "We conclude that the introduction of an operator t must fundamentally be abandoned and that the time t in quantum mechanics has to be regarded as an ordinary number ('c-number')".[2] What has gone wrong here is that energy eigenvalues have no lower limit and that therefore no system is stable, but the reason we cannot allow this does not lie in the formalism of the theory; rather it lies in our knowledge of dynamics.[3] We think that there are stable systems. Thus, as we shall see repeatedly, our knowledge of time is part of, and not prior to, our knowledge of dynamics. Here we are on Leibniz's side.

When physicists calculate they fill spacetime with a coordinate system, and they also make use of certain dynamical principles such as the conservation of energy and momentum that can be stated without coordinates. Usually the coordinate system plays a dual role. Equations of motion are written and solved in it. But also, the method of quantum mechanics requires that the apparatus that specifies an experiment and registers its result be described "classically", which means in practice that its layout and function can be expressed in terms of coordinates. If this is done, then the principle of indeterminacy says that we may not speak of its momentum, and consequently we may not use dynamical principles in describing its interaction with quantum systems. If we want to preserve dynamics we must untie the apparatus from the coordinate system defined by the laboratory and allow it to float freely. But then the places and times it defines have no clear relation to the walls of the room or the clock over the door, and we have merely postponed describing the experiment in classical terms.[4]

Equations of motion, whether quantum or classical, have nothing statistical about them; the statistical interpretation of quantum mechanics seems to be the price we have to pay for trying to combine incompatible modes of description. One way to say it is that since the equations of quantum mechanics nowhere contain the idea of an event and those of Newtonian mechanics do, events occur at random when the results of quantum mechanics are expressed in Newtonian terms. At any rate, a description of spacetime in terms of things and events is possible only in classical physics; if we ask for more than that we must invent some new kind of spacetime that hangs on quantum processes. Professor Wootters will tell you about his efforts in this direction, but, until the program is complete, spacetime as Leibniz defined it can be used only for classical physics. Thus we are back to Newton. Coordinates in his absolute space, far from situating everything in the created universe, have a meaning that for quantum mechanics is purely computational, but they have the great virtue that they allow a description in identical terms (if we are clever enough) of material systems ranging from galaxy clusters to the tissues of the central nervous system.

II. PROPER TIME

In 1941, E. G. C. Stueckelberg[5] noted that a U-shaped world line in the classical dynamics of a particle corresponds to the creation of a particle–antiparticle pair, so that antiparticles are just particles whose proper time runs backward. He tried to incorporate this idea into quantum mechanics, but there were difficulties of interpretation which still remain. Eight years later, Richard Feynman, at the cost of a radical reformulation of the theory,[6] was able to incorporate the antiparticle idea into quantum electrodynamics, and in the course of this work he obtained some results which suggest that there may exist a consistent quantum theory of proper-time dynamics.

Suppose we want to work out the quantum mechanics of a particle with internal degrees of freedom, for example, one that decays after a while. Classical physics provides an approach that is simple and clear: the internal dynamics should be considered in the particle's rest system, and the time of this dynamics is the particle's proper time. But proper time is not a variable in ordinary quantum theory, and so this approach cannot immediately be followed. Let us try to construct a version of the theory in which proper time occurs as a parameter.

For a single particle, the natural place to start is Hamilton's principle,

$$\delta \mathscr{W} = 0, \qquad \mathscr{W} = \int \mathscr{L}(x, \dot{x}) \, d\tau$$

where

$$\mathscr{L} = \tfrac{1}{2} m \eta_{\mu\nu} \dot{x}^{\mu} \dot{x}^{\nu} + e A_{\mu} \dot{x}^{\mu}, \qquad \dot{x}^{\mu} = dx^{\mu}/d\tau.$$

Here τ is the proper time and $\eta_{\mu\nu} = (1, -1, -1, -1)$. The momenta are $p_{\mu} = \partial \mathscr{L}/\partial \dot{x}^{\mu}$, and following the previous argument we arrive at a relation analogous to Eq. (2),

$$\tau(x, \bar{m}) = 2 \frac{\partial \mathscr{S}(x, \bar{m})}{\partial \bar{m}}, \qquad \mathscr{S}(x, m) = \int p_{\mu}(x, \bar{m}) \dot{x}^{\mu} \, d\tau \qquad (3)$$

If it was strange in deriving Eq. (2) to superpose states of closely neighboring energies in order to use the principle of constructive interference, the process by which Eq. (3) is derived is even stranger. It is, after all, possible for a particle in a given situation to have different values of energy, but if it is stable it has only one value of rest mass. For an unstable particle this is not a difficulty. We may think of stable particles as unstable with very long lifetimes. Perhaps this is what they are. But now let us see what the same theory looks like in quantum mechanics.

Assuming that as in nonrelativistic quantum mechanics $\mathcal{W}$ represents the phase of a wave and that the wave obeys Huyghens's principle, Feynman adapted his construction of the nonrelativistic wave equation to derive from the relations just given a proper-time wave equation[7]

$$i\frac{\partial\varphi}{\partial\tau} = \hat{\mathcal{H}}\varphi, \qquad \hat{\mathcal{H}} = \frac{1}{2m}(i\,\partial_\mu + eA_\mu)(i\,\partial^\mu + eA^\mu). \qquad (4)$$

Since the potentials A_μ depend on x^μ and not on τ, the equation as separable, and those states which vary as $\exp(-\tfrac{1}{2}im\tau)$ satisfy the Klein–Gordon equation in the remaining spacetime variables. If m is positive, the same argument as before shows that τ cannot be taken to be a q-number; the particle's proper time is a classical parameter.

The most natural interpretation for the wave function φ is that it is that the probability amplitude that a particle gets to the spacetime point x^μ with a given value of τ, as read on its little classical watch. In his first paper, Stueckelberg notes that with this assumption the expectation values of the coordinates x^μ satisfy the classical equations of motion. There is the difficulty that states of definite mass are not normalizable; this is analogous to the usual problem with plane waves and can be solved by considering states that are not exact eigenvalues of the mass.

Looking for a simple illustrative calculation I thought of finding the covariant version of the Aharonov–Bohm effect,[8] the shifting of the interference pattern formed by two electron beams when they pass on each side of a region of intense magnetic field without penetrating it at all. Viewed from a moving coordinate system, the situation would appear complicated by the varying electric field produced by the changing magnetic field, while of course the pattern would stay the same.

Consider a spacetime point x^μ on the trajectory of an electron at proper time τ. Let the electron's spacetime trajectory at proper times τ' prior to τ be given by $z^\mu(x, \tau')$, where $z^\mu(x, -\infty)$ is at spatial infinity and $z^\mu(x, \tau) = x^\mu(\tau)$. It is known[9] that if the quantum-mechanical wave function has its phase changed by an amount

$$-e \int_{-\infty}^{\tau} A_{\mu}[z(\tau')] \frac{\partial z^{\mu}}{\partial \tau'} d\tau'$$

then the resulting wave equation involves the fields, not the potentials. Since in our case the fields are zero, it is the equation for a free particle; thus the entire effect is contained in the phase just introduced. To calculate an interference pattern we must consider two beams which originate at the same point in spacetime and superpose them at the same point in spacetime on the screen. Consider these two points as fixed. Then the difference in phase between the two paths is given by

$$e \int A_{\mu}(z) \, dz^{\mu}$$

integrated around a loop in spacetime which encloses the region of the field. By the four-dimensional Stokes's theorem,[10] this is

$$e \int\int F_{\mu\nu} \, d\sigma^{\mu\nu}$$

where $F_{\mu\nu}$ is the electromagnetic field tensor and $d\sigma^{\mu\nu}$ is an element of two-dimensional surface. This is the covariant formula for the Aharonov–Bohm phase shift.

Even if proper time can be fitted into the conceptual framework of quantum mechanics,[11] I have found few papers in which it appears to be useful. An exception is Schwinger's[12] manifestly gauge-invariant analysis of vacuum polarization by an external field; that Schwinger's method is equivalent with the one sketched here has been shown recently.[13]

Coordinate time is a classical variable which looks out of place in quantum theory. Proper time looks as if it might be useful in the study of systems with internal dynamics, since a particle, so to speak, carries its own proper time with it as it moves, but it, too, can appear only as a c-number. It is time to see what quantum mechanics itself has to tell us about time.

III. QUANTUM MECHANICS

In starting this discussion we must remember that in the equations of motion the letter t denotes a coordinate and not a dynamical variable;

thus it stands outside the range of ordinary discussions of such subjects as measurability. In particle dynamics we are used to encountering the quantity x in two different roles: as a coordinate used to write equations of motion and as a dynamical variable, a q-number, used to denote the position of a particle. For t there is no such duality; it is only a coordinate. Nevertheless, t occurs in the "fourth uncertainty relation", and we must pause to see how it gets there.

Suppose a wave packet characterized by Δx and Δp moves through space past a certain point. Then the uncertainty in the time it moves past the point (or is registered in a detector placed at that point) satisfies

$$\Delta E \, \Delta t = \frac{p \, \Delta p}{m} \frac{m \, \Delta x}{p} = \Delta p \, \Delta x \geqslant \tfrac{1}{2}.$$

This is of course a genuine indeterminacy, but note that it does not reflect any supposed influence of the process of using a clock upon the phenomenon in question. Rather, it is an automatic consequence of the $\Delta x \, \Delta p$ relation which follows when one assume that the equations of motion are being obeyed. There is no question of the accuracy with which the numerical value of the letter t can be taken to denote a time. The "fourth uncertainty relation" is in this case parasitic on the other three.

Let us consider a slightly more sophisticated example. Suppose we have a device C which we wish to use as a clock to time an event in an apparatus A. To find out what time it is we read on C the value of an observable Q, which might be the angular position of a pointer. A and C are coupled dynamically and isolated from outside influences so that their combined energy is conserved. Let H be the Hamiltonian function of the entire system, and let the state of A and C be either an eigenstate of H or a superposition of states of so nearly the same energy that its variance is much less than the others to be considered. Then the expectation value of Q changes at a rate

$$\frac{d\langle Q \rangle}{dt} = i \langle [\hat{H}, \hat{Q}] \rangle$$

and according to a familiar identity this gives

$$\Delta E \, \Delta Q \geqslant \frac{1}{2} \left| \frac{d\langle Q \rangle}{dt} \right|$$

where ΔE represents the uncertainty in that part of the energy which depends on the variable Q: the clock and its interaction with the apparatus A. In order for the clock to function, $\langle Q \rangle$ must depend on time, and in order to read the clock the change in $\langle Q \rangle$ must be greater than the indeterminacy ΔQ. If the clock is observed for a time t throughout which C interacts with A, we have

$$t \, \Delta E \, \Delta Q \geqslant \frac{1}{2} t \left| \frac{d \langle Q \rangle}{dt} \right| \geqslant \frac{1}{2} \Delta Q$$

so that, if the clock is to serve its purpose, whatever Q may be, t must satisfy

$$t \, \Delta E \geqslant \tfrac{1}{2}. \tag{5}$$

Here, since A and C are coupled and otherwise isolated, ΔE represents the uncertainty in energy of either A or C.

Note that there is still no question of an indeterminacy in t. The clock with which we time the duration of the interaction is external to A and C. It is not coupled to them and belongs to Newtonian physics: that concept of the world in which time flows on "without regard to anything external". Time enters the calculation by way of the equation of motion satisfied by Q, since, as we have seen, the beginning and end of an interval can be related only in this way. It is true that as long as $|\Delta Q| > 0$ we do not find out exactly what time it is by reading C. The indeterminacy in Q prevents that, but I emphasize once more that it is the value of the Newtonian coordinate time t that the measurement leaves uncertain; we have still not come to grips with the central question, what ought time to mean in microphysics?

The question just asked actually contains a contradiction. When we ask what a physical concept means I take it that we are asking how it relates to our intuitive ideas of the world, based on language and everyday sensory experience. The question of meaning in microphysics contains a trap: the language in which we explain microphysics is the very language we are seeking to clarify, and we have no sensory experience to guide us except that relating to laboratory apparatus. Worse still, language and experience are often at odds, since language has developed in response to human situations which usually extend beyond the purely sensory sphere. A question to which we shall have to pay particular attention is the epistemological status—the reality, if one wants to use a simple word, of the past, the present, and the future.

In ordinary speech we behave as though the past were a fact which we know more or less perfectly, the present is experience itself, and the future has much less reality. Whatever reality is, we experience it in the present. I am not saying that I know the facts about the world around me; I may be completely confused, but at least I know what I am thinking, and that is real to me. I can with a certain degree of confidence predict the future—that is, what my experience will be, by using equations with initial conditions describing the situation at the present moment. To reconstruct the past I have my present memory of past events as well as certain pieces of material evidence—photographs, tabulations of data and the like, which I have learned to interpret more or less confidently. In doing so, however, I must always take account of the fact that Gibbs expressed so grandly: "While the probabilities of subsequent events may often be determined from the probabilities of prior events, it is rarely the case that the probabilities of prior events can be determined from those of subsequent events, for we are rarely justified in excluding the consideration of the antecedent probability of the prior events".[14] I intend, with John Wheeler[15] and many others, to consider the past and future as intellectual constructions, while the present, with its uncertainties, contains all the facts we know about the world. Such a conclusion was already forced on us by special relativity, according to which an event in the past of one observer may lie in the future of another. We became used to that long ago, but a continuing stream of literature warns us that not everybody has come to terms with related problems raised in the fateful collaboration of Einstein, Podolsky and Rosen.

To illustrate EPR, people usually discuss the experiment with the two spins, while to display the remaining mysteries of quantum mechanics they usually turn to the experiment with the two holes. Nobody claims to understand it, but there it is, and since it is also an EPR experiment I will stay with it.

Figure 1 shows a version in which the screen on which the interference pattern is formed contains a slit S, as wide as the pattern's central maximum and, behind it, a pair of detectors each of which can look at only one of the slits A and B. It is intended that when the rear slit is open it will be possible to tell by means of the detectors whether a given quantum passed through A or B. On the other hand, it is also easy to tell from the counters whether an interference pattern is being formed, because in that case they will count a total of about twice as many quanta as they would if the pattern were not being formed. Thus we hope to have our cake and eat it too. The proposed experiment is an EPR experiment because each of the two beams issuing from the source and passing

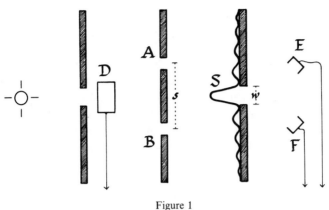

Figure 1

through A or B may be characterized by an occupation number or a phase. Suppose that first the detector D registers and then, a moment later, the detector E. We then know the occupation number of the beam through B and, by inference, that of the beam through A. Thus we cannot assign phases to the two beams and will not expect to see a diffraction pattern. If S is closed a diffraction pattern will form, and the counting of quanta by D will not harm it because from[16]

$$[\hat{n}, \sin \phi] = i \cos \phi, \qquad [\hat{n}, \cos \phi] = -i \sin \phi$$

it follows that

$$[\hat{n}_1 + \hat{n}_2, \sin(\phi_1 - \phi_2)] = 0.$$

This is only a minor variation on a familiar theme. Its relevance to the subject of time is that the field leaving the source must choose: either the beams through A and B have definite occupation numbers or they have a definite relative phase, and we can add a bit of spice to the experiment by arranging to open or close S only after a quantum has been registered at D and is on its way. We know from the general principles of quantum mechanics what ought to happen: as soon as we open S and it becomes possible to tell along which beam a quantum arrived, there can be no more diffraction pattern and the diminished total counting rate will reveal this at once. If we are going to give meaning to the experiment, to explain it according to familiar concepts and language, we must speak of a cause that operates subsequently to its effect and, furthermore, that if the apparatus violates the usual regulation which forbids us to know phase and intensity at the same time, we can observe visible proof of that reversal.

If the foregoing argument is correct then quantum time is indeed different from classical time, and we shall have to conclude that the mysteries of quantum mechanics include a temporal mystery. But of course the argument is not correct. The width of the central maximum of the diffraction pattern is given by $\lambda L/s$, where λ is the wavelength of the radiation. Slit S must be narrower than this:

$$w < \lambda L/s.$$

On the other hand, the angular discrimination possible for detector E or F looking through the slit of width w is λ/w, and at a distance L this translates into a linear resolving power $\lambda L/w$. In order to distinguish A from B at that distance this quantity must be less than s, so that

$$w > \lambda L/s.$$

Thus the proposed experiment, doomed from the start, cannot give a result which requires one to conclude that an effect can precede a cause, and it is hard for me to believe that the situation will be any different in more elaborate experiments.

Let me draw the moral of this little digression: our mental reconstruction of the past may indeed be affected by something that happens in the present; indeed this happens all the time. As long as the past is considered to be purely a mental construction, the question of what happens when it gets modified is not a question for physics. But if, as has occasionally been suggested in connection with the anthropic principle,[15] the reconstructed past can be shown to have observable consequences for us in the present, then physicists should be very much concerned. The reason why I embodied the EPR type of situation in which this effect might be expected in a two-slit experiment was to show in a simple instance how fundamentally the possibility of such a reconstruction conflicts with the laws of physics out of which it arose in the first place.

IV. A QUANTUM CLOCK

I have briefly mentioned quantum clocks in connection with the energy–time reciprocity and must now return to them. Two kinds must

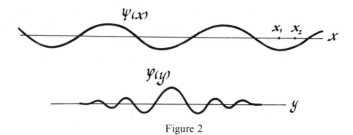

Figure 2

be carefully distinguished: those which register intervals of coordinate time by a quantum mechanism and those which produce a time reading which is recognizable as temporal but do it without reference to coordinate time. Quantum clocks which tell coordinate time were sketched long ago by Salecker and Wigner,[17] and more recently Peres[18] has worked out the details of a conceivable (though I think not very physical) model. Quantum clocks which define their own time are also present by implication in the Peres article and have been considered more fully by Page and Wootters.[19] The model I will discuss was suggested by one in Peres's paper.

Let us imagine the coupled system whose two parts are shown in Fig. 2. One part is a very long wave packet $\psi(x)$ describing a particle of mass m which passes two points x_1 and x_2 as it moves along the x axis. The other also is a wave packet $\varphi(y)$, of arbitrary configuration, coupled to the first one so that it moves only when the first particle is between x_1 and x_2. The motion and the coupling are defined by the Hamiltonian function

$$\hat{H} = \frac{p_x^2}{2m} + \frac{p_y^2}{2M} U(x), \qquad U(x) = \begin{cases} 1 & (x_1 < x < x_2) \\ 0 & \text{otherwise.} \end{cases} \qquad (6)$$

It is an accepted device of the quantum theory of scattering that a long wave packet may be treated as an energy eigenfunction, and this will make the work easier. We decompose the function $\varphi(y)$ into plane waves and write the wave function of the entire system as

$$\Psi(x, y) = \int \psi_k(x)\tilde{\varphi}(k)e^{iky}\,dk. \qquad (7)$$

The (approximately) time-independent wave equation for $\psi_k(x)$ is then

$$\left[\frac{p_x^2}{2m} + \frac{k^2}{2M} U(x) - E \right]\psi_k(x) = 0.$$

To the x-particle, it is as if it passed over a potential barrier of height $k^2/2M$ on its way along the x-axis. Suppose that M is large. By the time the wave packet has arrived at the other side of this barrier its form will be essentially unchanged but as a whole it will have undergone a phase shift

$$\Delta\phi = \left\{ \left[2m\left(E - \frac{k^2}{2M} \right) \right]^{1/2} - (2mE)^{1/2} \right\} L, \qquad L = x_2 - x_1$$

and if we write $(2E/m)^{1/2}$ as w, there is

$$\Delta\phi \approx -\frac{k^2}{2M}\frac{L}{w}.$$

Thus after the x-particle has passed through the region of interaction the wave function of the entire system will be

$$\Psi_{\text{after}} = \int \psi_k(x)\tilde{\varphi}(k) \exp i\left(ky - \frac{k^2}{2M}\frac{L}{w} \right) dk. \tag{8}$$

The wave function of the y-particle is once more frozen. The formula makes no mention of time, but the knowing eye will interpret w as the group velocity of $\psi(x)$ and see that the y-particle has expressed its own idea of time by moving forward a distance corresponding to our idea of the length of time during which it was in motion. Since we never knew where the x-particle was located, we have no idea when it passed through the region of interaction; but the motion of the y-particle tells what happened when it did go through. Thus quantum time is our time, at least in simple cases, and to the extent that the system can be taken to be in a stationary state, we may say that it defines a quantum time in the same way that a classical time is defined, also in terms of a long wave function, by Eq. (2). We should not be disappointed that it was impossible to construct an operator $\hat{T}$ to give coordinate time. That is Newtonian time. Now we are back with Leibniz.

I have conjectured earlier that probability enters quantum physics because of the conceptual mismatch at the boundary separating a theory that contains the idea of an event from one that does not. The example I have just given illustrates this point. If I were asked to say at what time the x-particle passed point x_1 I would have to answer in terms of probabilities. I can measure the time to go from x_1 to x_2 by putting counters there or by measuring how far the y-particle has moved; either course involves experimental uncertainties, but the quantity to be measured has been defined exactly in a calculation which, though it distinguishes early from late, makes no reference to any time coordinate.[20]

"Time", reads the epigraph to this paper, "is defined so that motion looks simple". My quantum time is registered simply, in terms of a distance, and I have to assume that we know what a distance is. If it becomes possible to define distance as well as time by quantum correlations, the problem of fitting a quantum description of a microphysical process into a classical description of the apparatus by which it is produced will have been solved, and when we have become used to thinking of phenomena in this new way our view of physics will be profoundly changed.

References

1. *Handbuch der Physik* (Springer, Berlin, 1933), 2nd ed., p. 140; *General Principles of Quantum Mechanics*, tr. P. Achuthan and K. Venkatesan (Springer-Verlag, N.Y., 1980), p. 63.
2. See also L. Susskind and J. Glogower, *Physics (N.Y.)* **1**, 49 (1964); P. Carruthers and M. M. Nieto, *Revs. Modern Phys.* **40**, 411 (1968).
3. Three papers by F. Engelmann and E. Fick: *Suppl. Nuovo Cim.* **12**, 63 (1959); *Z. Physik* **175**, 271 (1963); **178**, 551 (1964) discuss clocks to which the theorem of ref. 1 does not apply.
4. For a discussion of this point see N. Bohr, *Atomic Theory and the Description of Nature* (Cambridge University Press, 1935), p. 98. See also G. F. Chew, *Science Progress* **51**, 529 (1963).
5. E. G. C. Stueckelberg, *Helv. Phys. Acta* **14**, 322, 588 (1941); **15**, 23 (1942).
6. R. P. Feynman, *Phys. Rev.* **76**, 749 (1949).
7. R. P. Feynman, *Phys. Rev.* **80**, 440 (1950); **84**, 108 (1951).
8. Y. Aharonov and D. Bohm, *Phys. Rev.* **115**, 495 (1959); W. Ehrenberg and R. E. Siday, *Proc. Phys. Soc. (London)* **862**, 8 (1949).
9. B. S. DeWitt, *Phys. Rev.* **125**, 2189 (1962).
10. W. Pauli, *Relativitätstheorie* (Teubner, Leipzig, 1921), or *Theory of Relativity*, tr. G. Field (Pergamon, New York, 1958), Sec. 19.
11. R. E. Collins and J. R. Fanchi, *Nuovo Cim.* **48A**, 314 (1978).
12. J. Schwinger, *Phys. Rev.* **82**, 664 (1951).
13. F. Catara, M. Consoli and E. Eberle, *Nuovo Cim.* **70B**, 45 (1982).
14. J. W. Gibbs, *Elementary Principles in Statistical Mechanics* (Yale University Press, New Haven, 1902), p. 151.
15. J. A. Wheeler, in *Mathematical Foundations of Quantum Theory*, ed. A. R. Marlow (Academic Press, N.Y., 1978); also in *Problems in the Foundations of Physics*, ed. G. Toraldo di Francia (Varenna Course 72, North-Holland Pub. Co., N.Y., 1979).
16. P. Carruthers and M. M. Nieto, *Phys. Rev. Lett.* **14**, 387 (1965).
17. H. Salecker and E. P. Wigner, *Phys. Rev.* **109**, 571 (1958).
18. A. Peres, *Am. J. Phys.* **48**, 552 (1980).
19. D. Page and W. K. Wootters, *Phys. Rev.* **D27**, 2885 (1983); W. K. Wootters, *Found. Phys.*, to appear.

DAVID PARK

20. There is considerable literature on the possibility of interpreting the energy derivative
    of the scattering phase as a time delay, most of it stemming from the Princeton
    dissertation of L. Eisenbug in 1948. See E. P. Wigner, *Phys. Rev.* **98**, 145 (1955), and
    for modern work M. Büttiker, *Phys. Rev.* *B27*, 6178 (1983), and E. Pollak and W. H.
    Miller, *Phys. Rev. Lett.* **53**, 115 (1984). Here I have tried to construct a mechanism
    which provides its own interpretation.

IS SPACETIME A
BOOKKEEPING DEVICE FOR
QUANTUM CORRELATIONS?

William K. Wootters

Department of Physics and Astronomy, Williams College, Williamstown,
Massachusetts 01267

In this paper we consider the question whether the variables position *and* time *can be eliminated from the description of a system of particles, and replaced by correlations among internal variables such as spin. A specific proposal is given for effecting this reduction: View the state of the* internal *variables of a system of* N *localized particles from the vantage point of a specific location in spacetime; then translate the vantage point in both space and time, and superpose all the states thereby obtained. The result is a highly correlated state in the Hilbert space of the internal variables. By examining the correlations one can, at least in a simple example, reconstruct to some extent the original spacetime picture.*

I. INTRODUCTION

Nowadays it is not uncommon for physicists to speak of a wavefunction of the universe which does not depend on time:[1] Every instantaneous slice of every possible history of the universe is included as one "component" of a single timeless wavefunction. This idea was developed in the context of quantum gravity,[1-10] but in fact the possibility of condensing an entire history into one static wavefunction is a general property of quantum mechanics and has nothing particularly to do with gravity.[11,12] In the present paper, we present a few simple examples to illustrate how temporal information can be contained in a static wavefunction. We then ask the following question: If it is possible to describe temporal evolution by means of a state which does not depend on time, is it also possible to describe spatial relationships among particles by means of a state which does not depend on any spatial variables? We conclude that at least to some extent it is possible to do this. In particular, we consider a very simple example and a particular interpretation in which spatial information can be extracted from a state involving explicitly only spin variables.

II. TEMPORAL INFORMATION ENCODED IN QUANTUM CORRELATIONS

The simplest model universe in which one can see evolution within a stationary state is one consisting of two identical spin-$\frac{1}{2}$ particles in a uniform vertical (i.e., parallel to the z-axis) magnetic field. If the state of the spins of these particles at a given time is $|\psi\rangle = |+x\rangle|+x\rangle$, where $|+x\rangle$ is the state in which the spin is "pointing" in the positive x-direction, then there will be a genuine evolution: the spins will precess together around the vertical axis. On the other hand, if the state of the spins is initially

$$|\psi_s\rangle = \frac{1}{\sqrt{2}}(|\uparrow\rangle|\downarrow\rangle + |\downarrow\rangle|\uparrow\rangle), \tag{1}$$

then there will be no evolution in the usual sense, because the state is an eigenstate of the Hamiltonian.

Nevertheless, there is another sense in which one does find evolution in the stationary state $|\psi_s\rangle$. In real life, what do we mean when we say something is evolving? We mean that for different readings of a clock, the system of interest is in different states, or, in other words, that there is a correlation between the state of the clock and the state of the system. Such a correlation does exist in the above state $|\psi_s\rangle$, if we regard one of the particles as a clock. Let the spin states $|+x\rangle, |-y\rangle, |-x\rangle$ and $|+y\rangle$ of this "clock particle" be thought of as clock pointer positions labelled 12:00, 3:00, 6:00 and 9:00 respectively. (If the particle were ever in one of these states, it would progress through the rest of them in cyclical order and in equal intervals of time.) We can then ask questions such as this: if at 12:00 a measurement is made on the *other* particle to see whether its spin is in the $+x$ or $-x$ direction, what is the probability that it will be found in the $+x$ direction? That is, if the clock particle has been found pointing in the $+x$ direction, what is the probability that the other particle will also be found pointing in that direction? The answer is

$$\text{prob}(+x \mid +x_{\text{clock}}) = \frac{\text{prob (both in } +x \text{ direction)}}{\text{prob (clock in } +x \text{ direction)}}$$

$$= \frac{|(\langle +x|\langle +x|)|\psi_s\rangle|^2}{|(\langle +x|\langle +x|)|\psi_s\rangle|^2 + |(\langle -x|\langle +x|)|\psi_s\rangle|^2}$$

$$= \tfrac{1}{2}/\tfrac{1}{2} = 1. \tag{2}$$

Thus, at 12:00 the particle has a 100% chance of being found in the $+x$ direction, if the measurement "$+x$ vs. $-x$" is made. In a similar way one can show that at 3:00 the particle has a 100% chance of pointing in the $-y$ direction if the measurement "$+y$ vs. $-y$" is made, and so on. It is therefore reasonable to say that this particle is precessing with respect to clock time, where "clock time" is measured by the clock particle. This is the sense in which one can have evolution within a stationary state.

Clearly the state $|\psi_s\rangle$ is a special state whose correlations allow the above interpretation. Other stationary states, e.g., $|\uparrow\rangle|\downarrow\rangle$, have no correlations at all between the two spins. One can construct the special state $|\psi_s\rangle$ (and this is how I knew to choose it in the first place) by integrating an actually precessing state over time:

$$|\psi_s\rangle \propto \lim_{T\to\infty} \frac{1}{T} \int_{-T}^{T} e^{-iHt}|+x\rangle|+x\rangle \, dt, \tag{3}$$

where H is the Hamiltonian for the two particles in the vertical magnetic field. Thus, $|\psi_s\rangle$ is a superposition of all the instantaneous states of a pair of particles precessing together. It is therefore not surprising that it exhibits such strong correlations between the two particles.

In a similar way one can start with any evolving state of any system and "average" the state over all time. The resulting state is guaranteed to be stationary, and it will be a state in which the various subsystems are highly correlated. By examining the correlations one can often reconstruct the original evolution.

As another example of this construction consider a system of N particles each having only one coordinate and no internal degrees of freedom. The wavefunction of this system will be a function $\psi(x_1, \ldots, x_N)$, where x_j is the coordinate of the jth particle. Let us assume that the evolution of this system is governed by the unrealistic Hamiltonian

$$H - -i \sum_{j=1}^{N} v_j \frac{\partial}{\partial x_j}, \tag{4}$$

which says essentially that the jth particle always moves with velocity v_j. The state

$$\psi(x_1, \ldots, x_N) = \prod_{j=1}^{N} \delta(x_j) \tag{5}$$

is not a stationary state. Each particle will immediately start moving away from the origin with its particular velocity:

$$\psi(x_1, \ldots, x_N, t) = \prod_{j=1}^{N} \delta(x_j - v_j t). \tag{6}$$

On the other hand, the "averaged" state,

$$\psi_s(x_1, \ldots, x_N) \propto \int_{-\infty}^{\infty} \prod_{j=1}^{N} \delta(x_j - v_j t)\, dt$$

$$\propto \prod_{j=1}^{N-1} \delta\left(x_j - v_j \frac{x_N}{v_N}\right), \tag{7}$$

is stationary, and yet it contains correlations from which the original evolution can be reconstructed. Once the coordinate of one of the particles is specified, the wavefunction ψ_s determines the coordinates of the others: there is only one value for each of these other coordinates for which ψ_s is non-zero. Thus if we regard one of the particles as a clock—the hours are marked off at evenly spaced positions along its trajectory—then the stationary wavefunction ψ_s tells us how the other particles are evolving with respect to clock time.

The only thing ψ_s does not tell us is whether the clock particle really is moving uniformly with respect to "real time". But this is something one can never test experimentally. How do we know that all our clocks are running uniformly with respect to "real time"? We don't. All we know is that they are running uniformly with respect to each other. Thus the only information which was present in the original evolving state and which is absent in the stationary state ψ_s is information which is not accessible to us through observations.

In general, the correlations one obtains by averaging an evolving state over time will not necessarily be as perfect as they have been in the two above examples. Nevertheless, the point made in the preceding paragraph will always apply. One cannot observe "time"; one can only observe clocks. So, as far as observations are concerned, no information is ever lost if one replaces an evolving state by a well-chosen stationary state.[11]

One might argue that there is one thing that is always lost, namely, the *order* of events. In the case of the two spin-$\frac{1}{2}$ particles, for example, nothing in the stationary state $|\psi_s\rangle$ tells us that the clock particle proceeds from 12:00 to 3:00 to 6:00, etc.; it could just as well go from 12:00 to 6:00 to 3:00 to 9:00. If all one has to work with is a timeless wavefunction, how would one ever extract the answer to a question which depends crucially on the ordering of events? Such a question is this: If at 1:00 I measure the x-component of spin of a certain spin-$\frac{1}{2}$ particle and obtain the result $+\frac{1}{2}$, and at 2:00 I measure the y-component and obtain the result $+\frac{1}{2}$, then what is the probability that at 3:00 I will find the x-component to be $-\frac{1}{2}$? The answer to this question depends on whether 1:00 comes before or after 2:00, not to mention 3:00, and this is a matter on which the stationary state $|\psi_s\rangle$ seems to have nothing to say.

To resolve this problem, let us ask how in real life we determine the order of events. One way I can know that event B came after event A is to note that I remember remembering event A at the time when event B occurred. Or, if I don't trust my memory, I can keep a record on paper of the order of events. Then a question such as the one posed above, whose answer depends on the order of events, can be recast in terms of the state of the record: "If the results '$S_x = +\frac{1}{2}$' and '$S_y = +\frac{1}{2}$' are the only ones pertaining to the given particle which appear on the record, and if '$S_x = +\frac{1}{2}$' is higher on the list (i.e., that measurement happened first) then what is the probability that the x-component of the particle's spin is $-\frac{1}{2}$?" The stationary wavefunction, which is a function of the record's variables as well as the particle's variables, will give an unambiguous answer to that question.

We conclude that one is free to describe an apparently evolving world in terms of a non-evolving wavefunction. This is an option which is allowed quite generally in quantum mechanics, and one can take it or leave it as one pleases. However, in quantum gravity it is difficult to imagine any description *other than* that given by a single non-evolving wavefunction.

III. SPATIAL INFORMATION ENCODED IN QUANTUM CORRELATIONS

As we have said, the reason we can use a stationary state to describe evolution is that we never actually measure "time"; we only measure clocks. A stationary state can contain as much information about clocks and their correlations as a time-dependent state. We now ask whether we can do something similar with spatial information. We begin by saying, "we never actually measure distances; we only measure . . .". Whatever variables we use to fill in the ellipsis in that statement, those are the variables to which we will assign a quantum state. This state should be a highly correlated state from which information about distances can be extracted.

Our effort here is in the direction of pregeometry.[13] We are asking whether one can do away with "space" as a fundamental element of physics, and replace it with something more basic. To be consistent with this aim, we should choose as our basic variables quantities which are simpler than spatial variables. The simplest variables are binary variables such as the spin components of a spin-$\frac{1}{2}$ particle. Ideally, we would like to construct all of physics out of such variables. Needless to say, this hope will not be fulfilled in the present paper. For one thing, we will be using spin-j particles instead of spin-$\frac{1}{2}$ particles. Moreover, when we speak here of spatial information we will not mean to include the following: (i) information about the curvature of spacetime—here we will work entirely within the flat spacetime of special relativity; (ii) information about the spatial dependence of the wavefunction of a quantum particle. Rather, we will consider here only the spatial and temporal relationships among very massive objects whose trajectories can be treated classically. In the example we explore below, we endow each of these objects with a quantum mechanical spin. In the quantum state of the *spins* will be encoded spatial and temporal information about the objects themselves. Thus we are tentatively suggesting that the incomplete sentence in the preceding paragraph should be completed to read, "we never actually measure distances, we only measure spin". Unlike the analogous statement about time and clocks, this statement about distances and spins is not at all self-evident, and I would not want to stake my life on its validity. Our use of spin here is for the purpose of illustration. The essential feature of the following example is that spatial information is encoded in the correlations among simpler, non-spatial variables.

In the preceding section, we transformed an evolving state into a highly correlated "timeless" state as follows: we considered the evolving state from the point of view of what can be called an "instantaneous observer"; this gave us the state of the system at a particular time; we then superposed all these instantaneous views of the system to create a single state in which the points of view of all observers were equally represented. We will follow an analogous procedure here, but now an "observer" is specified by giving not only his location in time but also his location in space. We will take "his view" of the system to be the state of the system on the observer's past light cone. (The system will consist of a number of particles with spin, and when we speak of its state on the observer's past light cone, we mean just the state of the spins.) We will then superpose the views of all possible observers. The result of this superposition will be a highly correlated state of the spins. It does not depend on any spatial variables. Nevertheless, because it "contains" the points of view of observers at all positions and times, one can hope to reconstruct from this state the original spacetime picture which produced it. An analogous sort of reconstruction is done when we deduce the shape of a three-dimensional object from a set of photographs taken from different angles. Perhaps a better analogy is that of a hologram, which is a *single* photograph in which the three-dimensional information is more intricately encoded.

We will take as our example one of the simplest possible systems for which this procedure does not yield a trivial result. The system consists of two spin-j particles in one spatial dimension. Let the one dimension be labelled by the coordinate z, and let us picture it as vertical. In the frame in which we will analyze the problem the particles are at rest at the positions $z = +L/2$ and $z = -L/2$. There is also a uniform vertical magnetic field. At $t = 0$, the spin of each particle is pointing in the positive x-direction; that is, it is in an eigenstate of J_x with eigenvalue j. Because of the magnetic field, the two spins will precess together around the z-axis. (The particles have identical magnetic moments.)

That is the picture of what is happening in space and time. We now consider how the scene looks to various observers. An observer is specified by giving his location on the z-axis and his location in time. By assumption, the state which a particular observer sees is the state of the two spins when the particles are on the observer's past light cone. Thus each observer will see each particle pointing in some direction in the x–y plane, but the directions will be different depending on the observer's position in spacetime. If the observer's spatial coordinate is greater than

$L/2$, so that he is closer to the upper particle, he will see the lower particle lagging the upper particle by an angle ωL, where ω is the particles' precession frequency. (Here and in what follows we take the speed of light to be unity.) If the observer's spatial coordinate is less than $-L/2$, he will see the upper particle lagging the lower one by the same angle.

We now superpose the points of view of all observers to obtain our single correlated state $|\psi\rangle$, defined as follows:

$$|\psi\rangle \propto \lim_{Z \to \infty} \lim_{T \to \infty} \frac{1}{ZT} \int_{-Z}^{Z} dz \int_{-T}^{T} dt \, |z, t\rangle, \tag{8}$$

where $|z, t\rangle$ is the state of the two particles as seen by the observer at the spacetime point (z, t). The state $|z, t\rangle$ is written down explicitly below. We take the correlated state $|\psi\rangle$ to be *the* description of the system. The above spacetime picture we regard as merely a heuristic device which helped us to choose an interesting state $|\psi\rangle$. Ideally it should be possible to be presented with nothing but the state $|\psi\rangle$ and to determine whether it describes a history in space and time and, if so, what history.

When the limit in Eq. (8) is taken, the contribution from observers whose spatial coordinates lie between those of the two particles vanishes, because there are so many more observers either above or below both particles. We can do the integration by first expressing the state $|z, t\rangle$ in terms of eigenstates $|m\rangle$ of J_z ($J_z|m\rangle = m|m\rangle$):

$$|z, t\rangle = \begin{cases} |t - (z - L/2)\rangle_1 |t - (z + L/2)\rangle_2 & \text{if } z > L/2 \\ |t + (z - L/2)\rangle_1 |t + (z + L/2)\rangle_2 & \text{if } z < -L/2 \end{cases} \tag{9}$$

where

$$|\tau\rangle = 2^{-j} \sum_{m=-j}^{j} \left(\frac{2j}{m+j} \right)^{1/2} e^{-im\omega\tau} |m\rangle, \tag{10}$$

and the subscripts 1 and 2 refer to the particles at $z = L/2$ and $z = -L/2$ respectively. The result of the integration, up to a constant factor, is

$$|\psi\rangle \propto \sum_{m=-j}^{j} \cos(m\omega L) \left(\frac{2j}{m+j} \right) |m\rangle_1 |-m\rangle_2. \tag{11}$$

This is our state. Suppose now that we are presented with this mathematical expression and are not told from what spacetime picture it was constructed. Can we reconstruct the original picture?

There are many probabilities that can be computed from $|\psi\rangle$: the probability of finding particle #1 in the state $|m\rangle$; the probability of finding particle #1 in the state $|m\rangle$ given that particle #2 is pointing in the $+x$ direction, etc. If we were to start calculating such probabilities, we might eventually stumble upon a certain interesting coincidence, which is the answer to the following question: What is the probability of finding particle #2 in the state $|J_\phi = j\rangle$ given that particle #1 is found in the state $|J_x = j\rangle$? (Here J_ϕ is the component of angular momentum in the direction $(\cos\phi)\hat{x} + (\sin\phi)\hat{y}$, and the two states are eigenstates of the indicated operators.) In other words, what is the probability of finding particle #2 pointing in the direction $(\cos\phi)\hat{x} + (\sin\phi)\hat{y}$ given that particle #1 is found pointing in the direction $\hat{x}$? This conditional probability is computed as in Eq. (2). The result is

$$\text{prob}\,(J_\phi^{(2)} = j \,|\, J_x^{(1)} = j) = A \left| \sum_{m=-j}^{j} (e^{-im(\phi + \omega L)} + e^{-im(\phi - \omega L)}) \binom{2j}{m+j}^2 \right|^2,$$

(12)

where A is a constant independent of ϕ. For large j, this conditional probability is very small unless ϕ is very close either to $+\omega L$ or to $-\omega L$. Indeed, a plot of this probability as a function of ϕ has exactly two peaks, centered at $\phi = \pm\omega L$ and each having a width of order $\omega L/j$. A similar result obtains when particle #1 is found to be pointing not in the x direction but in some other horizontal direction. Again, particle #2 is most likely to be found pointing in a direction which makes an angle $\pm\omega L$ with the direction of particle #1.

By exploring the probabilities in this way, one finds that neither particle has a preferred horizontal direction, but that there is a strong tendency for them to have horizontal spin directions separated by an angle ωL. If one of the particles were always "ahead" of the other, we could interpret the state as describing the simple precession of one particle relative to the "clock time" kept by the other, as in Section II, with no need to introduce spatial concepts. But in fact there is an ambiguity as to which particle is ahead. A simple way of interpreting it is to say that the lack of synchronization comes from different light-travel-times to the observer, and that the ambiguity arises from the possibility of seeing the particle from different perspectives. If we adopt this strategy, we are led back to the original picture. We can even find a relation between the precession frequency and the particles' separation, namely, that the product of these two quantities is the ωL which appears in the expression given to us in Eq. (11).

Of course there are many alternative interpretations of the state in Eq. (11). If we did not know about space, we would certainly be able to find a non-spatial interpretation; e.g., we could say that the particles were constantly precessing except that at a certain instant the particle that was lagging suddenly jumped ahead of the other one. In no sense are we forced into a spatial interpretation. It seems likely, however, that for more complicated systems, in which the correlations can be more complex, a special state constructed as above may lead almost uniquely to a spatiotemporal picture. It could happen that the conditional probabilities inherent in such a state are very handily summarized by a picture of particles moving in space, so handily that one could almost be said to be forced to adopt that picture. Thus it is conceivable that intricate correlations among spins are indeed the correct pregeometry, i.e. the non-spatial "building material" out of which spatial and temporal structure is built.

On the other hand, there are a number of problems with this approach, including the following: (1) Spins do not always behave as conveniently as the ones in the above example do. One can invent other examples in which the particles follow non-trivial trajectories and yet the spins never change, so that all observers see the same spin state. In such a case the spin state constructed as above would contain no spatial information at all. (2) One would like to demand of a pregeometry that it provide some understanding of why the world has three spatial dimensions (or more precisely, three spatial dimensions which extend over an appreciable distance). The above ideas, taken by themselves, do not provide such an understanding. Even if the world had four spatial dimensions, one could still follow the procedure of "averaging" over observers, and one could presumably reconstruct the original picture from the resulting spin state. (3) As we have said, it is only a very special state that can be interpreted in terms of a spacetime picture. Therefore, even if the correlated spin state is the correct description of the world, one is still left with the question, why is the state of the universe one of those special states that allow a spacetime interpretation? (A similar question arises even in the case considered in Section II, where only temporal information is encoded in quantum correlations. In that case one finds that a *typical* state of the universe, even a typical stationary solution of the equation of motion, does not yield the kind of time that our universe has, namely, a time which is shared by many clocks.[12])

For these reasons we conclude that although it seems possible for a special correlated state of spins (or of other discrete variables) to have as its most natural interpretation a picture of events in spacetime, this possibility by itself does not argue strongly that such a state is the correct pregeometry.

If the possibility discussed here—that spatial information is to be extracted from correlations among simpler non-spatial variables—were in fact correct, this would have an interesting consequence for the interpretation of quantum mechanics. Consider, for example, a system of N two-orthogonal-state systems. Their Hilbert space has 2^N dimensions. If N is large, say 10^{23}, then the number of dimensions per particle would be so large that we could for many purposes regard it as infinite. This means that it would be possible in principle to encode, in the 2^N-dimensional state vector, enough information to determine a whole *wavefunction* for each particle. Thus, spatial quantum mechanics could conceivably come out of the quantum mechanics of spin. The encoded state could even have spatial correlations among two or three or one hundred of the particles—there is enough information for that. But, because of the limited dimension of the Hilbert space, there would be a limit to the number of particles which could have arbitrary correlations. Eventually the truth would come out: each particle really only has two orthogonal states. Such extensive correlations are never used in physics *except* when one considers the often branching state of the universe in the Everett theory.[14] In that theory, which is nothing but pure quantum mechanics with no collapse, vastly different states are superposed. If the sort of state considered here really were the case, such arbitrary superpositions would not be possible. This is the most likely way I can think of in which the Everett theory might possibly be wrong.

References

1. J. B. Hartle and S. W. Hawking, *Phys. Rev.* **D28**, 2960 (1983); J. B. Hartle, *Phys. Rev.* **D29**, 2730 (1984).
2. B. S. DeWitt, *Phys. Rev.* **160**, 1113 (1967).
3. R. F. Baierlein, D. H. Sharp and J. A. Wheeler, *Phys. Rev.* **126**, 1864 (1962).
4. R. Arnowitt, S. Deser and C. W. Misner, in *Gravitation: An Introduction to Current Research*, ed. L. Witten (Wiley, New York, 1962), pp. 227–265.
5. J. A. Wheeler, in *Battelle Rencontres: 1967 Lectures in Mathematics and Physics*, ed. C. M. DeWitt and J. A. Wheeler (Benjamin, New York, 1968), pp. 242–307.
6. A. Peres, *Phys. Rev.* **171**, 1335 (1968).
7. C. W. Misner, *Phys. Rev. Lett.* **22**, 1071 (1969); *Phys. Rev.* **186**, 1319 (1969); **186**, 1328 (1969).
8. K. Kuchař, *J. Math. Phys.* **11**, 3322 (1970); *Phys. Rev.* **D4**, 955 (1971); in *Quantum Gravity 2: A Second Oxford Symposium*, ed. C. J. Isham, R. Penrose and D. W. Sciama (Clarendon, Oxford, 1981), pp. 329–376.

9. J. W. York, *Phys. Rev. Lett.* **26**, 1656 (1971); **28**, 1082 (1972).
10. J. A. Wheeler, in *Problems in the Foundations of Physics*, proceedings of the International School of Physics "Enrico Fermi", course 72, ed. G. Toraldo di Francia (North-Holland, Amsterdam, 1979), pp. 395–497.
11. D. N. Page and W. K. Wootters, *Phys. Rev. D***27**, 2885 (1983).
12. W. K. Wootters, *Int. J. Theor. Phys.* **23** (1984).
13. C. W. Misner, K. S. Thorne and J. A. Wheeler, *Gravitation* (Freeman, San Francisco, 1971).
14. H. Everett III, *Rev. Mod. Phys.* **29**, 454 (1957).

PARTICLE PHYSICS AND THE INTERPRETATION OF QUANTUM THEORY

Max Dresden

Institute for Theoretical Physics, State University of New York at Stony Brook, Stony Brook, New York 11794

It is stressed that some of the most unusual experiments in particle physics all support the probability interpretation of quantum theory. The experiments on K–K̄ mixing and on weak-electromagnetic interferences directly verify the superposition and decomposition of states. A statistical discussion of a minimal quantum model based on just the superposition of states is presented. It is demonstrated that no classical statistical systems can reproduce all the results of these quantum systems.

I. BACKGROUND, GENERAL COMMENTS

Ever since its inception, the interpretation of quantum mechanics has been a difficult and controversial topic. Precisely because of the fabulous success of the quantum formalism and the—to classical conceptions—unfamiliar and even bizarre features, strong personal reactions resulted. It is important to recall that the basic elements of the current interpretation, such as wave functions defining probabilities, the expression of transition probabilities in terms of matrix elements, and the uncertainty relations, were all arrived at after lengthy struggles and drawn-out arguments. To many of the originators of quantum theory (Bohr, in particular), their successful liberation from the classical ideas was a continued source of inspiration in their further investigations in physics.[1] It is therefore not surprising and should not be judged too harshly if their interpretation—very loosely called the "Copenhagen Interpretation"—is sometimes a bit dogmatic and at times overbearing. But it is especially dangerous to adopt an equally dogmatic anti-Copenhagen interpretation and ignore or downgrade the significant successes these very interpretations have achieved.

Most of the articles on the traditional interpretation of quantum theory expressed the expectations (or perhaps even the hope) that the conventional probability interpretations of quantum mechanics would eventually break down. Many of the originators of quantum mechanics, such as Bohr, Pauli, Oppenheimer, Heisenberg, also anticipated (perhaps for different reasons) a breakdown or at least a major alteration in quantum mechanics in a number of areas. Interestingly enough, both critics and advocates of conventional quantum theory expected these major changes in quantum theory to occur in the same fields: in nuclear physics, relativistic quantum theory, high energy phenomena. Of course, their expectations ran in opposite directions; the quantum theorists expected an even further renunciation of classical concepts, an even more extended use of complementarity-type ideas. The critics expected and certainly hoped for a return to the trusted notions of classical physics, to causality and continuity with no reference to singularities, discontinuities, or probabilities. It is interesting and most surprising that the anticipated breakdown did not occur when physics moved into these new fields. Quite the contrary; the traditional formalism, technically somewhat refined, combined with the usual probability interpretation as in the formulation given by Feynman, scored some of its most impressive successes in the very areas in which it was supposed to fail. The extraordinary effectiveness of the quantum formalism and the probability interpretation in describing such diverse areas as neutron interferometry and $K-\bar{K}$ interference[2] does, of course, not prove that this interpretation is the only one or even the right one, but it puts severe constraints on the construction of competing schemes which legitimately must be expected to do equally well in those many fields. It remains a remarkable fact that even in quantum field theory, probably the most daring and possibly the most doubtful extension of the quantum theoretic notions, the basic structure and interpretation of quantum theory remains intact. The success of quantum field theory in quantum electrodynamics and recently in the electroweak Weinberg–Salam theory, culminating in the discovery of the W and Z particles, is such that the formalism and the interpretation of quantum field theory must be taken extremely seriously. Of course, this in no way means that any interpretation should be accepted without question, especially one as unintuitive and unusual as quantum theory. But it does mean that possible reinterpretations and certainly modifications should be considered only with the full knowledge and understanding of the accomplishments and power of the existing scheme.

The above experiments all show the great effectiveness of the probability interpretation of quantum theory. In all these experiments, the crucial feature is the superposition of states and, equally important, the decomposition of a state into component states. The assignment of probability amplitudes to transitions and their additivity emerges as the basic interpretative feature of quantum theory. This is formalized in Feynman's approach to quantum theory in which the path integral defining this transition amplitude in terms of the Lagrangian of the system is the most important ingredient. As could be expected, the various additional requirements such as relativistic invariance and gauge invariance are incorporated in the theory by imposing appropriate conditions on the Feynman path integral. As such, these physical principles are subordinated to the requirement of the additivity of the amplitudes, which remains the central element of all quantum theories in relativistic quantum theory as well as in field theory. The relativistic requirements are adapted to the additivity of the amplitudes. It is of interest that apart from the interpretive importance of superposition ideas in particle physics, recent experiments in neutron interferometry have provided direct experimental evidence for the superposition of Fermion spin amplitudes.[5] The importance of these experiments for the interpretation of quantum theory has especially been stressed by Greenberger.[6]

II. CONSTRUCTION OF A MINIMAL QUANTUM THEORY: HOW CLOSE A CLASSICAL COUNTERPART?

In spite of these unquestionably impressive achievements of quantum theory, it remains possible to inquire about the possibility of alterations or modifications which would retain the successes, have a coherent mathematical structure, have the same predictive power, but perhaps have a more intuitively congenial interpretation than quantum theory. To investigate such questions on a *concrete* level, it is of some interest to construct a "minimal quantum theory". Such a theory should contain the superposition principle and the mixing of states as a crucial ingredient and as little else as possible. It is clearly desirable to make the model simple enough to allow complete mathematical control of the system. Ideally, it would be desirable that the system be exactly solvable. The basic issue for the interpretation of quantum theory would then be whether or not it is possible to construct a classical model system which would reproduce all the observable results of this quantum system. Since the only genuine quantum features of the quantum system are the superposition principle and the mixing of states, these characteristics are specifically excluded from the *classical* model system. But the uses of classical probability notions, as in classical statistical mechanics, ensemble theory or the introduction of additional (hidden) variables are perfectly acceptable in the "classical" description of this system. Occasionally, the opinion is expressed that quantum features are intrinsically microscopic in character and that any averaging, smoothing, coarse graining will effectively destroy the characteristic quantum effects. In other words, *macroscopic* states, or macroscopic configurations, could not exhibit quantum properties. In that view, the description of the system via the quantum mechanical density matrix should not add any new physical features, it would merely provide an alternate but equivalent averaging procedure. All the characteristic quantum effects have presumably been obliterated by the averaging processes over microscopic variables and phases which are necessary to obtain the macroscopic description. It was stressed by Greenberger[7] that the typical quantum feature, that the outcome of an experiment depends on what measurement one chooses to make even for macroscopic distances, is, in spite of its unintuitive character, verified by the experiments in neutron interferometry. Even though the systems constructed (or contrived) here are not sufficiently complex to exhibit all the properties of the experimental arrangements of Greenberger,[7] it remains true even in these minimal quantum systems that the large-scale (macroscopic) or averaged properties of the quantum systems are surprisingly and systematically different from those of the classical systems.

The systems to be considered† are schematic and highly idealized versions of impurity scattering. The objects considered (particles, balls,

---

† These same systems, in a different context with different emphasis, are described in detail in references 8 and 9.

spins) are capable of just two states or two configurations (such as spin orientations, colors, strangeness). The objects in question move uniformly along a one-dimensional, discrete chain. The time variable is also taken as discrete, so all the basic variables are discrete. (This, of course, by itself does *not* make the model a quantum model!) The basic dynamical mechanism provides for a transition from one of the states to another. These transitions could be caused by impurities or local fields; the cause is not important. However, the interactions producing the changes of configuration are precisely localized. The lattice points along the chain are numbered by $p = 1, 2, \ldots, n$. Associate with each point a variable ε_p, which can be $+1$ or -1. $\varepsilon_p = 1$ indicates a point where no change of state occurs, while $\varepsilon_p = -1$ indicates a location where a change of configuration does occur. Clearly, the infinite set $\{\varepsilon_p\}$ specifies the complete dynamics of the system. (One could think of the locations where $\varepsilon_p = -1$, as magnetic impurities which would change the spin of a neutron as it moves past that impurity). The configuration of the system as a whole is described by the set of variables $\eta_p(t)$, $p = 1 \cdots n$. Each $\eta(t)$ can assume just two values, $\eta_p(t) = +1$, for one configuration of the object at p at time t, $\eta_p(t) = -1$ for the other configuration. ($\eta_p(t) = +1$ could denote a spin up or a color or any dichotomic property of the objects.) It is assumed that there is one object at each location p so that the configuration at time t is described by the set $\{\eta_p(t)\}$. Each successive time interval the objects move one lattice distance with or without changing their state depending on whether or not the point left is the location of an impurity or not. A measurement of *all* $\eta_p(t)$ would give complete information about the physical configuration of the system. This might be the result of a detailed microscopic experiment. To simulate the macroscopic features of the system, it is natural to assume that the locations of the impurities are only known in a probability sense. Thus, in general, it will not be known whether $\varepsilon_p = -1$, but just the probability that $\varepsilon_p = -1$ will be known. Macroscopic results will then be defined as averages over the impurity locations. If Q is a dynamical variable which, in general, will depend on the ε variables; the observed average value of Q, written as $\langle Q \rangle_\varepsilon$, is defined by

$$\langle Q \rangle_\varepsilon = \sum_{\varepsilon_1} \cdots \sum_{\varepsilon_n} \mathrm{Prob}(\varepsilon_1 \cdots \varepsilon_n) Q(\varepsilon_1 \cdots \varepsilon_n). \tag{1}$$

It should be stressed that this probability is a classical statistical probability; the averaging corresponds to an averaging over a set of impurities whose location is only incompletely known or known only in an average sense. The definition (1) has nothing to do with quantum interpretations, it just defines the manner in which macroscopic results are to be obtained from microscopic information. In particular, the averaging procedure is logically unrelated to the nature of the dynamics which describes the change of the state of the objects. The averaging process is further specified by the stipulations:

$$\text{Prob}(\varepsilon_j = -1) = \mu$$

$$\text{Prob}(\varepsilon_j = +1) = 1 - \mu. \tag{2}$$

Roughly speaking, if there are m impurities in a chain of length n, the ratio $\mu = m/n$ is a measure of the probability that an arbitrarily chosen lattice point shall contain an impurity. This is exactly what (2) expresses. Since the presence of an impurity at location p is unrelated and unaffected by the presence of an impurity at q, one has

$$\text{Prob}(\varepsilon_j \varepsilon_k) = (\text{Prob } \varepsilon_j)(\text{Prob } \varepsilon_k). \tag{3}$$

The formulae (1)–(3) define the statistical ensemble which specifies the macroscopic or statistical properties of the system. It must be reiterated again that this averaging process merely defines the macroscopic properties of the system; it has nothing to do with the underlying mechanics. The observables in this system are the ε or macroscopic averages of the one particle distributions, the correlation functions, etc. It is in the stipulation of the dynamics that the classical or quantum mechanical nature of the system shows up. Since the only interactions occur at the impurity sites, it is only necessary to specify the "interaction vertex" at the location of an impurity. Since the only effect of that interaction is the change of the configurations of the moving objects, it is only necessary to define the precise nature of that change. It is convenient for that purpose to describe these different states as *colors*. (B and W, blue and white.) A *classical* vertex has the effect of always changing the color with certainty:

$$\xrightarrow{\text{W}} \times \xrightarrow{\text{B}} \qquad \xrightarrow{\text{B}} \times \xrightarrow{\text{W}}$$

If $\eta_p(t) = 1$ for color W, $\eta_p(t) = -1$ for color B, the classical dynamics is completely contained in

$$\eta_{p+1}(t + 1) = \varepsilon_p \eta_p(t). \tag{4}$$

For a given set of ε_p, this is a completely deterministic dynamics. The statistical averages are obtained via an analysis of the Liouville equations which itself is still a deterministic equation describing the change of time of the ensemble density function $f^{(n)}$

$$f^{(n)}(\eta_1, \eta_2, \ldots \eta_n, t + 1) = f^{(n)}(\varepsilon_1 \eta_2, \varepsilon_2 \eta_3, \ldots \varepsilon_n \eta_1, t). \tag{5}$$

From the $f^{(n)}$, the contracted distributions are obtained in the usual manner

$$f_j^{(1)}(\alpha, t) = \sum_{\eta_1} \cdots \sum_{\eta_n} f^{(n)}(\eta_1, \eta_2, \ldots \eta_j = \alpha, \ldots \eta_n, t) \tag{6a}$$

$$f_{j,k}^{(2)}(\alpha, \beta, t) = \sum_{\eta_1} \cdots \sum_{\eta_n} f^{(n)}(\eta_1 \cdots \eta_j = \alpha, \ldots \eta_k = \beta, \ldots t) \tag{6b}$$

The quantities of macroscopic interest are the averages over ε, of $f^{(1)}$ and $f^{(2)}$. Thus $\langle f_j^{(1)} \rangle_\varepsilon$ is the classical macroscopic one particle distributions, $\langle f_{j,k}^{(2)} \rangle_\varepsilon$ is the classical correlation function. Both are, in principle, observable. The properties of this system, analyzed in detail in references 8 and 9, are neither obvious nor trivial. They are based directly and exclusively on the deterministic, mechanical equations (4). The Liouville equation (5) is a direct consequence of that equation, while the averaging over the impurities is a separate process. Schematically, the procedure can be outlined as follows:

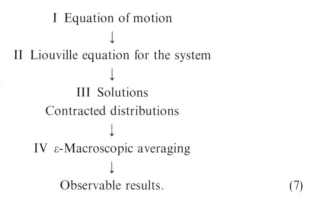

I Equation of motion

↓

II Liouville equation for the system

↓

III Solutions
Contracted distributions

↓

IV ε-Macroscopic averaging

↓

Observable results.   (7)

It is evident that this is a classical deterministic scheme; the microscopic variables η (the colors) are independently measurable and each measurement gives a unique, repeatable answer. Only the last step (IV) contains probability features, but these are classical, not quantum, probabilities.

The main problem now is to construct a parallel scheme where the underlying dynamics is quantum mechanics instead of classical mechanics. The rest of the scheme remains identical except for the obvious modifications, that the Liouville equation becomes the equation for the density matrix which has the usual quantum mechanical interpretation. (Its diagonal elements are probabilities.) The rest of the scheme (7) remains the same; in particular, the final step—the ε-macroscopic averaging—is identical in the two schemes. It was stressed repeatedly in Sections I and II that the single quantum feature, verified most often in particle physics experiments, is the mixing and superposition of states. To incorporate this feature in this dynamics, it is necessary to require that the action of an impurity on the state of an object will produce a *superposition* of colorstates from an initial eigenstate. Thus, a vertex should not change a W state in a B state (or conversely) but it should produce a superposition of W and B states from an initially pure W or B state. From an initially mixed state, the interaction will produce another but definite mixture. Stated succinctly, the interaction operator cannot be diagonal in the color representation. It is, of course, suggestive to use the usual two-component spinors to describe the color states. Thus, the W and B states are represented by spinors:

$$|W\rangle = \begin{pmatrix} 1 \\ 0 \end{pmatrix} \qquad |B\rangle = \begin{pmatrix} 0 \\ 1 \end{pmatrix}. \tag{8}$$

A general superposition (a general color state) is

$$|\eta\rangle = c_1|W\rangle + c_2|B\rangle \qquad |c_1|^2 + |c_2|^2 = 1. \tag{9}$$

The basic action of an impurity on the color eigenstates can then be represented by the diagrams as follows:

$$\xrightarrow{|W\rangle} \times \xrightarrow{c_1|W\rangle + c_2|B\rangle} \qquad \xrightarrow{|B\rangle} \times \xrightarrow{c_1'|W\rangle + c_2'|B\rangle}. \tag{10}$$

The coefficients c_1, c_1' are, by the usual quantum mechanical interpretation, the probabilities that, after the interactions, the object will be in a prescribed state. Thus $|c_1|^2$ gives the probability in (10) that the color will be white, $|c_2|^2$ that it will be blue, but the state itself is a superposition of both; a microscopic measurement could give either answer. In principle, $c_1\, c_2,\, c_1',\, c_2'$ all could depend on the location of the impurities. It also would not be necessary that there be a simple relation between c' and c. (The impurities could affect two different color states differently.) But the essential difference between quantum and classical systems is already evident in a simple situation. Assume that the action of an impurity in a pure color state produces a mixture in such a way that the probability for a *change* (i.e., from W to B or from B to W) is λ, while the probability for no change is $(1 - \lambda)$. Referring to the processes (10), this would mean:

$$|c_1|^2 = 1 - \lambda, \quad |c_2|^2 = \lambda \qquad |c_1'|^2 = \lambda, \quad |c_2'|^2 = 1 - \lambda. \qquad (11)$$

If $\lambda = 0$, this would mean *no* change ever; if $\lambda = 1$, this would imply always a change—in either case, no mixing of eigenstates would occur. Thus, $\lambda \neq 0, \lambda \neq 1$ is essential for the characteristic quantum mixing. To obtain the quantum equation of motion, it is best to introduce an operator E_p, which acts in the two-dimensional color subspace spanned by $|W\rangle$ and $|B\rangle$ (8).

$$E_p = \tfrac{1}{2}(1 + \varepsilon_p)1 + \tfrac{1}{2}(1 - \varepsilon_p)S_p. \qquad (12)$$

Here 1 is the unit matrix, while the stipulation (11), combined with (8), gives an explicit expression for S_p:

$$S_p = \begin{pmatrix} e^{i\alpha}\sqrt{1 - \lambda}, & e^{i\gamma}\sqrt{\lambda} \\ e^{i\beta}\sqrt{\lambda}, & e^{i\delta}\sqrt{1 - \lambda} \end{pmatrix}. \qquad (13)$$

Here $\alpha, \beta, \gamma, \delta$ are real numbers. The time evolution of the system is now determined by:

$$|\eta_{p+1}(t + 1)\rangle = E_p|\eta_p(t)\rangle. \qquad (14)$$

Starting with this equation, "the Schrödinger equation" of the system, the total scheme (7) can be carried out. This quantum theory is indeed a *minimal* quantum theory. The only quantum element contained in it is the continual mixing of states which the interactions with the impurities produces. It should be mentioned that if one requires that S_p, which defines the time, is unitary and that as $\lambda = 0$, S becomes the unit matrix (which is eminently reasonable since $\lambda = 0$ means that nothing happens), then only one *phase* remains in (13), so that it assumes the form:

$$S_p = \begin{pmatrix} \sqrt{1 - \lambda}, & e^{i\gamma}\sqrt{\lambda} \\ -e^{i\gamma}\sqrt{\lambda}, & \sqrt{1 - \lambda} \end{pmatrix}. \qquad (15)$$

With the Eqs. (14), (15) and the scheme (7), the quantum system is completely defined. But it cannot as yet be compared with any classical system. It should be noted that the dynamics, through E_p, depends on the ε variables over which, as in the classical theory, an average has to be carried out. This ε average is exactly the same as in the classical theory. However, the quantum theory, through S, contains a parameter λ, which defines the strength of the mixing or, more intuitively, the likelihood of a change of state. The classical theory considered so far contains no such parameter. Thus, if $\varepsilon_p = -1$, a change must occur, implying $\lambda = +1$, for which the quantum theory becomes trivial.

III. THE IDENTIFICATION OF THE ESSENTIAL QUANTUM FEATURES

It is possible and very instructive to construct a model which has the same dynamics and the same macroscopic averaging as the model just discussed but which is, nonetheless, classical. Again, consider for this purpose objects which can exist in two states, B and W, but these are now *classical* states. In particular, there is no such thing as an $\alpha B + \beta W$ state. These objects, when interacting with an impurity, will change their state with probability λ. Diagrammatically, one can illustrate this dynamics as:

$$\xrightarrow{\text{W}} \times \xrightarrow{\text{B}} \qquad \xrightarrow{\text{W}} \times \xrightarrow{\text{W}} \tag{16}$$

Thus, in this model, it is the *dynamics* which contains stochastic elements. The interpretation is perfectly classical, a state (both before and after interactions) is either a (pure) W state or a (pure) B state, but certainly *not* a superposition. However, which one of these definite states occurs is only known in a probability sense, determined by the parameter λ. This model will be called the *stochastic*, classical model. It is characteristically different from the quantum model. In the quantum model, the effect of the interaction on the state is completely definite; it produces a well-defined unambiguous mixture (given by (14), (12) and (15)) of states. However, the state itself has a probability interpretation. By contrast, the states in the classical, stochastic model have a well-defined unambiguous interpretation—they are either B or W states. However, the outcome of the interaction is not well defined; it only gives the probability for a well-defined state. Consequently, the proper description of the classical stochastic model is via the probability distribution functions. $f_j^{(1)}(\eta, t)$ is the (normalized) probability that the object at location j at time t shall have the color η. Clearly

$$f_j^{(1)}(\eta, t) + f_j^{(1)}(-\eta, t) = 1. \tag{17}$$

One can similarly introduce correlation functions $f_{jk}^{(2)}(\alpha, \beta, t)$, etc. The verbal description of the classical stochastic model leads directly to an equation for the one-particle distribution:

$$f_{j+1}^{(1)}(\alpha, t + 1) = \tfrac{1}{2}(1 + \varepsilon_j) f_j^{(1)}(\alpha, t)$$
$$+ \tfrac{1}{2}(1 - \varepsilon_j)[\lambda f_j^{(1)}(-\alpha, t) + (1 - \lambda) f_j^{(1)}(\alpha, t)]. \tag{18}$$

A similar, somewhat more complicated, equation can be obtained for $f_{jk}^{(2)}$ (see Ref. 9). As was to be expected, these equations all still depend on the ε variables, i.e., they depend on the impurity distributions. Thus, after solving (18), it is still necessary to average the solutions over the impurities. True averaged results are of physical relevance; the true comparison of the physical contents of the quantum model and the classical stochastic model must be carried out after the macroscopic average has been performed. These two models possess the identical interactions, the same (lattice) geometry, the physical results are obtained by the identical averaging process—just the interpretations of the *state* notions are different. Thus, a comparison of the macroscopic results of these two systems, would be most instructive in clarifying the physical differences resulting from difference in the interpretation of the state concept. To facilitate these comparisons it is helpful to summarize the model rules and the notations in the table on the next page. The table provides a convenient format to carry out the comparison between the classical stochastic and the quantum model. To make the results even easier to survey, it is best to assume that both classical and quantum system start from the identical initial state, which is conveniently picked as a state where all spins are up, $\eta_p(0) = 1$, or all balls are in the white (W) eigenstate. For the stochastic classical model, the probability that the object at time t, at location p is white ($\eta = 1$), is given by

$$\langle f_p^{(1)}(1, t) \rangle_\varepsilon = \tfrac{1}{2} + \tfrac{1}{2}(1 - 2\lambda\mu)^t. \tag{19}$$

On the other hand, the quantum model yields for this same probability the 1, 1 diagonal element of the density matrix.

Classical stochastic model

(1) Geometry, chain with impurities. $\varepsilon_p = +1$, no impurity, $\varepsilon_p = -1$. Yes.
(2) $\text{Prob}(\varepsilon) = \frac{1}{2} + \frac{1}{2}(1 - 2\mu)\varepsilon$,
$\text{Prob}(\varepsilon_i \cdot \varepsilon_j) = (\text{Prob } \varepsilon_i) \text{Prob}(\varepsilon_j)$
$\langle Q \rangle_\varepsilon = \sum_{\varepsilon_1} \cdots \sum_{\varepsilon_n} \text{Prob}(\varepsilon_1 \cdots \varepsilon_n)Q$
(3) States B, W, described by values $\eta = -1$, $\eta = \pm 1$ of a single variable. No superposition. No probability notion associated with the state.

(4) Interactions, when $\varepsilon_p = -1$, a state goes over into a state, λ is the probability of the new states being different; $1 - \lambda$, probability to be the same.
(5) Equation (18), for probabilities f.
(6) (4) and (5) imply an equation for the Liouville function $f^{(m)}$.
(7) From (6) the contracted distributions $f^{(1)}$, $f^{(2)}$ can be obtained. They depend on ε.
(8) Calculate the ε average using (2).

Quantum model

(1) Identical.

(2) Identical.
Identical
Identical.

(3) States $|B\rangle$, $|W\rangle$ described by two component objects, $B = \binom{0}{1}$, $W = \binom{1}{0}$. Yes, superposition. $|\eta\rangle = c_1|W\rangle + c_2|B\rangle$. State, amplitudes yield probabilities.
(4) When $\varepsilon_p = -1$, the interaction mixes the states in a prescribed fashion depending on λ.

(5) Equation (14) for the amplitudes $|\eta\rangle$.
(6) (4) and (5) imply an equation for the density matrix $\rho^{(m)}$.
(7) From (6), the reduced density matrix can be obtained. It depends on ε.

(8) Calculate the ε average using (2).

$\rightarrow$ Comparison $\leftarrow$

$$\langle 1|\rho^{(1)}(p, t)|1\rangle_\varepsilon = \frac{1}{2} + \frac{1}{2}\text{Re}[1 - 2\lambda\mu + 2i\mu\sqrt{\lambda - \lambda^2}]^t$$

$$= \frac{1}{2} + \frac{1}{2}R_0^t \cos v_0 t$$

$$R_0^2 = 1 - 4\mu\lambda(1 - \mu) \tag{21a}$$

and

$$\tan v_0 = \frac{2\mu}{1 - 2\mu\lambda}\sqrt{\lambda - \lambda^2}. \tag{21b}$$

It is easy to show that when $0 < \mu < 1$, and $0 < \lambda < 1$ (both required physically), that $|R_0| < 1$. A similar comparison can be made for pair correlation functions (and, indeed, for all the correlation functions, although the calculations become quite involved). The results for the classical two-point functions are:

$$\langle f_{q,p}^{(2)}(1, 1, t)\rangle_\varepsilon = \frac{1}{4} + \frac{1}{2}(1 - 2\lambda\mu)^t$$

$$+ \frac{1}{4}(1 - 2\lambda\mu)^{2t}, \qquad (t \leqslant p - q), \tag{22a}$$

$$\langle f_{q,p}^{(2)}(1, 1, t)\rangle_\varepsilon = \frac{1}{4} + \frac{1}{2}(1 - 2\lambda\mu)^t$$

$$+ \frac{1}{4}(1 - 2\lambda\mu)^{2(p - q)}[1 - 4\mu\lambda(1 - \lambda)]^{q - p + t},$$

$$(t > p - q). \tag{22b}$$

For the quantum model, it is best to calculate the correlation function proper rather than the two-particle density matrix. The definition of the correlation function in question is:

$$X(q, p, t) = \langle \rho_2(q, p, t) \rangle_\varepsilon - \langle \rho_1(q, t) \rangle_\varepsilon \otimes \langle \rho_1(p, t) \rangle_\varepsilon. \tag{23}$$

$\otimes$ is the tensor product. The quantum results can be expressed in various forms but, for the present purposes, the most useful is:

$$\langle X(q, p, t) \rangle = \tfrac{1}{4} R_0^{2(p-q)} - \tfrac{1}{4} R_0^{2t} - \tfrac{1}{4} R_0^{2t} \cos 2 v_0 t$$

$$+ R_0^{2(p-q)} R_1^{t-p+q} \cos[2(p-q) v_0 + (t - p + q) v_1]. \tag{24}$$

(24) is valid for times $t > p - q$. In (24), R_0 and v_0 are the same as in (20) given by (21a) and (21b). The new quantities R_1 and v_1 are also expressible in terms of the parameters μ and λ

$$R_1^2 = 1 - 16 \lambda \mu (1 - \mu)(1 - \lambda) \tag{25a}$$

$$\tan v_1 = \frac{4\mu(1 - 2\lambda)\sqrt{\lambda(1 - \lambda)}}{1 - 8\mu\lambda(1 - \lambda)}. \tag{25b}$$

It is, in principle, possible to calculate the higher-order distribution functions and correlation functions† but, for a first exploration of the interpretation question, the formulae quoted here are sufficient. There are two basic comparisons to be made. One is the comparison of the one-particle distribution function $\langle f_p^{(1)}(1, t) \rangle_\varepsilon$ (given by (19)) with the one-particle density matrix $\langle 1 | \rho^{(1)}(p, t) | 1 \rangle_\varepsilon$ (given by (18)). The other is between the two-particle distribution function $f_{q,p}^{(2)}$ (given by (22)) and the two-particle correlation function X given by (24). For a complete study, all distribution functions $f^{(s)}$ should be compared with all reduced density matrices. A direct inspection of the formulae for the one-particle functions, for stochastic classical and the quantum model shows some similarities: both $\langle f_p^{(1)} \rangle_\varepsilon$ and $\langle \rho_p \rangle_\varepsilon$ are independent of p—this expresses in both cases the translational invariance of the model, thus this was to be expected. Both, in the limit that $t \to \infty$, approach a uniform state. This follows from (20) by observing that $R_0 < 1$. It is also easy to check that, in general (not only for the special initial state chosen), the limit $t \to \infty$ obliterates any dependence on the initial state. However, where the

---

† These results are very far from obvious. They take some considerable calculations starting from the formulae in reference 9.

classical stochastic model approaches equilibrium (the value 1/2) *monotonically*, the quantum approach is oscillatory with frequency v_0 (given by (21b)). This is a striking difference between classical and quantum systems which is a general property of all correlation functions. It should be noted that the oscillatory behavior of ρ_0 contains just one single frequency v_0. The differences between the two point functions $\langle f^{(2)} \rangle_\varepsilon$ of the stochastic model and the correlation functions X of the quantum model are pronounced. The two-point function approaches its expected value $\frac{1}{4}$ as $t \to \infty$, it does so monotonically. On the other hand, the correlation function contains two distinct frequencies, $2v_0$, double the frequency of the $\rho^{(1)}$, and v_1, a distinct frequency given by (25). There is also a phase shift which depends on $(p - q)$. The distinction between the monotonic approach to equilibrium of the classical system and the oscillatory approach of the quantum model occurs again on the $f^{(2)}$ and $\rho^{(2)}$ levels—with additional frequency v_1. But even more striking is that the limit of X at $t \to \infty$ is not zero. (The terms $+\frac{1}{4}R_0^{2(p-q)}$ in (25b) remain.) This, expressed in terms of the density matrix, means that:

$$\lim_{t \to \infty} \langle 1, 1 | \rho_2(q, p, t) | 1, 1 \rangle_\varepsilon = \tfrac{1}{4} + \tfrac{1}{4}[1 - 4\lambda\mu(1 - \mu)]^{p-q}. \quad (26)$$

In other words, there are *persistent* correlations which in fact depend on the initial data.

Thus, the intrinsic quantum features are:

(a) An oscillatory approach to equilibrium for the one particle density matrix, with a fixed frequency v_0.

(b) For the higher correlation functions, an oscillatory approach containing several frequencies to a limiting state.

(c) There are *persistent correlations* in $\langle \rho^{(2)} \rangle_{\varepsilon 1} \langle \rho^{(3)} \rangle_{\varepsilon 1} \cdots$ that is why the limiting state is *not* the equilibrium state. (26) is an example.

(d) There is a memory effect. Apart from $\langle \rho^{(1)} \rangle_\varepsilon$, the higher correlation functions, all retain a memory of the initial configurations.

IV. COMMENTS, CONCLUSIONS

The specific quantum features, (a)–(d), are all direct consequences of the equation (14) for the amplitudes $|\eta\rangle$. This basic equation describes how the interaction mixes the eigenstates. This is what makes the model quantum mechanical. The question of fundamental significance for alternate interpretations is now whether one can construct classical models leading to exactly the same results for the observable probabilities, such as $\langle 1|S_p^{(1)}|1\rangle_c$ and $X(p, q, t)$. One could, in constructing such classical systems, introduce more types of impurities or objects with more than two colors. The physical observables in these classical theorems could then emerge upon averaging over the additional colors or upon averaging over the new type of impurities. One also could give the additional impurity elements different statistical properties. Such procedures would, of course, confer a different microscopic dynamics and different statistical mechanics on the classical substructure from the quantum dynamics. Yet another possibility would be to invoke a classical dynamics where the probability that an object upon interaction retains its color would become *position* dependent. The equation for the classical one-particle distributions (18) then would contain λ_j instead of λ. The "observable" λ then might be taken as:

$$\lambda = \frac{1}{\sum \lambda_j} \sum p_j \lambda_j \tag{27}$$

where p_j is, in turn, a probability.

Many other schemes are conceivable; one might, for example, guess that if the quantum results are averaged over the phase γ left undetermined in the expression for S_p (15), that the classical results would emerge. This, however, is not the case; averaging over γ has no effect at all. The minimum quantum system presented here has actually some maximal properties. It appears to be completely unrelated to any kind of classical system or classical model. Although all these suggestions for the construction of classical systems with additional variable and additional states have a certain plausibility, it has so far been impossible to construct any classical system which reproduces the properties of the quantum model. In particular, it has been impossible to devise a classical dynamics whose macroscopic properties show the *persistent* correlation as $t \to \infty$, the memory effects, the oscillatory approach to equilibrium and the multiple frequencies of the higher correlation functions. It is true that some rather contrived models can reproduce *some* of these features but *none* can reproduce them *all*. Certainly the qualitative relation between v_0, v_1 (see (25) and (21b)), cannot be reproduced. For example, one can investigate a model where the η variables can assume three values—$\alpha_-, \alpha_0, \alpha_+$. The dynamics is described by a transition matrix, a 3×3 extension of the matrix (15). The calculations of $\langle f_j^{(1)}(\eta, t) \rangle_\varepsilon$ and $\langle f_{j,k}^{(2)}(\eta_1, \eta_2, t) \rangle_\varepsilon$ can be carried out for this model.[9] Both $f^{(1)}$ and $f^{(2)}$ approach their respective equilibrium values $1/3$ and $1/9$. There is no memory effect nor are there persistent correlations. One unusual feature is that for large enough μ the solution decaying to the equilibrium value does have an oscillatory behavior with a single frequency. However, *no* averaging or elimination of the additional color variable can reproduce all the quantum results. It is possible to consider more extensive elaboration of these classical models. For example, one can consider not only systems where the η variables can have three values—in addition, one can consider two types of impurities, so that the ε variables can assume three values—say the three cube roots of unity. In this case, the *mechanical* equation is again:

$$\eta_{p+1}(t+1) = \varepsilon_p \eta_p(t) \tag{28}$$

with

$$\eta_p = \alpha_+ = e^{2\pi i/3} \quad red \quad \text{at } p$$

$$= \alpha_0 = 1 \qquad \text{white at } p$$

$$= \alpha_- = e^{4\pi i/3} \quad \text{blue at } 1 \tag{29a}$$

$$\varepsilon_p = \varepsilon_+ = e^{2\pi i/3} \quad \text{impurity of type 1}$$

$$\varepsilon_p = \varepsilon_0 = 1 \qquad \text{no impurity}$$

$$\varepsilon_p = \varepsilon_- = e^{4\pi i/3} \quad \text{impurity of type 2.} \tag{29b}$$

The analysis of this model can be carried out on the classical level (see Ref. 10). (The quantum level has not yet been solved.) This classical model does show a rather involved frequency spectrum which, however, is *not* identical with the quantum spectrum of the model discussed here. Although this more complex system shows oscillations, no averaging over the third color or the second type of impurity reproduces the quantum spectrum for all correlation functions. Furthermore, this classical model (as all classical models) is altogether incapable of producing either persistent correlations or memory effect, which are so characteristic for quantum theory. Thus, even these generalized classical schemes are unable to produce *all* the quantum effects of a minimal quantum system based solely on the superposition of states. The results reported here do not show and do not purport to show that a classical reconstruction of the results of this minimal quantum scheme is impossible. It is not altogether excluded that a very complex, very contrived classical scheme could be devised which has the same content as the minimal quantum scheme presented here, but the evidence points in the opposite direction. The rather artificial even byzantine features which appear when one introduces more states, more objects†, more types of impurities in the system, in the hope that integration or averaging on these new elements will replace the quantum effects, seem to guarantee that such systems are complex and contrived. Their physical properties, even though they might be classical and non-quantum are, nevertheless, so contrived that they do little to provide the physical or pictorial intuition so necessary for the effective description of physical systems. The general moral is rather that the physical interpretation in quantum theory (and probably in general) should *follow* the formalism and procedures and not precede it. There are numerous instances where mathematical structure and formal beauty have been more reliable guides towards progress in physics than a too-literal adherence to common visualizability. This strongly suggests that it might be advantageous to investigate interpretations of quantum theory starting from the Feynman path integral formulation. That method incorporates the superposition and decomposition idea from the very beginning. That would appear like an excellent beginning which might provide physical insights in an effective formal scheme. In the rather unlikely event that a classical or pseudo-classical interpretation is still possible, it would seem that in that case one recovers physical visualizability only at the expense of mathematical pathologies.

References

1. A. Pais, in *Niels Bohr: His Life and Work as Seen by His Friends and Colleagues*, ed. S. Rozenthal (North-Holland, Amsterdam, 1967).
2. M. Gell-Mann and A. Pais, *Phys. Rev.* **97**, 1387 (1955); for readable account, see John D. McGervey, *Introduction to Modern Physics* (Academic Press, New York, 1971), p. 583.
3. M. A. Bouchiat and C. Bouchiat, *Phys. Lett.* **48B**, 111 (1974).
4. C. Y. Prescott, in *International Symposium on High Energy Physics with Polarized Beams and Polarized Targets* (American Institute of Physics Proceedings, 1979).
5. G. Summhammer, H. Badurek, N. Rauch, U. Kischko and D. Zeilinger, "Direct Observation of Fermion Spin Superposition by Neutron Interferometry", *Phys. Rev. D*, in press.
6. D. Greenberger, *Rev. Mod. Phys.* **55**, 875 (1983).
7. D. Greenberger, paper in this volume.
8. M. Dresden, in *Studies in Statistical Mechanics*, vol. I, ed. J. de Boer and G. F. Uhlenbeck (North-Holland, Amsterdam, 1962).
9. M. Dresden and F. Feiock, *J. Stat. Phys.* **4**, 111 (1972).
10. Y. Chinn, thesis, SUNY–Stony Brook (1983).

---

† One has to go to complex representations of the ε, η variables. The fs, ρs, all become complex. The moduli or real parts are quite complicated.

PHILOSOPHICAL PROBLEMS IN THE INTERPRETATION OF QUANTUM MECHANICS

K. D. Irani

Department of Philosophy, The City College of the City University of New York, New York, NY 10031

A philosophical interpretation of a scientific theory seeks to formulate a description of reality entailed by the theory. Usually one first attempts to extract the conceptual framework of the theory. The Conceptual Framework is specified by two principles: 1) Principle of Existence, i.e. that which is taken, in the postulates of the theory, to be ultimately existing; and 2) Principle of Explanation, described in the concepts of the theory, is explained. Such an approach is eminently successful in theories of classical physics and chemistry which have realistic frameworks. Analysis of the conditions of measurement and description show that a realistic interpretation of Quantum Mechanics is impossible. Exploring possible alternatives, one may notice two approaches. Reality may be viewed as constituted of Observable Events and the theory as a scheme for achieving maximum prediction. Or, we may construe reality to be a hierarchy, each level relative to the condition (or energy) of interaction.

Quantum Mechanics is a highly successful theory in physics. In its later development including Quantum Field Theory, it has led to a variety of new predictions of great accuracy. The problem with this theory, since its inception in the nineteen-twenties, has been that of interpretation, that is, the description of reality contained in the theory. What we shall explore here is the idea of interpretation. This will require a philosophical analysis of scientific theories directed toward extracting their conceptual frameworks, for that is where the interpretations of theories are most clearly disclosed. We shall notice how Quantum

Mechanics differs from Classical Physics in clearly identifiable ways. The possibilities of obtaining an interpretation of Quantum Mechanics and their limits will be discussed at the end. This is essentially a philosophical exploration attempting to relate the empirical findings of Quantum Mechanics to our conception of reality.

A scientific theory is a set of interrelated postulates, from which one can deduce consequences. These consequences are expressed in the terms which appear in the theory and are not normally directly connected with items of observation. Therefore the theoretical terms have to be related by special rules with items of observation. When this is done, the deduced consequences of the theory can be determined to be observationally true or false. A sufficient number of true predictions leads to the acceptance of the theory. Of an accepted theory one asks the question: what according to this theory is the nature of that aspect of reality to which it refers? The account that one obtains from such an inquiry is the interpretation of the theory. In most cases we obtain this by extracting the conceptual framework of the theory.

The Conceptual Framework of a theory consists of two (sets of) propositions: i) the *postulate of existence* and ii) the *postulate of explanation*. The content of the postulate of existence of a particular theory is the answer to the question: What, according to the theory, is taken to be the ultimate objects or entities in existence? The content of the postulate of explanation of a particular theory is the answer to the question: What is an acceptable form of the explanation of an event in the theory? For example, in Classical Mechanics the existence postulate asserts that all matter consists of corpuscles which may be represented as points in space at any given time and which have specifiable masses; and the explanation postulate asserts that all change occurs as a result of the action of force either by impact or gravitation, the magnitude of the change in motion being proportional to the magnitude of the force.

Other examples may be mentioned briefly. In early theory of electricity there were, in addition to the framework of mechanics, objects which had the property of being charged electrically, which produced forces acting upon each other which led to change. In atomic theory of the nineteenth century there existed atoms of 92 different elements. They had special properties, such as atomic weight, atomic number, affinities and valences. Their behavior was explained by the rules of combination according to which they combined and interacted.

In every case what a theory explains is an event, usually an event which is the result of a change. The change is always an alteration in the properties or their magnitudes of the fundamental entities. The postulate of explanation states how the change occurs. In a well-constructed theory what the theory provides is a description of the continuous transformation of the state of the system in space and time from one state to another. Any state at any point is explained by being shown to be a specific spatio-temporal cross-section of the process. Explanation is considered to be adequately provided when events were recognized as elements of such process-descriptions. This is the requirement of causality. To show how an event is caused is to show that it is a cross-section of a spatio-temporally *continuous process description*. The differential equations of classical physics provided the descriptions of such processes and thus satisfied, what was taken in classical physics, as the requirement of causality.

To repeat, the interpretation of a theory was given by the conceptual framework which was a qualitative conception of reality according to the theory. There were some theories, particularly thermodynamics and electromagnetism, which did not yield a conceptual framework in a direct and clear way. Thus there emerged a distinction between two types of theories: Realistic Theories and Instrumentalist Theories. These terms are relatively modern, but the distinction in different terminologies arose in the second half of the last century.

The theories considered realistic were those whose conceptual frameworks referred to entities and processes existing in nature, thereby making the claim that they were descriptions of events in nature, even though the phenomena were unobserved. Whereas theories considered instrumentalistic were essentially mathematical schemes which, given the initial values of the variables of a system, enabled one to calculate, and thereby predict, the values at a later time. The theory makes no descriptive claim, it is merely a convenient and reliable predictive instrument. There was a somewhat radical position taken in philosophy of science using the instrumentalist notion. It argued that all theories should be treated as instruments. The realist interpretation, when given, should be viewed only as a convenient heuristic gloss having perhaps some psychological benefit for those reflecting on the theory, but of no ontological significance. For one holding such a view the problem of interpretation does not arise. However, a realistic interpretation of science has been a demand in our thinking about the world and need not be abandoned by a radical philosophical change in perspective on the nature of scientific knowledge.

The general philosophical position in science and common sense is

Realism, the view that there is a world existing independently of our knowledge of it and capable of being known. This position can be made a little more precise as one which asserts the following propositions:

1) Objects exist and are conceived of as existing in space and time.

2) Objects are known through their properties

a) Some properties belong to the objects themselves, whether these properties are observed or not. These are called primary qualities.

b) Some properties of the objects are given in observation, but do not belong to the object, they are effects the objects produce on the observing system.

3) An Event is a particular combination of properties of one or more objects at some time in a location in space.

4) The primary properties change in space and time. The relations of the regularities of the changes are formulated in the Laws of Nature. If the properties are unobservable, they are formulated in the postulates of theories.

5) The laws and theories enable us to make predictions. If the predicted event is a segment of a continuous process description, it is thereby explained as being part of a causal chain.

The correct description of the world is not determined merely by observation, even aided by instruments; it is that description which enables successful prediction and explanation. The last point here is a subject of dispute. It focuses on the demand for an explanation by way of constructing a continuous process description, over and above achieving prediction. We may therefore be justified in identifying two realist theses: The weaker, which requires the realistic interpretation to provide successful prediction; and the stronger, which requires, in addition, the formulation of prediction through a continuous process description.

In order to see how this works in practice, we may consider the epistemology of an experiment. Consider some system S which has many variables, some related to the others and some not. This system with its interacting sub-systems is changing in space and time. This process is the one we wish to study experimentally. Let this be called the *Substantive Process*. Assume that we have a theory, or a set of theories, which describe the process. How shall we verify it? We will make an observation on the substantive process at time t_1 which gives us a set of observations o_1, from which we infer the state of the system at t_1, call it s_1. It is clear that s_1 is some appropriate combination of the variables of the system, usually called the state-variables. If our theory is adequate, as we assume, it will provide us with an equation of state, such that, by substituting the value of the variables forming state s_1 at time t_1, we shall be able to derive the values of future states at future times. We make such an observation at time t_2 and obtain s_2 which was predictable from the theory, and which made us anticipate the result of the observation o_2 at time t_2. One must note that there were two *Observation Processes* interacting with the *Substantive Process* at two moments, t_1 and t_2. An observation process is a physical interaction with the elements of the substantive process in order for it to give us observations which provide information regarding the state of the substantive process. Of course, the observation process disturbs the flow of the substantive process, but we assume that we can diminish the disturbance to a negligible amount and thus our predictions are not measurably affected.

We are now in a position to see how the structure of Quantum Mechanical Theory diverges from this classical picture described above.

Since energy exchange has a lower limit, the quantum, interaction between the substantive process and the observation process cannot be indefinitely reduced in quantum phenomena. There is thus an intrinsic limit to the accuracy of observation here. At one time it was believed that this was the implication of the Heisenberg uncertainty principle. But this is not so. If that were the case, one could conceivably have a theoretical description of the substantive process. Of course, the observation process would grossly disturb, and therefore the results would only have statistical validity. What we have in quantum mechanical theory is not an intrinsic inability to make a non-disturbing observation upon a microscopic process, but a theory which renders a substantive process that can be realistically described impossible.

This situation is forced upon us by the wave-particle duality. Again, it was thought at one time that the corpuscular and wave aspects of matter and energy could be separated as manifestations in different physical situations. Energy is emitted and absorbed in quanta, thus manifesting corpuscular properties, but in its transmission from one place to another it proceeds as a wave as manifested by interference phenomena. But this description fails to deal satisfactorily with the single-slit and double-slit experiments. Particularly the result of obtaining interference patterns when only single photons or single particles were transmitted through double-slits resulting in an interference pattern, even though their absorption on the screen was at point localizations and hence corpuscular.

If we are limited in our list of existent things to particles or waves, and here we include field variables propagated as waves, we get the following result: The observations made at the points of the observation processes may lead you to describe these events in terms of particles or waves, the substantive process in the spatio-temporal interval between the two observations cannot be described in those terms. There is no realistic description of quantum substantive process which can be used for all observations, because that would entail the process being both a corpuscular as well as a wave transmission. These two descriptions are mutually incompatible. This problem is treated by fiat in Bohr's Principle of Complementarity, where an appropriate measurement of a variable is to be obtained by an appropriate experimental arrangement and permits only one kind of description of the system. In that situation the theory forbids a measurable value of a non-commuting variable, which could have been obtained by another experimental arrangement, incompatible with the first, and which would have required the other, hence complementary, description of the system.

The theoretical postulate here is the Schrödinger equation, together with the requirements of how the observation data are to be incorporated into the initial state of the equation, and the way the function is to be operated upon. The theory does provide an inference from observation o_1 of state s_1 at time t_1 to a later state at time t_2. This is the transformation in the Schrödinger function, ψ, as given in the time-dependent Schrödinger equation. However, the value of ψ at time t_2 does not, in general, predict a specific state s_2. It predicts a set of possible states with a probability distribution; a distribution which can be verified by performing the experiment a large number of times. If, following Reichenbach, we call the variables we observe in the observation process, the *phenomena*, and the variables in terms of which we describe the substantive process between the observations as *inter-phenomena*, then we find that in quantum mechanics the phenomena and inter-phenomena are necessarily different. A description of reality in a homogeneous set of variables is impossible in quantum mechanics.

We can now formulate the difference between the classical and quantum conceptions of explanation. In quantum mechanical descriptions, a continuous process description is impossible. Even a purely theoretical state-description which is homogeneous with observable states is impossible. When a description is attempted in one conceptual form, it requires that another incompatible description of the same situation, but measuring a different variable, be recognized as possible. Thus explanation in the form of causality is impossible in quantum mechanics. The only explanation available in quantum mechanics is prediction. It is a statistical prediction which gives us a probability distribution of possible outcomes. This was viewed as an inadequacy in the theory—a point made by several pioneers of quantum theory, such as Einstein, De Broglie and Schrödinger. It is not that the theory fails to provide an explanatory scheme in the strong sense; it goes further, and makes the providing of such a scheme by an extension of the theory impossible. This was the impact of Von Neumann's theorem against the extension of the theory by hidden variables.

The thought-experiment devised by Einstein, Podolsky and Rosen to demonstrate the incompleteness of quantum mechanics essentially commits itself to the existence of an underlying substantive process, which, though unavoidable in a realistic process mode of thinking, is not compatible with quantum mechanics. The Bell theorem draws consequences from quantum mechanics which have been experimentally confirmed, and which make a process-descriptive approach to quantum phenomena impossible. When one might think that an order in the form of a process is impossible, the delayed-choice experiments, suggested by Wheeler, seem to indicate an order which we cannot as yet express, in spite of the heroic analogic attempts by Bohm in his notion of implicate order.

It may well be that we fail to express the order because we are operating with inadequate concepts. For that, we should turn to the postulate of existence in quantum mechanics. First let us see where we stand. Quantum mechanics provides a predictive theory, but not a process-descriptive system. The latter being a requirement of a traditional realist interpretation, makes quantum mechanics, as far as the postulate of explanation is concerned, a non-realistic theory. What is in the postulate of existence in contemporary quantum mechanics? Although we speak of particles and waves, these cannot be the objects of ultimate existence. The uncertainty principle prevents us from ascribing continuously existing properties to particles and waves which their definitions demand. What exist are corpuscle-like and wave-like events. They exist when they are observed. Two questions arise at this point: 1) Are these corpuscle-like and wave-like events independent of observation? 2) What can be taken to exist in quantum physics?

It may be that our perceptual mechanism and conceptual repertory can grasp the events of the external world as it impinges upon our perceptual system in a limited way. And we are ultimately limited to construe the perceived world in those concepts. The theories we construct are the rational adjustments the paucity of our conceptual repertory is forced to make in exploring an indescribably rich and varied reality. To assume that the events we experience, as we experience them, constitute reality seems to me a somewhat excessive commitment to realism. But the alternative is not the total abandonment of realism.

In the advanced theories of physics what do we take reality to be? They are entities defined as a set of attributes which co-exist in a space–time location. Existence is the invariance of those attributes under conditions of change, i.e. energy transfer. The various properties we postulate, and the symmetries we ascribe to their structural schemes, tell us what is stable under what conditions; the rest changes. We also postulate the range of change. In any particular context, the unchanging, the invariant, is the existent.

This position relativises the concept of existence. That which exists is relative to the energy with which it interacts with its environment. In a particular energy environment the structure which is stable is the existent. The condition of quantization gives relative stability and similarity of stable patterns. This is Weisskopf's idea of the Quantum Ladder. For example, the various substances we know on earth exist because they are stable at the level of energy exchange on the surface of the earth. At high levels of energy the molecules decompose and the atoms become the existents. What is more, because of the quantum rules they have systematically stable forms. All gold atoms are alike, they do not have slight variations. At higher energies still the atoms break and what exists are sub-atomic particles. At the other end, some molecules combine to form reasonably stable DNA molecules and thus, at that level, the animal species exist.

Our ability to describe that which exists is dependent upon their having common and stable characteristics. That is what quantum mechanics explains. But since stable forms are relative to energy exchange, and all information is communicated by energy exchange, existence is justifiably relativised. There is no single existence postulate in Quantum Mechanics, they vary with energy levels. And though we have explanatory rules of invariance and symmetry we realize that there are underlying principles of order which are unknown to us. Thus though predictive explanation is abundant, a deeper principle of order is still to be formulated. Traditional realism has become untenable, but a total espousal of instrumentalism is not taking place.

We cannot say clearly what we are looking for. We recognize a new realism with a different principle of existence and a new, more extended, principle of explanation, even the vague form of which we do not know.

I am reminded of a friend who remarked to Croce that the philosophers arrive at the conclusion that they know little or nothing after exhausting toil; while he knew nothing without any effort at all, simply as a generous gift of nature.

BUDDHIST MADYAMIKA PHILOSOPHY AS THERAPY FOR REALISTIC VIEWS

Laura M. Roth

Department of Physics, State University of New York at Albany, Albany, NY 12222

Parallels between quantum paradoxes and eastern philosophies have often been noticed. We describe here the Madyamika or middle way philosophy of Nagarjuna based on emptiness or no-thing-ness. The view that outer objects and the self are real and solid is regarded by Buddhism as being a kind of sickness in need of treatment. Nagarjuna's dialectic was aimed at curing realistic views by cutting away concepts of self-existence or own-being of things, and showing their interdependence. According to the Madyamika view, nothing exists on its own; everything depends upon something else for its existence. This dependent arising seems to be closely related to what quantum theory tells us, especially in Bell's theorem and the measurement process. Buddhism goes further and says that the inner mind and outer world also arise in dependence, which implies that ultimately our inner world needs to be subjected to the same objective scrutiny as the outer world, discomforting though this may seem.

The similarity between eastern world views and quantum puzzles and paradoxes has often been cited. Thus Bohr said "For a parallel to the lesson of atomic theory ... [we must turn] to those kinds of epistemological problems with which already thinkers like the Buddha and Lao Tzu have been confronted, when trying to harmonize our position as spectators and actors in the great drama of existence".[1] A popular exponent of these ideas in recent years has been Fritjov Capra in his celebrated book, "The Tao of Physics".[2] Capra has a chapter in his book entitled "Form and Emptiness' which relates to the very profound

and little understood Buddhist concept of emptiness or sunyata, which
we will find has a very great relevance to fundamental questions in
quantum mechanics. The source of the line "Form is Emptiness" is the
Heart Sutra,[3,4] which is part of a group of Mahayana or great vehicle
Buddhist Scriptures called the Sutras on the Perfection of Wisdom. It is
these scriptures on which the Madyamika or middle way philosophy of
emptiness is based. A major exponent of the Madyamika view was the
philosopher Nagarjuna who lived in the second century A.D., about 600
years after the Buddha's time. The Heart Sutra is discussed in a recent
paper by Schumacher and Anderson,[5] who pointed out a
correspondence with the recent thinking of David Bohm.[6,7] Bohm has
interpreted quantum mechanics, and especially the Bell's theorem
arguments,[8] as implying a quantum interconnectedness of space and
time, and has proposed a holographic model for reality in which every
region of space–time contains the totality. These arguments are shown
by Schumacher and Anderson to be closely related to the Heart Sutra,
and indeed we will find interconnectedness to be the essence of
emptiness.

In this paper I will discuss the correspondence between the
Madyamika view of emptiness and quantum physics. It would be a
wonderful thing if one could apply the Madyamika view to help give a
coherent foundation and understanding of quantum mechanics.
Unfortunately I, for one, am unable to go as far as that at this time,
although it is most hopefully a goal for the future. A more modest goal is
to ask what the Madyamika view has to say about various questions
which we have heard discussed in this conference. Two such questions
are: 1. Is a realistic foundation for quantum mechanics possible? and 2.
Can quantum mechanics be interpreted without reference to the human
consciousness?

THE MADYAMIKA VIEW

In order to see the answers the Madyamika view gives to these questions
I wish here to set forth as clearly as possible just what the Madyamika
view is, because it is a little subtle, hard to hold on to, and easily
misinterpreted. In the above-mentioned chapter, Capra compares
emptiness with the quantum field. "Thus the void of the Eastern mystic

can easily be compared to the quantum field of subatomic physics. Like the quantum field, it gives birth to an infinite variety of forms which it sustains, and eventually reabsorbs".[9] This is an interesting analogy, and indeed the idea of the physical vacuum with its potentiality for creation of innumerable particles and hence all physical form seems likewise to be a good analogy for form being emptiness. However, we will find below that there is more to emptiness than that, and that a correct understanding of the concept of sunyata will show more clearly its relation to quantum mechanics. In discussing the Madyamika view, I will draw on both the written and oral traditions. Nagarjuna's works are beginning to be available in translation,[10] and a particularly valuable source is the "Prasanapada"[11] or "Clear Words" by Cantrakirti, an eighth-century commentator on Nagarjuna's Middle Way verses. Also the Tibetan Buddhist schools place a very great emphasis on the view of emptiness, and the recent availability of Tibetan Buddhist teachers and interpreters provides a rich source of this oral and written tradition.[12-14]

As implied by the title of this paper, realism, especially naive realism, is viewed by Buddhism as a kind of pathological condition requiring treatment. There is a kind of wrong seeing, the illusion that there are real objects in the world and that our egos exist in a substantial way. This is part of a collection of neurotic mental patterns to which ordinary beings are subject and which leads to suffering. These patterns include selfishness, attachments and aggression.[12,15] The Buddhist path which is also the middle way between indulgence and austerity works toward transforming the neurotic patterns into wholesome patterns such as generosity, compassion, and wisdom or correct seeing. The path involves meditation, study, virtuous action, and so forth, and through this there can be awakening, satori, peak experience, nirvana, whatever you want to call it. John Wheeler[16] talks about the universe lighting up to itself, and Buddhism says that the mind can awaken to itself. So really we are talking about a religious path, and the relation to the scientific approach, which is based on the wish to know about the world, is that the awakening experience is described as cognitive. It is a knowing and a seeing; in fact, it is seeing the world correctly, according to the tradition.

Briefly, the Madyamika view is that while things appear to be separate, permanent, and independent, actually nothing exists on its own. Everything is dependent for its existence on something else. The Madyamika view is thus a kind of cosmic ecology, stating that things arise in dependence. This is called dependent origination or co-dependent co-origination. The Madyamika emptiness is an emptiness of self-existence.

C. C. Chang, in his "Buddhist Teaching of Totality",[4] describes three main arguments for the line in the sutra "Form is emptiness".

> "First, when we observe the momentary and constant changes of all things in the world we can conclude that all beings are empty. Second, when we contemplate the fact that all things are produced through the principle of dependent arising and are therefore devoid of selfhood or own being we can conclude that all things are empty. Third, since the external world is the sum total of collective karma and the projection of one's own mind, subject to change and extinction we can conclude that all things are empty".[17]

The first and last of these arguments are considered to be useful approximations but the middle one is considered to be the ultimate view of emptiness.

The first argument relates to the Hinayana or small vehicle of Buddhism which emphasizes analysis or abhidharma. In the Buddhist system the person or the psychophysical continuum is analyzed first into the five aggregates: form, feeling, perception, volition or mental formations, and consciousness,[18] and these are further broken down into atomic physical and mental elements or dharmas. These elements arise, abide briefly, and decay from moment to moment, so that there is no permanence at all. Furthermore, if one examines this stream of impermanence, no self or I or ego can be found. "There is no unmoving mover behind the movement. It is only movement. . . . thought itself is the thinker".[19] If we do not examine ourselves we experience the sense of I, but if we try to look for this I, for example in the body, from head to toe we cannot find it. The purpose of the analysis is not quite the same as a scientific analysis, but rather the examination is a therapy for attachment and egoism. The ego which we spend so much energy protecting, is unfindable under analysis. This is so shocking when one gets an inkling of it that the notion is rather quickly forgotten. The analysis extends to the physical realm. All objects are made of atoms, in that if one imagines breaking something up into smaller and smaller bits, there must be a smallest bit. Thus there was a Buddhist atomism and, though the Buddhist theory of atoms[20,21] is rather primitive by modern standards, the arguments hold just as well for our modern atoms. Here again the Buddhist purpose is less to establish a theory than to lessen attachment. A captivating object is just a collection of tiny unseeable atoms, so why are we so attached to it? One's body is also made of many small parts— organs, molecules, atoms, particles, so why be so attached to it?

Hinayana Buddhism emphasizes moral practices and meditation with the aim toward freeing oneself from attachment and liberating oneself from samsara. Mahayana or great vehicle Buddhism places a greater emphasis on developing compassion toward others. The ideal is the Bodhisattva, who vows to liberate not only himself but all beings. At a philosophical level, there is a problem with analysis in that one becomes attached to the elements of analysis. Mahayana Buddhism says that the aggregates and even the dharmas are themselves empty of self-existence. They are interdependent. This is what "Form is emptiness" is all about in the Heart Sutra. I should remark here that the Mahayanist is not saying that one should not analyze but that self-existence will not be found. Everything is interdependent. After analyzing one must recognize the emptiness of the elements of analysis.

At a seminar at Harvard in 1981, His Holiness the Dalai Lama gave a refutation of partless atoms. He said that if atoms were just points without extension nothing could be made of them, but if they have extension then atoms of something must touch each other at various places (traditionally on six sides), and so must have parts. This argument is more than a thousand years old. It is interesting to see that our current analysis of atoms into smaller and smaller parts has led to quarks which cannot exist outside of the particles they compose.

The third argument for "Form is emptiness", that the world is a projection of one's mind, is idealistic. In the words of Eddington, "the stuff of the world is mind stuff". This is a phase of thinking which can be a useful therapy against a materialistic view but is not the final view because there is a tendency to regard the consciousness or intelligence as real. The argument is given by the Tibetan Schools against this mind only view that if the mind and the world are the same substance then the eye could see itself or the finger touch itself.[12,14] The mind and objects are interdependent, but not the same. The mind only view is described by many scholars as the third turning of the wheel of dharma, but the latter designation more accurately concerns the teaching of the philosopher Asanga regarding the Buddha nature, or the potential for awakening that all beings are said to possess. There is some controversy among the Tibetan schools over whether emptiness or Buddha nature is more fundamental, but exploring this would be getting a little too specialized.

NAGARJUNA'S ARGUMENTS

I wish to discuss now some of the Madyamika arguments and I will begin with an argument from the oral tradition.[12] We see objects in the world with properties such as color which appear to be independent of us. But where is the color of something? Is it really in the object? For there to be an experience of color we must have the light on, our eyes open, and we must have vision. Is the color in the light waves? (Here I am modernizing the argument a little!) Actually it is interesting to note that we don't really see light waves, but we see objects. Is the color in the eye, or in the brain? If we examine carefully, the color is not in any one of these places. It is an interdependent situation.

Nagarjuna's "Verses on the Middle Way"[10] gives many such arguments. He propounded a dialectic which is designed to refute any realistic view his opponent could propose. An example related to physics is his refutation of motion.[11,12] The dialectic begins by asking where on the path is the motion, on the part already traversed, the part to be traversed, or the part being traversed? We think of someone walking along. The motion is clearly not in front and not behind. Nagarjuna says it is also not on the part being traversed, and his commentator explains that the foot is made of atoms, and a point under the foot is behind for an atom in the toe, and in front for an atom in the heel and so one cannot find a place where there is motion. The opponent argues that the motion is in the mover, and Nagarjuna here invokes at some length a different kind of argument in saying that one cannot attribute motion to a mover, because then you could take away motion from the mover and have a mover without motion, which is absurd, or else there are two motions, and then two movers. He goes on to argue that it is unintelligible to suppose that a non-mover commences motion, or a mover comes to rest. These arguments have been compared with Zeno's denial of motion.[23] Zeno was trying to prove that motion and change are illusory, while Nagarjuna was denying the self-existence of things and leading his opponent toward a radically relativistic view. Certainly our present concept of motion is relative, as even classically one body moves relative to another.

Another matter taken up by Nagarjuna is the question of causation. At an ordinary level we understand that there is cause and effect. The seed grows into a sprout when the proper conditions come together; the soil, water, sun, et cetera. It is wrong to say that the sprout pre-exists in the seed, but it is equally wrong to say that the sprout is totally different from the seed. Causation is of course related to time. Nagarjuna had a refutation of time based on the division of time into past, present and future. Nagarjuna said that if present and future depend on the past they are in the past which no longer exists, but if they do not depend on the past, relative to what are they the present and future? A related argument comes from the oral tradition. If a cause in one moment produces an effect in the next moment, is the first moment over before the next begins? If so, how can the cause which no longer exists produce an effect? But if the first and second moments occur together, how can the cause and effect be distinguished?

These arguments are meant to refute our notions that things are real in the sense of self-existent or independent. Nothing exists on its own. Everything depends on something else. One important distinction is the two truths—ordinary truth or the everyday world, and ultimate truth. At an ordinary level things exist and are effective. The follower of the great vehicle in Buddhism develops compassion for others and helps them and accumulates merit or positive qualities. Ultimately there is emptiness. There is no doer, no deed, no recipient of the action. Yet if one were to say because of the emptiness and insubstantiality of things it does not matter what one does, one would fall into error. This is the error of nihilism. There are two extreme views: Externalism is the view that things exist in a permanent fixed way, and nihilism is the view that things do not exist. The middle way is a path between these extreme views. It is not that things do not exist, but just how they exist is the issue. Things exist interdependently.

MIND AND EXTERNAL OBJECTS

An important aspect of this interdependence is between the mind and external objects. It is said that without external objects there would be no mind, and without mind there would be no objects. Let me quote from a Tibetan text, "The prayer of Mahamudra", by the thirteenth-century teacher, Ranjung Dorje:

All the dharmas are an illusory projection of the mind;
The mind is no mind, its nature is void.
Although it is void, everything is continuously appearing from it.
Through the careful observation, may the root be understood.[24]

It is finally said, however, that the two truths, the relative and ultimate truths, are actually the same. Nirvana or enlightenment is samsara, or the round of existence. Form is emptiness. Compassion and emptiness are inseparable. Quoting further from the same source,

> There is no existence, Buddha himself didn't see any.
> All is not non-existent because the basis of samsara and nirvana exists.
> In the unity of these statements, neither in agreement nor in contradiction, lies the middle path.
> May this dharma essence of the mind, beyond limitations, be understood.
>
> If someone says, "this is it", there is nothing to show for it.
> If someone says, "this is not it", this cannot be denied.
> The uncreated Dharma nature is beyond thinking.
> May the actual and final meaning of this truth be realized with certainty.[24]

There are various analogies about how things exist such as a magical show, a rainbow, a reflection in a mirror, or an echo. A Buddha, according to the tradition, sees things the way they are. The Hwa Yen sutra has some wonderful descriptions of how a Buddha sees things:

> The infinite lands that cannot be described
> Gather in the tip of one hair [of the Buddha]
> They neither crowd nor press
> Nor does the hair tip swell . . .
> In it all lands remain
> Just as they were before . . .
> How these lands enter the hair . . .
> The huge vastness of the realm . . .
>
> . . . When a Bodhisattva obtains the ten wisdoms, he can then perform the ten universal enterings . . . to being all the universes into one hair, and one hair into all the universes; to bring all sentient beings' bodies into one body, and one body into all sentient beings' body; to bring inconceivable aeons into one moment, and one moment into inconceivable aeons . . . to make all thoughts into one thought, and one thought into all thoughts . . . to make all the Three Times into one time, and one time into all the Three Times . . .
>
> He enters Samadhi in one moment, and emerges from it in billions of aeons, enters in billions of aeons and emerges in one moment . . . enters in the present, and emerges in the past . . . enters in the past, and emerges in the future . . .[25]

IMPLICATIONS FOR QUANTUM MECHANICS

Can we use this to help with quantum mechanics? The news is I think both good and bad. The purpose of the dialectic and Buddhist meditation and practice in general is to lead one to a direct experience of how things are. Thus in a sense the world is said to be understandable. But of course Buddhahood is difficult to achieve, and the mental awakening promised as a result of spiritual practice takes a long time. Meanwhile we scientists have the more modest goal of finding a description of the world which is both instrumental and satisfying in some way.

If we feel that we need separate real things to be satisfied, Madyamika would help us give up such ideas, and the Buddhist answer to my first question is that realism in that sense is illusory. The particle in the Bell's theorem experiment with its definite instructions on how to respond to a given detector setting seems to have the kind of self-existence that the Madyamika arguments refute, and of course quantum mechanics refutes it also. Wave functions are certainly insubstantial and rainbow like, describing as Capra says tendencies to exist in places. But the main characteristic of quantum mechanics with which the Madyamika view resonates is interdependence. Quantum correlations evidently occur over macroscopic distances. Reduction of wave packets requires in most versions an assist from the macroscopic level to define microscopic things which are supposedly constituting the macroscopic object. "Form is emptiness", however, also says that the world as perceived exists in some sense, which is encouraging, but how the world exists is the question.

But the answer to the second question, can quantum mechanics be interpreted without reference to human consciousness, is more discomforting. The Madyamika answer here is quite clearly no. The argument is that mind and object or subject and object are interdependent. Buddhism is actually mainly concerned with the mind, whereas we physicists have specialized to studying the external world. One of the difficulties with the interpretation of quantum mechanics is that we have an ambivalence as to whether we, in the sense of our conscious awareness, are in the system or not. Somehow we want to assume that the external world is independent of us and that we are not in the system. Most discussions of the measurement problem concern ways to effect reduction of the wave packet without the intervention of the consciousness.

This preference for not including the mental aspect is probably based on the way things appear to us. Objects appear to be independent of us. According to the Madyamika view, this appearance is illusory. Colors and shapes appear to be external but there is, as we have argued, an interdependence between mind and objects. Mind and objects appear together, according to the Madyamika view. Here the whole objective view of science is in question, and this is a serious problem. The word objective can be thought of in two ways. One is as part of the subject–object duality. In this sense we can say that looking just at the external world necessarily gives an incomplete picture. It is likely that the probabilistic feature of quantum mechanics and its paradoxical nature are related to this incompleteness. It has often been said that science is limited to looking at the external world. However, there is another way of construing the word objective, and that is as dispassionate. Looking at it in this way, if we allow ourselves to be in the system this necessitates subjecting our mental processes to the same objective scrutiny as the external world. If this is discomforting it is mainly because it is unfamiliar to us. Psychology may have many of the answers, in particular perception. It is interesting that several papers in this conference, most notably that of John Schumacher, are beginning to turn in this direction. It may be that we will not understand quantum mechanics, or the external world, until we understand the mind.

References

1. N. Bohr, Atomic Physics and Human Knowledge (Wiley, New York, 1958), p. 20, as quoted in Ref. 2.
2. F. Capra, *The Tao of Physics* (Shambala, Berkeley, 1975).
3. E. Conze, *Buddhist Texts Through the Ages* (Harper, New York, 1975).
4. C. C. Chang, *The Buddhist Teaching of Totality* (Penn State Press, University Park, Pa., 1971), p. 65.
5. J. A. Schumacher and R. M. Anderson, *Philosophy East and West* **29**, 73 (1979).
6. D. Bohm, *Wholeness and the Implicate Order* (Routledge and Kagen Paul, London, 1980).
7. D. Bohm, *Foundations of Physics* **1**, 359 (1971); **3**, 139 (1973).
8. N. Merman, this conference.
9. F. Capra, Ref. 2, p. 208.
10. K. K. Inada, *Nagarjuna* (Hokuseido, Tokyo, 1970).
11. M. Sprung, *Lucid Exposition of the Middle Way* (Prajna Press, Boudler, 1979).
12. Khenpo Karthar Rinpoche, Seminars at Karma Triyana Dharmachakra Tibetan Buddhist Monastery, Woodstock, N.Y.
13. P. M. Hopkins, "Meditation on Emptiness", thesis, University of Wisconsin, 1973.
14. Trangu Rinpoche, *The Open Door to Emptiness* (Tara, Manila, 1983).
15. R. H. Robinson and W. L. Johnson, *The Buddhist Religion* (Wadsworth, Belmont, Cal., 1982).
16. J. A. Wheeler and W. H. Zurek, *Quantum Theory and Measurement* (Princeton, N.J., 1983), p. 209.
17. C. C. Chang, Ref. 4, p. 69.
18. W. Rahula, *What the Buddha Taught* (Grove Press, New York, 1974).
19. Ibid, p. 26.
20. F. Th. Stcherbatsky, *Buddhist Logic* (Dover, New York, 1962), p. 79.
21. K. V. Ramanan, *Nagarjuna's Philosophy* (Motilal Banarsidass, Delhi, 1975), p. 209.
22. T. R. V. Murti, *The Central Philosophy of Buddhism* (George Allen & Unwin, London, 1955).

23. M. Siderits and J. B. O'Brien, *Philosophy East and West* **26**, 281 (1976).
24. Akong Tarap Rinpoche, *Refuge, Part 2* (Karma Drubgyud Darjay Ling, Dumfriesshire, Scotland, 1977), p. 37.
25. C. C. Chang, Ref. 4, p. 11.

SUMMARY OF CONFERENCE

Abner Shimony

Since I has been asked to summarize the Conference, I have the impossible task to go and catch a falling star. In thinking about how to summarize, let me list five different scenarios for conferences, and then we can see where ours fits. The five scenarios are by Joseph Conrad, the Book of Daniel, Don Marquis, Plato, and Robert Frost. In my opinion, each has caught one aspect or another of our Conference.

What is Joseph Conrad's scenario? If you recall, the typical setting at the beginning of a Conrad story is the deck of a sailing ship, on a sultry summer night. Cigars are lit, and the travellers and sailors are remembering the strange lands they have visited and the strange human beings they have encountered, and they have come back to exchange tales about them. Not all of the tales are believable, but they are all fascinating. This scenario partly describes our Conference, if we recollect the reports of the experimentalists and of the investigators of superconductivity and ferromagnetism, who told us some very strange things, for the most part credible.

The second scenario is that of the Book of Daniel: many things are said, mainly irresponsible, and at the end a hand appears and writes on the wall.

"Mene, Mene, Tekel, Upharsin,"

and what does that mean? It means,

"Measurements have been made, your words have been weighed, and your theories are torn to bits."

That hand did not appear to us. I have attended one conference at which something similar happened. It was in Erice in 1975, perched on a high cliff in Sicily. Fog and rain held the conference bound for four straight days, and as soon as it ended the fog lifted and the sun came out. I think that on that occasion someone was telling the conferees something.

331

Although nothing like this happened to us during the last few days, some of the auditors of lectures took it upon themselves to make comments like those in the Book of Daniel. Hence, the second scenario applied also.

Don Marquis is the author of the great Archie and Mehitabel poems. In one of my favorites, Archie meets a flea and has a long and inconclusive argument. Archie reports, "and so we parted, each feeling superior to the other, which after all is one of the desiderata of social intercourse." One is reminded of elastic collisions: they come, they meet, they talk, and their internal states are unchanged. Some of that happened here.

There is the scenario of Plato, particularly in *The Republic*, in which we are shown how the increase of knowledge goes hand in hand with the clarification of concepts. Happily, something of this process happened here. There were some examinations of concepts, and I think that at the conclusion of the discussions, the concepts were clearer than before. To some extent Plato's scenario was realized.

And then I think the Frost scenario is very important. I have in mind a poem of Frost which is only two lines long, which goes approximately,

We dance round in a ring and suppose,
But the Secret sits in the middle and knows.

Of course, the poem is based on a children's guessing game. Many of us still feel that even though we have explored various aspects of quantum mechanics, the Secret is still there and we don't know it. It remains a fascinating game, and I am not sure that we want it to end too quickly.

Finally, then, I think that our Conference had some elements of all of these scenarios. And I want to thank the three organizers—Laura Roth, Akira Inomata, and John Kimball, for providing a many-sided occasion. We are very grateful to them for a most stimulating Conference.